Broadband Communications

A Professional's Guide to
ATM, Frame Relay, SMDS, SONET, and BISDN

Other Books in the McGraw-Hill
Series on Computer Communications

In order to receive additional information on these or any other McGraw-Hill titles, in the United States please call 1-800-822-8158. In other countries, contact your local McGraw-Hill representative.

Broadband Communications

**A Professional's Guide to
ATM, Frame Relay, SMDS, SONET, and BISDN**

Balaji Kumar

McGraw-Hill, Inc.

New York San Francisco Washington, D.C. Auckland Bogotá
Caracas Lisbon London Madrid Mexico City Milan
Montreal New Delhi San Juan Singapore
Sydney Tokyo Toronto

Library of Congress Cataloging-in-Publication Data

Kumar, Balaji.
  Broadband communications : a professional's guide to ATM, frame
relay, SMDS, SONET, and BISDN / Balaji Kumar.
    p. cm.
  Includes index.
  ISBN 0-07-035968-7
  1. Broadband communication systems. I. Title.
TK5103.4.K86   1994
004.6'6—dc20   94-33621
                                                      CIP

  3 4 5 6 7 8 9 0 DOC/DOC 0 9 8 7 6 5

ISBN 0-07-035968-7

*The editor of this book was Theresa Burke, the managing editor was
Susan W. Kagey, and the director of production was Katherine G.
Brown. This book was set in ITC Century Light. It was composed in Blue
Ridge Summit, Pa.*

Printed and bound by R. R. Donnelly & Sons.

Author's Disclaimer

Every effort has been made to include the latest information available at the time of writing. Much of the information that was at a draft stage at the time of writing might have become standard by the time of publication. I have made every effort to include the reader who has little background knowledge. Finally, please excuse any personal biases that might have crept into the text because of my background or work environment.

This book does not reflect any policy or position of Bell Northern Research (BNR), Northern Telecom, DSC, or MCI. This work was not funded or supported financially by BNR, Northern Telecom, DSC, or MCI. Ideas expressed here are my own. Information provided here is available in the public domain.

This is dedicated to my parents, Joola and Sahasranaman; my wife, Durga; and my brother, Vijay.

Acknowledgments

Many people have helped me prepare this book. They provided valuable critique, information, and other services. I thank all of them. In particular, I am grateful to Rick McLean, Fred Homayoun, Mohan Lakshminarayan, Kalyan Basu, Brent Earley, Sri Nathan, Giri Giridharagopal, Joe Buckhoff, Pete Bernadian, Stephen Flemming, Tom Steffens, Eric Chern, Zhara Ghasamian, Lisa Casto, Ian MacDonald, and Spencer Dawkins.

In addition, I would like to mention some of my friends who have helped me in many different ways, namely Durga, Sridhar, Jey, Karl, Chitra, Biswajit, Magesh, Anu, Jay Ranade, Bob Hawkins, Ram, Achala, Beena, and Joylyn Quintina.

Contents

Introduction

The 1990s have inaugurated the second revolution of telecommunications. Changes have already occurred so rapidly in the telecommunications and computer environment that it is hard to believe more is to come during this decade. This book gives the reader a snapshot of different technologies that will drive the future of telecommunications. Welcome to the new world!

The primary objective of this book is to present a comprehensive view of one aspect of next-generation telecommunications technology—broadband communications, which encompasses multimedia applications where voice, video, and data are integrated. The reader learns the standards, technology, services, architecture, and protocols of broadband communications. Among the different broadband technologies mentioned, the most important, asynchronous transfer mode, or ATM, is covered in detail. Here, the different environments that ATM can affect are covered, including local area networks (LANs), wide area networks (WANs), public networks, and cable television networks. Details are also given of the commitment by the equipment vendors and service providers toward participating in ATM trials.

The focus is in the ATM aspects of broadband communications that will be applicable in the next few years. Although ATM is capable of handling all types of traffic (voice, video, and data), data traffic will most likely be the initial target for ATM, with other traffic to follow later. Because broadband communications is a new technology and its applications are still unfolding around the world, it is not possible to cover every aspect. Within the context of data traffic, experts argue about which environment (LAN or WAN, etc.) ATM will penetrate first. The penetration will most likely depend on which vendor(s) design the right equipment at the right time. Some say ATM's first environment will be LAN, some say WAN, and others say public backbone (long distance). This book discusses some equipment vendors' ATM products that target the different environments.

Intended Audience

This book covers the basics of broadband technologies, emphasizing the broadband switching technology (ATM) and the transmission technology (SONET/SDH). As this area is relatively new, the book is organized to address audiences who have some understanding of communication (voice or data) as well as those in any other professional field. For those in other fields, sufficient background and history are provided. In addition, this book can be used as course material for a senior or graduate-level communications class. Every effort has been made to ensure that this book can remain useful as a reference guide for a long time.

Organization of the Book

This book is organized into seven parts, with each part containing related chapters.

Part 1 provides the background of communications as a whole, including a history of telephone and computer networks. We then introduce the broadband concept and discuss possible evolution of different types of networks.

Part 2 describes the different broadband technologies. Here, FDDI, DQDB, frame relay, SMDS, ATM, and SONET are covered. All these are switching technologies, except SONET, which is a broadband transmission technology.

Part 3 provides the broadband architecture as defined by the ITU-T. Here, the BISDN architecture protocol layers are covered in detail, along with other aspects of BISDN.

Part 4 covers broadband switching and transmission. The basic architecture and functional components of broadband switching and transmission systems are described here.

Part 5 covers the different environments applicable to ATM switching. Here, ATM in LAN, WAN, public network, and CATV networks are covered.

Part 6 discusses how to design a broadband network from any existing network facilities. The emphasis here is not on the simulation or modeling of traffic characteristics, but on the process of designing a broadband network.

Part 7 covers topics that do not fit in any of the above categories. These chapters add value in terms of understanding ATM in the real world and where we go from the existing ATM/SONET systems. Figure I.1 illustrates the organization of the book.

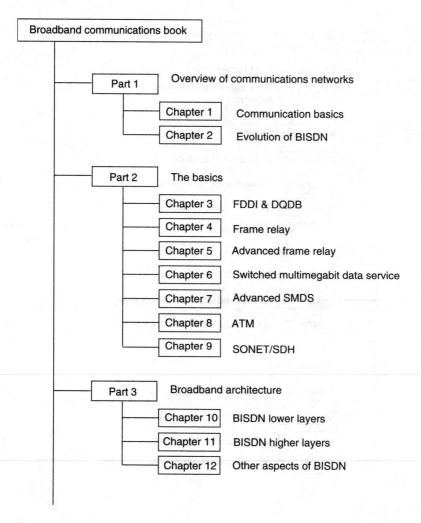

Figure I.1. Book organization.

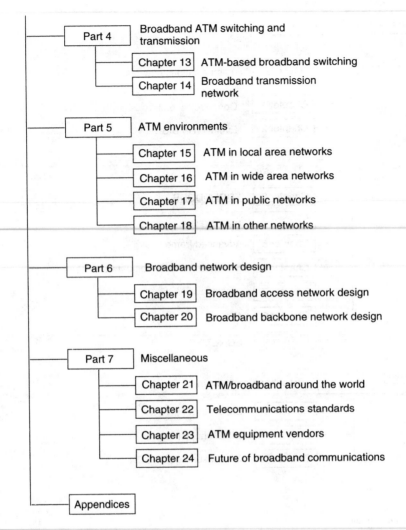

Figure I.1. Continued.

Acronyms and Abbreviations

AAL	ATM adaptation layer
ADSL	asymmetrical digital subscriber loop
AN	access node
APS	automatic protection switching
ATM	asynchronous transfer mode
BECN	backward explicit congestion notification
BER	bit error rate
BH	busy hour
BISDN	broadband ISDN
BOM	beginning of message
CAC	connection admission control
CAD	computer-aided design
CAE	computer-aided engineering
CAM	computer-aided manufacturing
CATV	cable television or community antenna television
CBR	continuous bit rate, or constant bit rate
CBDS	constant bit rate data service
CFM	configuration management
CIR	committed information rate
CLLM	consolidated link layer management
CLP	cell loss priority
CMISE	common management information service elements
CMT	connection management

CO	central office
COI	community of interest
COM	continuation of message
CPE	customer premises equipment
CPN	customer premises node
CRC	cyclic redundancy check
CS	convergence sublayer
CSU/DSU	channel service unit/data service unit
DAS	dual attachment stations
DCC	data communications channels
DCE	data communications equipment
DE	discard eligibility
DLCI	data link connection identifier
DQDB	distributed queue dual bus
DSP	digital signal processors
DTE	data terminal equipment
DTP	data transport protocol
DTPM	data transport protocol machine
EA	extended address
ECM	coordination management
ECN	explicit congestion notification
EO	end office
EOM	end of message
FCS	frame check sequence
FDDI	fiber distributed data interface
FDM	frequency division multiplexing
FECN	forward explicit congestion notification
FEP	front-end processor
FR	frame relay
FRI	frame relay interface
FSK	frequency shift keying
FTAM	file transfer access and management
GAN	global area network
GFC	generic flow control
HDLC	high-level data link control
HE	header extension
HEC	header error control

HOB	head of bus
HIPPI	high-performance parallel interface
HRC	hybrid ring control
HSSI	high-speed serial interface
I/O	input/output
IAO	intraoffice optical interface
IC	integrated circuit
ICI	intercarrier interface
ICIP	intercarrier interface protocol
IEC	interexchange carriers
IIME	intranetwork SMDS mapping entity
IN	intelligent network
INTUG	international trade and user groups
IP	intelligent peripheral
ISDN	integrated services digital network
ISO	international organization for standardization
ISSI	interswitching system interface
IWU	internetworking unit
JPEG	joint photographic experts group
JTM	job transfer and manipulation
LAN	local area network
LAP-B	link access protocol-B
LATA	local-access transport area
LEC	local exchange carriers
LED	light-emitting diodes
LEN	local exchange node
LLC	logical link control
LME	layer management entity
LMP	layer management protocol
LOH	line overhead
LTE	line terminating equipment
MAC	media access control
MAN	metropolitan area network
MFJ	modified final judgment
MHS	message handling system
MIB	management information base
MMF	multimode fiber

MPEG	motion picture experts group
NIF	neighborhood information frame
NISDN	narrowband ISDN
NME	network management entity
NNI	network-network interface
OAM	operations, administration and maintenance
OAMP	operations, administration, maintenance, and provisioning
OC	optical carrier
OCI	optical carrier interface
ONI	optical network interface
OSI	open systems interconnection
OSS	operations systems
PA	prearbitrated
PCS	personal communications services
PDH	plesiochronous digital hierarchy
PDU	protocol data unit
PFM	parameter frame management
PHY	physical layer protocol
PLPC	physical layer convergence protocol
PM	physical medium
PMD	physical layer medium dependent
POH	path overhead
POP	point of presence
POTS	plain old telephone service
PPL	phase locked loop
PRM	protocol reference model
PSTN	public switched telephone network
PT	payload type
PTE	path terminating equipment
PTM	packet transfer mode
PTT	post, telegraph, and telephone
PVC	permanent virtual circuit
QA	queued arbitrated
QOS	quality of service
RBOC	regional Bell operating company
RME	routing management entity
RMN	remote multiplexer node

RMP	routing management protocol
RMT	ring management
SAP	service access point
SAR	segmentation and reassembly sublayer
SAS	single attachment stations
SCM	subcarrier multiplexing
SCP	service control point
SDH	synchronous digital hierarchy
SDU	service data unit
SIF	status information frame
SIP	SMDS interface protocol
SMDS	switched multimegabit data service
SMF	single mode fiber
SMS	service management system
SMT	station management
SNA	system network architecture
SNI	subscriber network interface
SOH	section overhead
SONET	synchronous optical network
SPE	synchronous payload envelope
SRF	status report frame
SS7	signaling system number 7
SSP	service switching point
STM	synchronous transfer mode
STP	shielded twisted pair
STP	signal transfer point
STS	synchronous transport signal
SVC	switched virtual circuit or signaling virtual circuit
TC	transmission convergence
TDM	time division multiplexing
TEN	transit exchange node
THT	token holding timer
TNN	transport network node
TOH	transport overhead
TP	transaction processing
TRT	token rotation timer
TTRT	target token rotation time

TVX	valid transmission timer
UNI	user-network interface
UTP	unshielded twisted pair
VBR	variable bit rate
VCI	virtual channel identifier
VOD	video on demand
VPI	virtual path identifier
VT	virtual tributaries
WAN	wide area network
XAME	exchange access SMDS mapping entity
XC	cross connect
WDM	wavelength division multiplexing

1

Overview of Communications Networks

Part 1 of this book introduces you to the history and evolution of communications networks. The reader learns about existing telephone and computer networks and follows the technological paths these networks are taking to reach the target network of the future: an ATM/SONET-based broadband network. Many reasons exist for the migration toward this type of integrated network, the most important being that no other existing network has the ability to handle all types of communications traffic (voice, data, and video). Each of these types of traffic has different characteristics, and some of these are described in chapter 2.

1

Communication Basics

You've just driven home from work. You get out of the car and walk up to the front door. You don't need to reach into your pocket to get your keys because your eyes open the door for you. As you enter your home and walk through the living room, you have the following dialog with your computer:

You: Do I have any messages?

Computer: Yes, you have three new messages. Do you want to play them?

You: Yes, I want them on my bedroom screen (they are video messages).

The computer displays the video messages on an enormous flat screen on the bedroom wall. (Additional screens are in other rooms of the house.)

After dinner, you would like to help your child with his or her homework. The assignment is to write a paper on the pygmy tribes of Africa. Although you don't have a clue about the topic, you don't worry. First, you ask the computer to gather enough information about the pygmies to complete the paper. The computer searches all available libraries and video images to compile the data. In a matter of minutes the computer asks, "Are you ready to view the data?" Once you are ready, the computer then displays the information on the flat screen.

Now that your child is busy writing the report, you decide to relax and watch a movie. You ask the computer for a list of the latest PG-13 ones. It brings up a list, you select one, and ask the computer to play it. The computer extracts the movie from the database at a video store nearby and plays it as if it were being played from your home VCR, the only difference being that the VCR is at a remote location.

Suddenly, the computer interrupts you from viewing the movie to announce that a video call is coming in. You ask the computer to put the movie on pause and answer the video call. When you complete the call, the computer automatically switches the movie from pause to play so you can continue to view it from the point where you stopped before the video call.

The above example illustrates the range of services available through broadband technologies. They include the following:

- *video voice mail*: voice-activated remote video answering machine

- *interactive video phone*: voice-activated, video-based telephone conversation

- *video on demand*: the ability to access remotely located VCRs on demand

- *data and video transfer*: voice-activated accessing of information from remote text and video libraries across the country

In addition to these, numerous other applications are being provided by broadband communications. The objective of the example, however, is to illustrate the integration of voice (a video call of voice and video), data (accessing information such as text, graphics, and video from a library) and video (viewing a movie played at a remote location). This integration is not possible with today's segregated communication systems.

The example is not a science-fiction movie, and yes, one day you will be living in this type of world. The technology that is going to bring you this type of service is right here. Broadband communications is made up of three components:

- Synchronous optical network (SONET)

- asynchronous transfer mode (ATM)

- intelligent networks (IN)

These are the transmission, switching, and intelligent network architecture technologies of the future.

This book provides you with a clear understanding of these and other related technologies that are the backbone of broadband communications. This chapter reviews the basics of communications from a standpoint of both the computer and telecommunications. The future of communications is an integrated communications system that can handle both voice from telephone networks and data from computer networks. As a result, the future network will not be distinguished solely by the applications of the network, unlike today's telephone networks based on voice and computer networks based on data traffic.

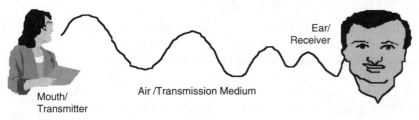

Figure 1.1 Basic form of communication.

Before we go into detail, let's first understand the basics of communications. The word *communication* is derived from Latin *communicare*, which means to share. Communication is a process of representing, transforming, interpreting, or processing information among persons, places, or machines. This process involves a sender, receiver, and transmission medium over which the information flows.

Figure 1.1 illustrates the simplest of all communications: a basic conversation between two people, where the person speaking transfers information by talking, thus changing the pressure of the air that impinges on the other person's eardrum.

The problem with this type of communication is distance. Once the two persons move away from each other, the information carried in the air loses its strength before it reaches the other person; thus, the need for a better transmission medium arises. The earliest form of telecommunications was the telephone. To communicate, both persons need similar equipment. A basic telephone conversation uses electricity in a wire instead of air as the transmission medium, allowing the transfer of information between two persons to be carried over a longer distance. Figure 1.2 shows a simple telephone connection. This connection is not adequate if more than two people need to communicate with each other, which is where a communications network comes into existence. A communications network provides the means of linking many people efficiently using links and nodes.

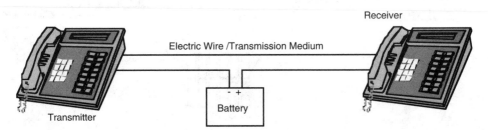

Figure 1.2 Simple telephone connection.

1.1   What Is a Communications Network?

A *communications network* is a system of interconnected facilities designed to carry traffic from a variety of telecommunication sources. A network consists of nodes and links. *Nodes* represent switching offices, junction pairs, or both. *Links* represent cable, terminating equipment, etc. *Traffic* is the information within the network that flows via these nodes and links. Three characteristics influence the design of a communications network:

- traffic carried over large geographic areas
- variation of traffic distribution pattern
- ability to exchange information with negligible delays

A communications network is a common resource shared by numerous end users who need to communicate with other users at remote locations. Not every user uses the network all the time, so it is logical to share this important resource. Sharing is where the concept of switching comes into play. Suppose no switching exists in a network. Then, in a network of m locations, you would have the following:

- the number of links required is $m \, (m-1) \, / \, 2$
- the number of telephones needed is $m \, (m-1)$

A *central switch* reduces the number of links and telephones required to m. As shown in Figure 1.3, a communications network consists of three major categories of components:

- station equipment or customer premises equipment (CPE)
- switching
- transmission facilities or links

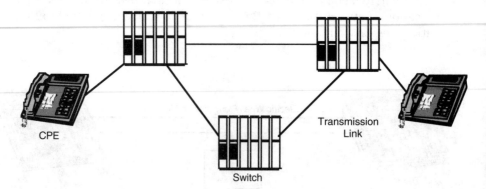

Figure 1.3 Interaction between switch, links, and CPE.

1.1.1 Station equipment

Station equipment is generally located at the user's premises. Its function is to transmit and receive user information (traffic) and exchange control information with the network to place calls and access services from the network. The information is converted into electrical signals and transmitted to the other end, where it is converted back to its original form. The CPE equipment can be a telephone, computer terminal, or fax machine.

1.1.2 Switching

Switching systems are generally called nodes. These systems interconnect the transmission facilities at various locations and route the traffic through the network. In a typical node, the number of incoming lines is greater than the number of outgoing lines (trunks). This statement is true because of multiplexing, which results in improved utilization in the use of trunk facilities. Multiplexing is possible because the utilization on the incoming, or access, side is very low—usually less than ten percent.

1.1.3 Transmission facilities

Transmission facilities provide communications paths to carry the user's voice and network control information between nodes in a network. In general, transmission facilities consist of a medium such as air, copper wires, coaxial cables, or fiber-optic cables, along with various electronic equipment located in the path. This equipment amplifies the signal so it can be carried over a longer distance without losing its strength. The most difficult part of building a communications network is laying the transmission network because of the cost and time involved in deploying cable across the country or even around the globe.

Having explained the basis of a communications network, we can now categorize this communications network based on its dominant applications. A network used for voice communication is called *telecommunications network*, and one used for data communication among computers is called a *computer communications network* or *data communications network*. This separation does not mean that data cannot be carried over the telecommunication (voice) network or vice versa. It only means that the network is designed and optimized for its specific application, and transporting any other information results in the inefficient use of network resources.

1.2 Telecommunications Network

This section describes a voice-oriented network, starting with the history and then the specific architecture of the network.

1.2.1 Telecommunications history

Alexander Graham Bell invented the telephone in 1876 and obtained a patent for his invention. Following this great invention, the U.S. public telephone network became almost the exclusive property of AT&T, but is now fragmented among a number of companies. To summarize the history of telecommunications, Table 1.1 highlights the history of the telephone network in chronological order.

Historically and today, telephone traffic (voice) has been the major user of communications facilities. Over the last 25 years, however, a stable growth of facilities transmitting information other than voice from the telephone has occurred. To address those and other new services, new technologies are needed that adapt easily to the new services.

TABLE 1.1 Highlights in the History of Telecommunications

1844	Morse sends the first public telegraph message
1876	Telephone patent issued to Alexander Graham Bell
1877	First telephone in private home
1881	First long-distance line, from Boston, MA, to Providence, RI
1889	A.B. Stowger invents telephone switch, dial telephone
1891	Undersea telephone cable, England to France
1915	First transcontinental telephone call in U.S.
1929	Coaxial cable invented; Herbert Hoover becomes the first president with a phone on his desk
1947	Transistor invented
1951	Direct long-distance dialing
1956	First transatlantic-repeated telephone cable
1960	First test of electronic switch
1963	Touch-tone service introduced
1965	First trial offers for reversing telephone charges (collect call)
1970	Laser invented
1976	First digital electronic switch installed
1977	First light wave system installed
1984	Divestiture of AT&T (Ma Bell and the baby bells)
1988	First transatlantic optical fiber cable
1989	First fiber-optic cable to the home field trial, Cerritos, CA
1990	Demonstration of 2000-km link using optical amplifiers without repeaters

1.2.2 Telephone network architecture

Before we go further, we need to understand the U.S. public telephone network architecture as the benchmark for the development of networks around the world.

The U.S. network consists primarily of local services offered by Regional Bell Operating Companies (RBOCs), which were a part of AT&T before the 1984 divestiture, and long-distance service still offered by AT&T. A growing number of providers have sprung into being both in local and long distance service to compete against Ma Bell (AT&T) and the baby bells (RBOCs). In this section, we look at the architecture of the network formed by local exchange carriers (LEC) and interexchange carriers (IEC), which interact with each other to complete an inter-LATA (local-access transport area) telephone call.[1] LEC services are provided by RBOCs, whereas IEC services (long-distance services) are provided by long-distance companies such as AT&T, MCI, and Sprint.

Although both routing and specific architecture of the telephone network have evolved since the AT&T divestiture, the overall architecture can still be described with the basic components of a communications network. In a telephone network, each subscriber is connected via the local loop to a switching center known as an end office (EO) or central office (CO). Typically, an end office can support thousands of subscribers in a localized area. About 25,000 central offices exist in the U.S. today. Clearly, it is impractical for each CO to be connected with a direct link. If that were the case, we would need 3×10^8 links. Therefore, intermediate switching nodes are used. These intermediate nodes provide traffic aggregation and reduce the number of links required to connect the central offices together. The intermediate nodes are called *access tandems*. Each of these nodes in the network has an average of 10 to 15 central office switches connected to it. Thus, in the U.S., about 1,200 access tandem switches exist. Traffic routing in the LEC network is based on how the access tandem and central offices are connected. The switching centers are connected by links called trunks. These trunks are designed to carry multiple voice frequency circuits using frequency division multiplexing (FDM) or synchronous time-division multiplexing (TDM) or wavelength division multiplexing (WDM) for fiber-optics.

Figure 1.4 shows a segment of an LEC and IEC interconnected by a 1984 Modified Final Judgment (MFJ) act. Without detailing MFJ, post-MFJ telecommunication functions, service, responsibilities, and restrictions can be described by tracing an inter-LATA call via an LEC and IEC public switched telephone network (PSTN).

---

[1] LATAs are logical boundaries defined by regulatory bodies to define the scope of service for local exchange carriers and long-distance carriers.

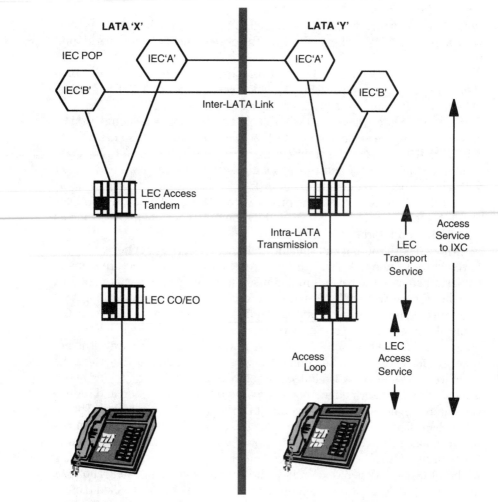

Figure 1.4 Relationship between IEC and LEC in U.S.

In Figure 1.4 a call originates in LATA X. If the call terminates within the same LATA, the traffic goes no further than the access tandem node. The traffic is routed to the appropriate central office and on to the destination phone within the same access tandem serving area. If the call has a destination address in a different LATA (LATA Y), the call is handed to the appropriate IEC (depending on which carrier the call originator has subscribed to), which switches the call to the appropriate LATA. At the LATA, the call is handed to the local exchange carrier serving the destination address. This LEC then routes the call appropriately to the destination. In most countries around the world, the telecommunications network is owned by the country's government under Post Telegraph and Telephone (PTT) authorities. As a result, these networks are not as segmented as in the U.S. The hierarchy mentioned here, however, is now adopted by most countries. With the privatization of PTTs around the world, competition now thrives in every segment of telecommunications, and U.S. telecommunications stands as an example for other countries. The entry of other carriers to compete against Telecom Canada in the Canadian long-distance market for example is a good starting point. The Canadians are learning from U.S. experiences in long-distance. In a telephone network, which primarily carries voice traffic, circuit switching is the preferred switching technology. Communication via circuit switching implies a dedicated communication path between two terminals. The path is a connected segment of links between network nodes. On each physical link, a channel is dedicated to the connection. Communication via circuit switching involves three phases:

- circuit establishment
- signal transfer
- circuit termination

Note that the circuit path is established before data transmission begins. Thus, the channel capacity must be reserved between each pair of nodes in the path, and each node must have sufficient internal switching capacity to handle the requested connection. The switches must be intelligent switches to make these allocations and route the call through the network. Some of the requirements for circuit switching are the following:

- establishing/maintaining and terminating calls on subscribers' request
- providing a transparent full-duplex signal
- limiting acceptable delays for call setup (≤ 0.5 sec)
- providing adequate quality for the voice connection
- limiting blocking probability

1.3 Computer Communications Network

This section discusses the data or computer communications network to explain the differences between telecommunication and data communications networks.

1.3.1 Computer network history

The unprecedented technology revolution involving computers did not begin until the later part of this century. It is now predicted that, by the turn of the century, the traffic generated from computers will dominate communications networks, compared to today's voice-dominated traffic.

The communications requirement for a computer of the 1940s and 1950s was batch-based and minimal. The processor communicated with its peripheral via input/output (I/O) devices over short distances at a very low speed. The 1960s brought the concept of *timesharing*, where users were connected to computers via a dumb terminal, as shown in Figure 1.5.

The 1970s saw the development of integrated circuits (IC) technology and the microprocessor, making it possible to bring personal computers to an individual's desk. This development drastically changed the way people viewed computers. The growth of local area networks (LANs) in the 1980s provided personal computers the technology to communicate with each other, thus enabling the migration from centralized computing to distributed. Figures 1.6 and 1.7 show centralized and distributed computing, respectively.

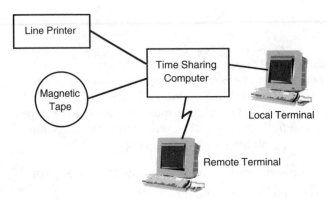

Figure 1.5 Time-shared computer system.

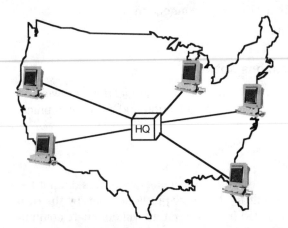

Figure 1.6 Centralized computing.

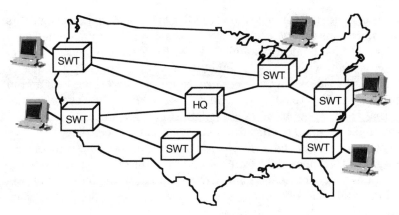

Figure 1.7 Distributed computing.

In a centralized computing environment, the main system or processor is centrally located, and all remote locations are connected via a direct link. All the information, i.e., the entire database, is located at the central system.

In a distributed computing environment, the processors, or main computers, are distributed at different locations, with each node having a complete copy or a portion of the database. The user accesses the information from the nearest processor, which keeps current by periodically updating the information in its databases.

The future trend is a migration toward a distributed environment because of both the advances in processor or switching connectivity and the intelligence available at the CPE level.

1.3.2 Computer network reference model

Before discussing the different types of computer networks, one must understand the basics of computer interactions, or the protocols on which all computer communications systems work. A data communications network requires a high degree of compatibility and interoperability among network elements, particularly with respect to its physical and logical interfaces and controls. So, to address such issues, the International Organization for Standardization (ISO) established a subcommittee (Technical Committee 97) in 1977 to develop a standard architecture to achieve the long-term goal of open systems interconnection (OSI). This committee established the current data communications standards.

The term *OSI* denotes the standards for the exchange of information among systems "open" to one another by virtue of incorporating the ISO standards. The fact that a system is open does not imply any particular system's implementation, technology, or interconnections, but refers to the system's compliance with applicable standards.

With this in mind, ISO has specified an OSI reference model that segments the communications functions into seven layers. Each layer is assigned a related subset of communications functions, which are implemented in data terminal equipment (DTE) that communicate with other DTEs. Each layer relies on the next lower layer to perform more primitive functions and in turn provides services to support the next higher layer. The layers are defined so that changes in one layer do not affect other layers.

Information exchange occurs when corresponding (peer) layers in two systems communicate using a set of rules as their protocols. Protocols define the *syntax* (arrangements, formats, and patterns of bits and bytes) and *semantics* (system control and information context, or the meaning of patterns of bits and bytes) of exchanged data, as well as numerous other characteristics such as data rates and timing.

Defining the details of seven layers of protocols for data communications is an enormously complex task. To understand the concepts and objectives of layering, we use an example. Figure 1.8 illustrates the three layers of communications between two philosophers, one in China and one in India. The exchange of ideas between the two philosophers represents layer 3, peer-to-peer communication.

As the philosophers do not share the same language, they each engage the services of a translator. In Figure 1.8, the exchange in Latin represents layer 2, peer-to-peer communications. Each translator must then engage the services of an engineer to transmit the ideas by letter, telegram, telephone, computer network, or other means. Just as the translators had to agree on a common language, layer 1, peer-to-peer communications requires that an agreement on the physical transmission media must be made between the two engineers.

Note that the translators could use English to speak to each other without affecting either the layer 1 or layer 3 process. Similarly, neither message integrity at layer 3 nor the translation process of layer 2 is affected should the engineers change layer 1 physical medium choices. Performance aspects, such as message delivery time, however, could be drastically different if the postal service were substituted for a real-time form of communications.

1.3.3 ISO reference model for OSI

Figure 1.9 shows the ISO reference model for OSI. The objective is to solve the problem of heterogeneous DTE and data communications equipment (DCE). The OSI model is not a product blueprint; two companies can build computers consistent with the model, but unable to exchange information. The model is only a framework meant to be implemented with standards developed for each layer. Standards must define services provided by each layer, as well as the protocols between the layers. Standards do not dictate

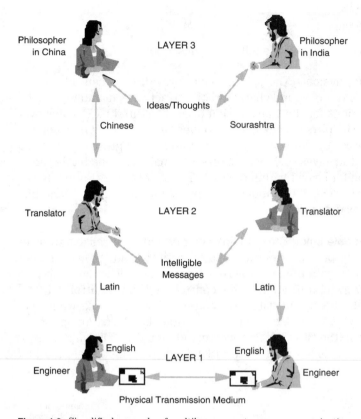

Figure 1.8 Simplified example of multilayer peer-to-peer communication.

how the functions and services are implemented in either hardware or software, so these can differ from system to system.

In Figure 1.9, the protocol stacks to the left and right represent two DTEs connected by a communications subnetwork shown in the middle. Each layer has a specific name and job function. There are seven layers in the OSI model:

1. physical layer
2. data link layer
3. network layer
4. transport layer
5. session layer

6. presentation layer

7. application layer

Each layer is described in the following subsections.

1.3.3.1 Layer 1: physical. The physical layer provides mechanical, electrical, functional, and procedural characteristics to activate, maintain, and deactivate connections for the transmission of unstructured bit streams over a physical link. The physical link can be connectors and wiring between the DTE and the DCE at a network access point and fiber-optic cable within a network. Layer 1 involves such parameters as signal levels and bit duration. X.21 is an option for layer 1 within the CCITT X.25 recommendation. In the United States, the RS-232C standard is generally used at layer 1, and bits are exchanged as data units.

1.3.3.2 Layer 2: data link. The data link layer provides for reliable transfer of data across the physical link. It maps data units from the next higher (network) layer to data frames to permit transmission. The data link provides necessary synchronization, error control, and flow control. Layer 2 provides for multiplexing one data link into several physical links when necessary. A single physical link can support multiple data links. An option for this layer exists in the CCITT X.25 recommendation called link access protocol-B (LAP-B). It is a subset of the ISO-developed high-level data link control (HDLC).

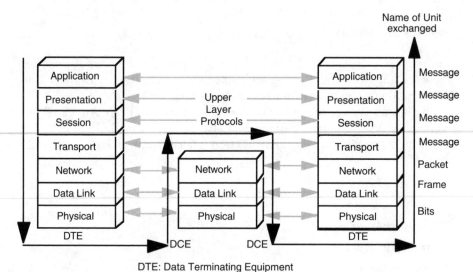

DTE: Data Terminating Equipment
DCE: Data Communications Equipment

Figure 1.9 OSI reference model.

1.3.3.3 Layer 3: network. Layer 3 provides higher-level layers with independence from the routing and switching associated with establishing a network connection. Functions include addressing, end-point identification, and service selection when different services are available. An example of a layer 3 protocol is the CCITT X.25 recommendation.

1.3.3.4 Layer 4: transport. In conjunction with the underlying network, data link, and physical layers, the transport layer provides end-to-end (station-to-station) control of transmitted data and optimizes the use of network resources. This layer provides transparent data transfer between layer 5 session entities. In ISO terminology, an *entity* is the network processing capability (hardware, software, or both) that implements functions in a particular layer.

Transport-layer services are provided to upper layers to establish, maintain, and release transparent data connections over two-way, simultaneous data transmission paths between pairs of transport addresses. The transport protocol capabilities needed depend on the quality of the underlying layer services.

When used with reliable, error-free virtual circuit network service, a minimal transport layer is required. If the lower layers provide unreliable datagram service, the transport protocol must implement error detection, recovery, and other functions.

1.3.3.5 Layer 5: session. *Session* is a connection between stations that allows them to communicate. For example, a host processor might need to establish multiple sessions simultaneously with remote terminals to accomplish file transfers with each station. File transfer access and management (FTAM) is an example of an ISO application-layer standard for network file transfer and remote file access.

The purpose of the session layer is to enable two presentation entities at remote stations to establish and use transport connections by organizing and synchronizing their dialogue and managing the exchange of data.

1.3.3.6 Layer 6: presentation. The presentation layer delivers information to communicating application entities to resolve syntax differences but preserve meaning. Toward this objective, layer 6 can provide data transformation (data compression and encryption), formatting, and syntax selection. Virtual terminal protocol, a layer 6 protocol, hides differences in remote terminals from application entities by making the terminals all appear as generic or virtual ones. When two remote host processors use virtual terminal protocols, terminals appear as locally attached to their host.

1.3.3.7 Layer 7: application. The application layer enables the application process to access the OSI environment. It serves as the passageway between application processes that use OSI to exchange information. All services directly usable by the application process are provided by this layer. Services include the following:

- identifying intended communications partners
- determining the current availability of intended partners
- establishing of the authority to communicate
- agreeing on responsibility for error recovery
- agreeing on procedures to maintain data integrity

Protocols currently being defined or enhanced for this layer include FTAM, transaction processing (TP), directory services (ISO 9595, CCITT X.500), and job transfer and manipulation (JTM). The CCITT X.400 message handling system (MHS) protocols issued in 1984 were substantially revised in 1988 as application-layer protocols.

In a communications environment, the information flow starts at the application layer at one end and terminates at the application layer at the other end. This process is shown in Figure 1.9 by a line in the middle of the protocol stacks. At the intermediate node, the information goes up to layer 3 in case of an X.25-based network. The protocols of the future (described in this book) will only require that the information go to layer 2 or even layer 1, rather than layer 3.

1.3.4 Types of computer networks

Computer networks can be classified based on their geographic scope. The four major categories are

- LANs (local area networks)
- MANs (metropolitan area networks)
- WANs (wide area networks)
- GANs (global area networks)

One can infer the geographical reach of the network from the name itself.

1.3.4.1 Local area networks. LANs are typically used to interconnect computers and PCs within a relatively small area, such as within a building, office, or campus. A LAN typically operates at speeds ranging from 10 Mbps to 100 Mbps, connecting several hundred devices over a distance of up to 5 or 10 km. LANs became popular because they allow many users to share scarce resources, such as mainframes, file servers, high-speed printers, and other expensive devices.

Figure 1.10 shows two different types of LAN connectivity. One is con-

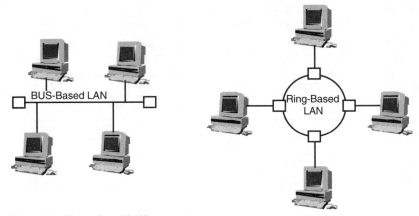

Figure 1.10 Examples of LANs.

nected via a bus architecture, in which physical medium is shared by users connected to the bus. A protocol called carrier sense, multiple access high collision detection (CSMA/CD) is used in this network. The next topology is called the *ring topology*, in which the token protocol is used. In the ring topology, the user who has control of the token transmits data on the ring.

1.3.4.2 Metropolitan area networks. MAN, as the name implies, is a network covering a metropolitan city. The MAN connects many LANs located at different office buildings. A MAN has a larger geographical scope compared to a LAN and can range from 10 km to a few hundred km in length. A typical MAN operates at a speed of 1.5 to 150 Mbps. Figure 1.11 illustrates a MAN.

1.3.4.3 Wide area networks. A WAN is designed to interconnect computer systems over very large geographic scopes, such as from a city to another

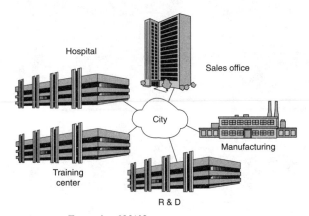

Figure 1.11 Example of MAN.

city within a country. A WAN can range from 100 km to 1,000 km, and the speed between the cities can vary from 1.5 Mbps to 2.4 Gbps. In a WAN, the cost of transmission is very high, and the network is usually owned and operated by a public network. Businesses lease a transmission system from the public networks to connect their geographically diverse sites. Figure 1.12 illustrates an example of a WAN.

1.3.4.4 Global area networks. As the name implies, GANs are network connections between countries around the globe. A good example of such a network is the Internet in the United States, which has a connection to similar networks in other countries. A GAN's speed ranges from 1.5 Mbps to 100 Gbps and its reach is several thousands of kilometers. Figure 1.13 illustrates an example of GAN.

Figure 1.14 shows how the different networks interact. In this type of network, the traffic leaving the network is usually small, but this fact is changing daily. As more and more companies do business globally, the constant need to communicate within the company or with international customers is increasing drastically, thus making the GAN appear as a LAN in terms of access. The interfaces between these networks are usually via international or regional standard interfaces, enabling equipment from different vendors to be interconnected.

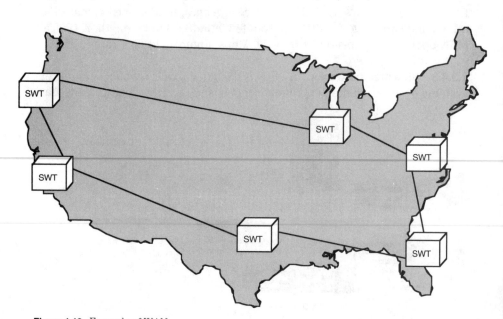

Figure 1.12 Example of WAN.

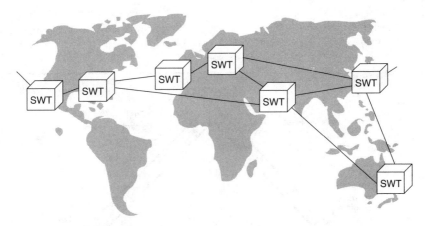

Figure 1.13 Example of GAN.

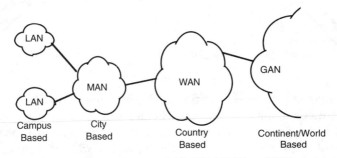

Figure 1.14 Interaction between LAN, MAN, WAN, and GAN.

1.4 Switching Technologies

In this section, we look into the different switching technologies in use and why some were successful and why others are now becoming more prevalent. Figure 1.15 shows the spectrum of switching technologies available, ranging from traditional circuit switching on one end to packet switching on the other.

One of the first uses of switching technologies used in telecommunications was telegrams sent via "packets" or "messages" from one relay station to another. This packet contained the source and destination address along with the contents of the message.

The next switching technology, introduced at the end of the last century, was *circuit switching*. It is used even today for plain old telephone service (POTS), the classical telephone service. In this application, a circuit must be established and dedicated for the duration of the call (i.e., the connection remains intact).

Next came the requirement to interconnect a computer and terminals. Because the need for data interconnection was more stringent than voice

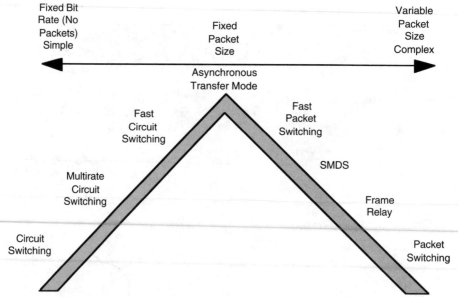

Figure 1.15 Different switching techniques.

(large bursts for a short duration and error-free transmission) circuit switching was not suitable. Circuit switching was thus replaced by *packet switching*. In a packet-switched network, the data stream is broken into a sequence of packets, and each packet is appended with the source and destination address, just like the address in a postal letter. These packets are received, stored, processed, and retransmitted at each node until they reach the matching destination address on the packet. In this technique, many users can share network resources because no one can reserve the rights for a channel at any given time. The network can therefore use resources more efficiently. To help your understanding of packets, compare a packet to a postal envelope, as shown in Figure 1.16.

The following information is in the header of a packet:

- source address
- destination address
- size of the packet

There are two types of packet switching: datagram (connectionless) and virtual circuit (connection-oriented). Table 1.2 shows the difference between circuit switching, datagram packet switching, and virtual-circuit packet switching.

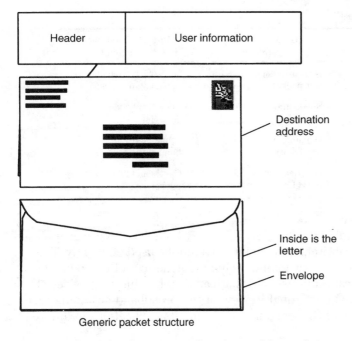

Generic packet structure

Figure 1.16 Comparison between postal envelope and data packet.

TABLE 1.2 Differences Between Switching Types

Circuit switching	Datagram packet switching	Virtual-circuit packet switching
Dedicated transmission path	No dedicated path	Dedicated path
Continuous transmission of data	Transmission of packets	Transmission of packets
Fast enough for interactive	Fast enough for interactive	Fast enough for interactive
Message not stored	Packets stored until delivered	Packets stored until delivered
Path is established for entire conversation	No path is established	No path is established
Call setup delay	Packet transmission delay	Call setup delay
Busy signal if the party is busy	Sender can be notified if the packet is not delivered	Sender is notified of the connection denial
Overload can block call	Overload increases packet delay due to queuing	Overload might block call setup, increases packet delay

TABLE 1.2 Continued

Circuit switching	Datagram packet switching	Virtual-circuit packet switching
User responsible for lost message	Network responsible for lost packets	Network responsible for packet sequences
Usually no speed or code conversion	Speed and code conversion	Speed and code conversion
Fixed bandwidth transmission	Dynamic use of bandwidth	Dynamic use of bandwidth
No overhead bits after call setup	Overhead bits in each packet	Overhead bits in each packet

For future broadband networks that satisfy broadband types of applications, the existing switching techniques cannot be used efficiently. Therefore, we must devise a switching technique generic and independent of the type of application in use and that can adapt to future services. The following section is a thorough review of the broadband switching technology that can meet these requirements and its applications in different environments.

1.5 What Is Broadband?

CCITT defines broadband service as the following:

> A service requiring transmission channels capable of supporting rates greater than 1.5 Mbps or primary rate in ISDN or T1 or DS1 in digital terminology.

Many other definitions exist, but we follow this one for now. From the telecommunications perspective, broadband communications evolved from current integrated services digital network technology (ISDN), which is now called narrowband ISDN. The broadband communications network described in this book is known as BISDN, for broadband integrated services digital network.

As shown in Figure 1.17, BISDN is not a technology but a platform supported by different technologies, including ATM, SONET, and IN (intelligent network). We address each of these technologies in detail.

1.6 The Need for Broadband

One of the most important factors driving the need for broadband communications is changing user needs and demands. Prior to the late 1970s and

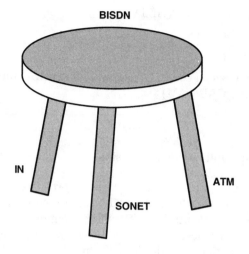

BISDN	Broadband ISDN
ATM	Asynchronous Transfer Mode
SONET	Synchronous Optical Network
IN	Intelligent Network

Figure 1.17 BISDN framework.

early 1980s, public network needs were almost entirely driven by telephony (voice). Data traffic has been growing slowly until recently, indicating the changing of user needs and demands. Many conditions led to this sudden drastic change, some of which are the following:

- lower relative telecommunications cost because of lower cost/higher performance of integrated circuits (ICs) and microprocessors used in telecommunications switches
- competition in the telecommunications sector
- introduction of low-cost, reliable fiber-optics transport in the network
- proliferation of PCs at homes and LANs that connect the PCs at the work place

Together, lower cost and increased processing power of computers enabled a number of users to appear in the public network, but their applications and needs are completely different. To name a few needs:

- video telephony
- low-cost video conferencing
- imaging

- High-Definition Television (HDTV)
- hi-fi audio distribution
- LAN Interconnect
- computer-aided design/computer-aided engineering/computer-aided manufacturing (CAD/CAE/CAM)
- visualization
- multimedia
- supercomputing and channel extension

Each of these is a potential application for broadband communications because they all have very high bandwidth requirements. Figure 1.18 shows the bandwidth requirements for different applications, including the traditional application of voice. One can see that the applications mentioned here require very high bandwidth in real-time.

In addition to these new applications, other factors are also driving the need for broadband around the world. They include the following:

- The service demand has changed in nature. Customers want mobility (both terminal and personal), bandwidth on demand, access to management functions, flexibility in establishing connections, end-to-end connectivity, and management.
- A competitive market generally operates more efficiently than a regulated one.
- Equipment and components for the processing and transport of information are becoming increasingly inexpensive.

We address these trends and drivers in detail in chapter 2.

1.7 Overview of Broadband Technologies

In this book, we explain the technologies that drive the components of a broadband network. Before we go into any of these technologies, it is very useful to understand where and how they fit with each other. Figure 1.19 shows a broadband network that can be implemented with the technologies. The information (voice, data, and video) that originates at the source node is carried to a destination through a switched network. The origination and destination are typically LAN based. The fiber distributed data interface (FDDI), distributed queue dual bus (DQDB), and asynchronous transfer mode (ATM) are three emerging LAN protocols that coordinate information flow at the source and destination node. These protocols replace the current standards of Ethernet and token ring.

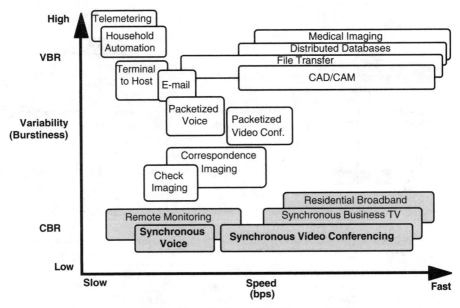

Figure 1.18 Characteristics of potential applications. *(Rick McLean, BNR)*

The data that flows from the originating node into the switched network is routed using frame relay, SMDS, ATM, etc. All these technologies can be used together, and each can use any physical transmission medium, such as fiber, coax, etc. Each physical medium has its own standard for implementation. In this book, we discuss a standard for fiber called synchronous optical network (SONET), which is used in the United States and Canada, and synchronous digital hierarchy (SDH), which is used in Europe and other countries. Other global standards, such as ATM, are also discussed in detail. When ATM is combined with SONET/SDH, we get BISDN, whose protocol is also discussed.

1.8 Summary

There are two fundamental types of networks: computer and telecommunications. Each is designed for specific applications using the most suitable technology for the application. With the emergence of new services and the need for integrated voice, data and video, none of the existing networks can be used to handle or provide optimum cost-effective services for the future. In recent years, new network technologies have been developed that provide the backbone for BISDN.

The following chapters provide a better understanding of different broadband communication technologies. The reasons why ATM and SONET have become the backbone technology for BISDN are explained, and a detailed explanation of BISDN's principles, standards, and architecture is given.

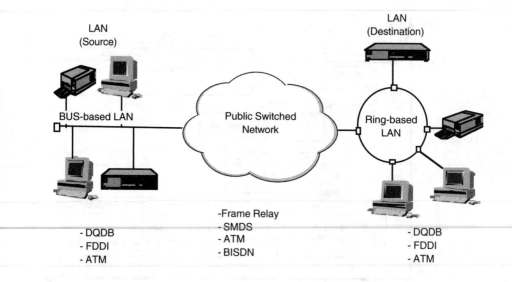

SMDS	Switched Multimegabit Data Service
ATM	Asynchronous Transfer Mode
BISDN	Broadband Integrated Services Digital Network
DQDB	Distributed Queue Dual Bus
FDDI	Fiber Distributed Data Interface

Figure 1.19 Interaction of different broadband technologies.

2

Evolution of BISDN

2.1 Overview

This chapter addresses the drivers leading to the evolution of the BISDN network from existing networks. No single factor is driving BISDN, but rather a combination of technology, standards, industries/business, and applications. All these factors are converging to create a new paradigm called *broadband networks* or *broadband communications network* or simply *BISDN*. Figure 2.1 illustrates this convergence. In this chapter, the different potential applications are explained and the third technology of BISDN, intelligent network (IN), is also described.

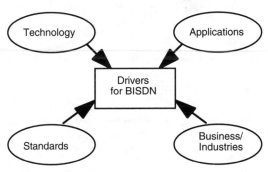

Figure 2.1 Drivers for BISDN.

2.2 Drivers for BISDN

This section addresses the various drivers for BISDN and discusses BISDN trends.

2.2.1 Technology

Some of the technologies behind the evolution of BISDN are

- semiconductor
- fiber-optic
- computer

Each are described in the following subsections.

2.2.1.1 Semiconductor technology. The cost of semiconductors is dropping every year, yet its complexity is simultaneously increasing (Figure 2.2). For example, when IBM invented the PC, it had a 64K memory, two floppy drives running on Intel's 8080 microprocessor, and a monochrome monitor. If you tell your kids about this machine today, they will laugh and say, "Do you mean that such a primitive computer existed?" Don't worry! Some day your children's computers will be obsolete because of the ever-changing technology. Regarding cost, the early, primitive IBM PC cost $2,000 to $3,000. Today, most complex IBM compatibles cost less than $3,000.

2.2.1.2 Fiber-optics. The evolution of fiber-optics technology is shown in Figure 2.3. The amount of data carried on a single strand of fiber has increased drastically, while at the same time, the cost per bit carried on fiber-

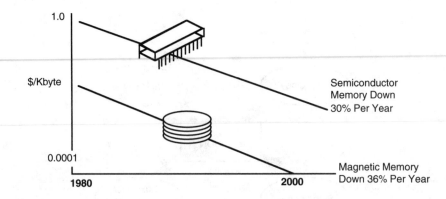

Source: Adaptive Corp.

Figure 2.2 Cost of semiconductors vs. complexity.

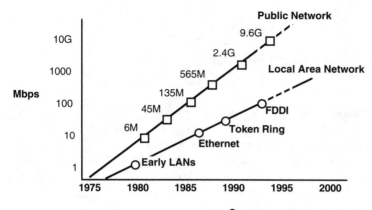

Source: IEEE Communications

Figure 2.3 Fiber-optics technology.

optics has decreased drastically. Today, fiber-optics carry about 2.4 Gbps of traffic over a single strand of fiber, and 9.6-Gbps fiber terminals are expected to be on the market by 1995. These high-speed fibers are typically used in public backbone networks.

Recently, fiber has begun to penetrate local area networks (LANs) with fiber distributed data interface (FDDI) technology. In the past, fiber cost much more when compared with coaxial cable limiting its penetration into LANs. Today, the cost of fiber is comparable to that of coaxial, thus enabling fiber to appear in the LAN environment.

2.2.1.3 Computing technology. Another factor driving broadband is the cost of computing equipment. The price of a home computer has dropped so much that it has become a commodity item available in every electronics retail outlet. Today, about 40 percent of U.S. homes have a personal computer. Figure 2.4 shows the comparative cost decrease in computing equipment. Cost is calculated based on cost of processing information.

2.2.2 Applications

So far, we have seen the technological progress that has enabled the evolution toward BISDN. In this section, we address the applications of BISDN in detail by explaining their characteristics. Most of the protocols addressed in this book use these applications as their drivers. By understanding the characteristics of the applications, we can better understand the use of the protocols. The applications discussed in this section are the following:

- LAN interconnect
- computer-aided design (CAD)/computer-aided engineering (CAE)/computer-aided manufacturing (CAM)

- visualization
- imaging
- supercomputing and channel extension
- multimedia

2.2.2.1 LAN interconnect. LAN interconnect evolved in the late 1980s to connect computers within a building or campus. Initially, a LAN's backbone capacity was 4 Mbps, then 10 Mbps, 16 Mbps, and recently 100 Mbps. Each one of these LANs is based on the IEEE 802.x protocol (x varies from 1 to 6). The number of LANs also has increased drastically from 30K in 1986 to 600K in 1993 and is projected to reach 1.5M by 1995. The reason behind this expansion is the exponentially increasing number of PCs connected to LANs. Figure 2.5 shows how two LANs can be interconnected by a network.

Table 2.1 describes some applications that run on LAN and their characteristics, in terms of data rate. These applications are not the only ones that run on a LAN; they are just the potential applications that can impact LAN traffic because of their data rate or because they are more frequently used.

2.2.2.2 CAD/CAE/CAM. Figure 2.6 shows a typical CAD environment. Typical traffic sources are engineering, computing, manufacturing, and design. It is essential to compare local traffic with remote to characterize CAD traffic. Today, CAD traffic is primarily local and involves file transfers between workstations and the mainframe, all located at the same site.

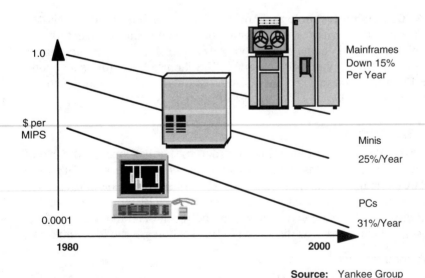

Source: Yankee Group

Figure 2.4 Computing technology cost.

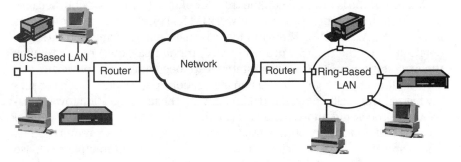

Figure 2.5 Typical LAN interconnect network.

Table 2-1. LAN Applications and Their Data Rates

LAN application	Typical data rate in bits per second
E-mail	9.6K to 56K
File transfer	56K to 1.5M
X-Windows system	56K to 1.5M
Remote database access	1.5M to 10M
Imaging	1.5M to 45M
Multimedia	3M to 45M

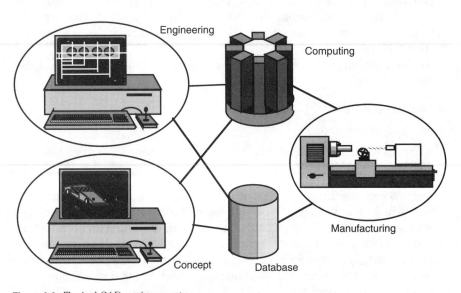

Figure 2.6 Typical CAD environment.

Remote traffic is primarily offline transfer of files (offline possibly because of the limitations imposed by the network). The key characteristic of today's interactive CAD/CAM/CAE environment is that peak transmission rates are determined by the transfer protocol (X.25, frame relay, etc.) rather than by the workstation or PC. CAD workstations are becoming more powerful, and the cost for supporting higher million instructions per second (MIPs) or simply higher processing power, is declining. With the fast migration toward distributed processing, more traffic is also imposed on the network. Speeds of 56 kbps to 1.5 Mbps are sufficient to meet today's standalone requirements. These speeds are not sufficient, however, to meet the interactive response time required for most CAD/CAM/CAE applications.

2.2.2.3 Visualization. Visualization is one of the upcoming applications where graphics workstations and vector processors are used to process large amounts of data and display them graphically, such that the human eye can process the important properties of information. Operating on the premise that "seeing is believing," visualization differs from CAD technology in the following two ways:

- No known accurate physical model exists in visualization.
- The graphics display is an abstraction of the object.

A good example of visualization is the modeling of molecular structures. Because molecules are invisible to humans, the models are simply a useful but abstract way of visualizing what we know about molecular structure. For a visualization application to be effective, it depends on the reality of the presentation. High resolution must be displayed on a large viewing area. The abstraction is then structured to allow trends or patterns in a data set to be perceived.

Let's look at an example where the basic requirements for such a display are the following:

- display using 1280×1024 pixels
- pixel of 24 bits of data
- pictures with alterations occurring 18 times per second or once every 55 milliseconds

With these specifications, the required data rate is approximately 23 Mbps, which is obviously a very high bandwidth application that requires high-speed access. As a result, the economics of designing an architecture for visualization are not very favorable. The visualization system also requires a supercomputer for processing, which makes the system even more expensive. Even with all these restrictions, about 100,000 visualization-capable systems are available. By 1995, about one million systems will exist.

Because the data rate requirement for visualization is 23 Mbps, there is also a need for a high-speed communication backbone to enable users residing at remote sites to access a supercomputer simultaneously shared by other users. Currently, visualization is a LAN-based solution or application, but, with the availability of cost-effective, high-speed links, the application will go into MAN, WAN, and GAN.

2.2.2.4 Imaging. Imaging is defined as the process that digitizes and stores or retrieves documents, drawings, photographs, and other information in bit-mapped format. Systems that digitize, compress, and store this information are called *imaging systems*. The imaging application typically falls under three categories:

1. *Item capture applications.* Items such as checks, drafts, credit card slips, etc., are characterized by the need to process a large number of items quickly. These items are stored and only retrieved when the need arises. The applications are classified as *capture and store-intensive*. American Express, for example, uses imaging to capture all credit card receipts.

2. *Records management applications.* Claim forms, loan documents, etc., are characterized by a low volume of items, small storage needs, but high frequency and the need for nationwide retrieval. These applications are classified as *retrieval and distribution intensive*. For example, in the insurance industry, a claim could be processed electronically by imaging the claims documents, thus enabling quick transfer for proper authorization. For banking, loan applications could be processed quickly if they were processed electronically.

3. *Medical and scientific applications.* X-rays, scans, seismic data, satellite photos, etc., are characterized by a unique, near one-to-one capture-to-access ratio with little need for storage. These applications are characterized as *resolution and distribution-intensive*. These items are stored, but they are usually stored in a tape backup system that can be retrieved in the future if the need arises. For example, a patient's X-rays could be imaged and electronically transferred from the X-ray lab to the doctor's office.

To better understand why imaging is a broadband application, let's look at the size of the imaging files.

Say you are scanning a page at 300 dots per inch (dpi), using approximately 700K bytes of memory. If compressed, the page is between 80K and 250K bytes (depending on page content and compression ratio), which is about 0.25M to 2M bits (each byte is 8 bits). Using a 16-level gray scale, the size of the file is four times larger. With true color, 24 times larger. In other words, a true-color image can involve a file of 40MB (compressed). This application is currently LAN-based, but once it appears in MANs and WANs, it will become one of the predominant applications from user perspective because it uses most of the available bandwidth.

2.2.2.5 Supercomputing and channel extension. Figure 2.7 shows a typical supercomputer/channel extension environment. The supercomputer is at a remote site, connected via a channel extender. The channel extender enables the supercomputer's input/output (I/O) bus to be extended, generally over a metropolitan area or campus, to support locations that have significant performance limitations or simply do not have a supercomputer.

Let's first address supercomputing in general. Many industries routinely use computers with enormous processing power. These high-powered computers are used for two types of applications:

1. *Data reduction.* In today's business, a vast array of data, from market facts to actual statistics, are increasingly encountered. Supercomputers manipulate and reduce such data to usable format. This application is categorized as *input-intensive.*

2. *Modeling and simulation.* Models of business, scientific, or engineering processes are exercised repeatedly with varying parameters. Supercomputers accomplish thousands of exercises in reasonable amounts of elapsed time. These applications are typically *output-intensive.* In business models, supercomputers simulate a business environment. In a scientific environment, data collected from the Hubble telescope, for example, undergoes numerous processing before being interpreted. In engineering, many designs must be attempted using different parameters. In designing a new car, for example, the supercomputer could experiment with different shapes and sizes to meet design requirements set by the user.

Having addressed supercomputing, let's now look into channel extension. Channel extension enables data transfer between mainframes/peripherals in one of the following ways:

- by being directly attached to the I/O channel
- by bypassing the front-end processor (FEP)
- by serializing the information for transport on a communication link

Figure 2.7 shows a typical environment of a supercomputing channel extender. The channel extender is connected to the supercomputer and peripherals, enabling the peripheral to access the mainframe and bypass the FEP.

In the 1970s and 1980s a small number of market segments used channel extenders at low speeds in a *host-to-host communications environment.* From 1989 and into the 1990s, however, the characteristics of this application have rapidly changed. Current characteristics are summarized as follows:

- specialized systems
- increasing need of higher bandwidth
- media dependent
- large market

In recent years, the broad acceptance of T1 digital technology has been the most significant factor in the success of channel-extender applications. Host-to-host and host-to-peripheral applications, which are T1-capable, have been extremely successful. In addition, the installation of fiber by service providers in the public network has reduced the cost of high-bandwidth pipes.

2.2.2.6 Multimedia. *Multimedia* is the combination of multiple forms of media in the communication of information between the user and machine. A multimedia application is an application that uses different forms of communication as one application. For example, information can be displayed on a computer using voice, text, and video simultaneously. A multimedia terminal allows text, graphics, and audio to be displayed simultaneously. Whether these applications are implemented as a standalone or shared system, they are usually bit-intensive, real-time, and very demanding on the networks. In this type of application, even in standalone systems, the

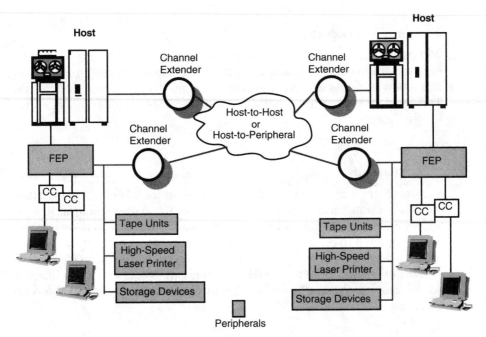

Figure 2.7 Example of supercomputing/channel extension environment.

source might be a location remote from the workstation, such as used when teaching a customer how to set up a new product. The customer can read relevant material, see a full-motion video demo of the product, and listen to a warning about possible hazards; all can be accessed from a remote location within the vendor's premises.

To support audio (voice) and video, a workstation must support vast amounts of data. In the case of video, 10 minutes of uncompressed full-motion video consumes 22GB of memory. Obviously, even a broadband network cannot handle this amount of data. One of the driving forces of multimedia is compression technology and multimedia standards. Some of the compression standards available today are joint photographic experts group (JPEG) and motion picture experts group (MPEG). Today, a compression ratio of 10:1 is available and is expected to reach 50:1.

Compressing converts a regular stream of data, called continuous bit rate (CBR), to variable bit rate (VBR) traffic. The problem with VBR is that an uneven bit stream arises because of how the compression is accomplished. The compression algorithm is usually performed on a frame-by-frame basis. The content of the first frame is transmitted first, and then the change from the preceding frame is transmitted next. This variation in frame size results in the uneven flow of data. In this book, we address the technology capable of handling this type of traffic.

2.2.3 Business/industry

With today's global economy, powerhouse companies averaging three percent growth in existing markets are looking for new markets to penetrate. Many new entrants are vying for partnerships that provide new services. Thus, new industries are forming. A recent example is MCI's alliance with British Telecom.

2.2.4 Standards

Standards are very important in the success of a technology. Many standards organizations around the world are working on the standardization of telecommunications networks and protocols so that any vendor can develop a product to be used in the worldwide network. A list of the standards organizations is given in Table 2.2. This list can only give you an idea on the number of organizations involved in the standards-making process. Chapter 22 describes some of the major international organizations and the relationships between them.

TABLE 2.2 Global Standards Organizations

AFNOR	Association Francaise de Normalisation
ANSI	American National Standards Institute
AOW	Asian-Oceanic Workshop
ARC	Administrative Radio Conference
BCS	British Computer Society
BSI	British Standards Institute
CCIR	International Radio Consultative Committee
CCITT	International Telegraph and Telephone Consultative Committee
CEN/Cenelec	Comite Europeane de Normalisation Electronique
CEPT	European Conference of Postal and Telecommunication Administrations
COS	Corporation for Open System International
COSINE	Cooperation for Open System Interconnection Networking in Europe
DIN	Deutsches Institut für Normung
DOD-ADA	U.S. Department of Defense—ADA Joint Program Office
ECMA	European Computer Manufacturers Association
ECSA	Exchange Carriers Standards Association
EDIFACT	Western European Electronic Data Interchange for Administration, Commerce, and Transportation
EMUG	MAP/TOP User Group
ETSI	European Telecommunications Standards Institute
EWOS	European Open Systems Workshop
GOST	USSR State Committee for Standards
IEC	International Electrotechnical Commission
IEEE	Institute of Electrical and Electronics Engineers
IAB/EITF	Internet Activities Board/Internet Engineering Task Force
ISA	Integrated Systems Architectures
ISO	International Organization for Standardization
ITU-T	International Telecommunications Union—Telecommunications
ITRC	Information Technology Requirements Council
JISC	Japan Industrial Standards Association
JSA	Japan Standards Association
JTC1	Joint Technical Committee 1—Information Technology

TABLE 2.2 Continued

NIST	National Institute for Standards and Technology
NNI	Nederlands Normalisatie—institut
OSF	Open Software Foundation
POSI	Pacific OSI Group
SAA	Standards Association of Australia
SCC	Standards Council of Canada
SIGMA	Unix Open Applications Group—Japan
SIS	Standardiseringskommissionen i Sverige
SMPTE	Society of Motion Picture and Television Engineers
SNV	Swiss Association for Standardization
SPAG	European Standards Promotion and Applications Group
T1	Standards Committee T1—Telecommunications
TTA	Telecommunications Technology Association of Korea
TTC	Telecommunications Technology Council
UI	Unix International
UAOS	Users Association for Open Systems
URSI	Union Radioscientifique Internationale
VESA	Video Equipment Standards Association
X/OPEN	Unix Open Applications Group

2.3 Building Blocks of BISDN

As mentioned earlier, BISDN is not a technology but a combination of multiple technologies. The three basic technologies are ATM, SONET and IN. ATM and SONET form the major topics of this book and are addressed in detail in chapters 8 and 9, respectively. A discussion of IN follows.

2.3.1 Intelligent network

IN is not a technology or a set of services but rather a concept for a new approach for the development and deployment of telecommunications and information services. An IN permits functionality and capabilities to be distributed flexibly at a variety of network entities and allows the service provisioning via a service-independent control architecture.

IN was developed with the following objectives in mind:

- rapid development and deployment of services
- ubiquitous service access
- multimedia services support, i.e., integrating voice, data, and video
- vendor independence
- cost-effective service delivery environment
- compatibility with existing networks

IN's long-term goal is the ability to introduce new services or change existing services quickly without having to adapt service switching point (SSP) software.

Having defined the objective of IN, let's now look at the IN architecture. Figure 2.8 shows a high-level view of IN architecture and how the different components interact with each other.

The following are the elements in the IN network:

- service management system (SMS)
- service control point (SCP)
- signal transfer point (STP)
- service switching point (SSP)
- intelligent peripheral (IP)

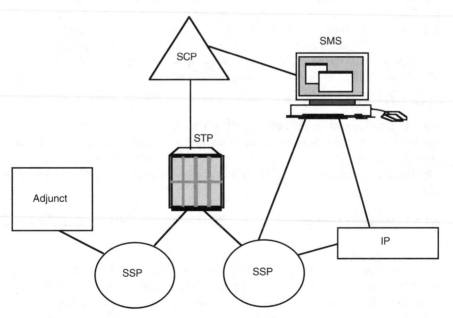

Figure 2.8 IN architecture.

The functions of these IN network elements are described in the following subsection.

2.3.1.1 Service management system. The SMS is usually owned by the network operators. It updates the SCPs with new data or programs and collects statistics from them. SMS also enables the service subscriber to control his or her own service parameters via a terminal linked to the SMS. For example, the subscriber can define the day and time when a number should be routed to a specific route. This modification is filtered or validated by the network operator. SMS is a commercial computer, usually a mainframe, that provides a development environment for new services.

2.3.1.2 Service control point. SCP is used when new IN services are introduced into the network and activated. If a service is based on functional components, the functional components are elected using a service logic interpreter. Some SCP services require large amounts of data that must reside on direct-access storage devices such as disks. The service programs and data are updated from the SMS. SCP is a commercial computer or modified switch. It should be able to access databases efficiently and reliably and provide a software platform for rapid service creation.

2.3.1.3 Signal transfer point. STP is part of the common channel signaling no. 7 network, or simply SS7. SS7 is a standardized communication interface through which the goal of the multivendor SCP and SSP can be achieved. STP switches SS7 messages to different SS7 nodes. The use of standalone or integrated STP depends on specific network configurations. STP is normally produced by traditional switch manufacturers such as Northern Telecom or AT&T.

2.3.1.4 Service switching point. SSP serves as an access point for service users and executes heavily used services. It is produced by traditional switch manufacturers. The higher-layer protocol of SS7, namely transaction capabilities application part (TCAP), is used to communicate between the SSPs and the SCPs.

2.3.1.5 Intelligent peripheral. IP provides enhanced services controlled by an SCP or SSP. It is more economical for several users to share an IP because the capabilities in IP are either not in SSP or too expensive to pull in all SSPs. Typical examples of IP functions are customized announcements, speech synthesizing, voice messaging, speech recognition, or database information that is made accessible to the end user. By making IP physically separate from SSP but connected via a standard interface enables any vendor to own and operate an IP, thus enabling the IP provider to provide value-added services to its users.

2.4   Progress in Residential
and Business Broadband Networks

The evolution toward a BISDN network will allow broadband services to be provided. The goal of broadband services is thus to provide an integrated network that offers service to everyone, from business customers to residential. Providing services both to business and residential customers over the same network poses numerous problems, however, because of the different needs of each customer. Figure 2.9 shows the typical business and residential services provided today.

Three major categories of services exist: data, voice, and video. Figure 2.9 shows that both residential and business customers use voice, but data services are used only by business, and video services only by residential users. This categorization does not mean that business users do not use video services or residential users do not use data services; it only represents the majority. For example, residential video services are provided via cable television (CATV) network. This service is the most dominant one used by the residential users, unlike data.

In the broadband world, all services are adapted to a single protocol, thus making the distinction impossible from a network point of view. For a service provider or an end user, however, the distinction is required to assign priority, quality-of-service, and allocation of bandwidth. To make these assignments, the service providers and end user must understand the characteristics and usage pattern of each service. For example, most business customers' traffic is voice-based and its peak usage is during working hours

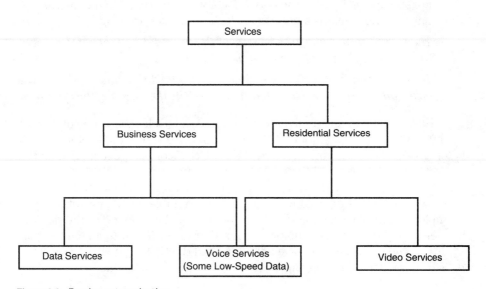

Figure 2.9 Service categorization.

(8 AM to 5 PM), whereas the residential customer's peak time is during evenings and weekends.

Although future BISDN promises to deliver all the above-mentioned services on a single unified network, in reality, it is a lot more difficult to achieve such objectives. The reason is that today's networks are designed to provide only cost-effective specific services. Therefore, it is cumbersome for a new network such as BISDN to provide an overall cost-effective service compared to the existing specialized networks. So, to achieve this goal of providing all services as a single network, an evolution process must occur. New capabilities that have the best chance for success must be implemented in the BISDN network, then allow other services to slowly migrate onto the BISDN network on a justification basis. The evolution toward BISDN can be defined in the following manner:

1. *Provide limited new services on an existing platform (network).* For example, CATV with its existing network cannot provide switched video. So, as a first step, CATV providers are providing pay-per-view, where the user dials the number, and the CATV provider unscrambles the channel for that specific decoder assigned to that particular home.

2. *Replace or enhance the platform, maintain backward compatibility with existing services and revenue streams while expanding the new service growth potential.* For example, analog fibers can be first used in the network and then CATV providers can migrate to digital fiber by replacing the terminal equipment. One of the reasons that CATV providers do not migrate to digital systems currently is because TV sets are still analog. Even if the network converts the signal to digital, the signal must be converted back to analog for ultimate viewing.

3. *Increase the volume of services with minimal operational impact once critical network capacity is achieved.*

4. *Interwork with other networking services and systems to fill service gaps temporarily or permanently as best fits the needs of the business.*

Having defined a generic evolution strategy, let's look at the evolution of different networks mentioned above. Figure 2.10 shows the evolution path of the different networks toward the BISDN vision, i.e., using ATM as the switching platform and SONET/SDH as the transmission platform. Figure 2.10 illustrates how the different existing networks can lead to the target network of BISDN by taking different paths suitable for the evolution from their current base. The figure does not specify the timeline of the deployment of the different technologies but rather the different technologies that enable the evolution. We have not addressed the CATV network here, but it is discussed in detail in chapter 18.

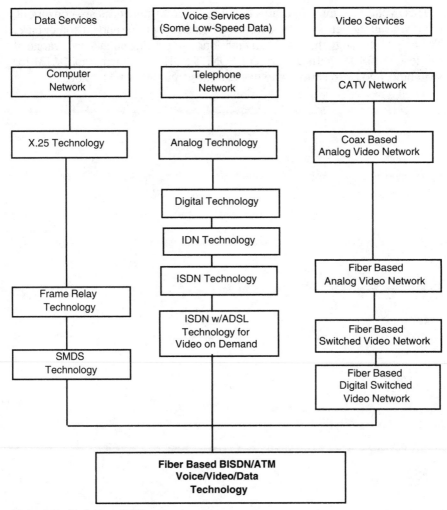

Figure 2.10 Evolution of different networks.

2.5 Summary

This chapter addresses the various factors leading to the evolution toward BISDN. The four basic categories that are driving the evolution for BISDN are technology, application, standards around the world, and businesses/industries. The three main building blocks of BISDN are ATM, SONET, and IN. This chapter discussed the generic architecture of one of the BISDN building blocks, IN. Finally, we looked at the evolution path taken by voice, data and video networks to achieve the target network of BISDN.

This evolution simply indicates how the different networks are converging to provide all the services, leveraging their existing service and customer base. The only technology that has been envisioned to provide such a wide range of services is BISDN, with the help of SONET/SDH for transmission, ATM for switching, and IN to control and manage these and other technologies.

The Basics

In this part we provide an overview of different broadband technologies. The different technologies covered are FDDI, DQDB, FR, SMDS, ATM, and SONET. Every technology is a switching technology except SONET, which is a transmission technology.

Chapter

3

FDDI & DQDB

3.1 Overview

This chapter explains the basic working standards and protocols of fiber distributed data interface (FDDI) and distributed queue dual bus (DQDB) technologies.

FDDI is included because, as per our definition, any service requiring more than 1.5 Mbps is considered broadband. Because FDDI is a LAN technology operating at 100 Mbps, it therefore falls under our definition. In chapter 15, we also see how asynchronous transfer mode (ATM) technology competes against FDDI as the technology of choice in the LAN environment. The evolution strategies for migrating from FDDI to an ATM-based network architecture are also covered in that chapter.

Both FDDI and DQDB are quite similar. Both use fiber as the transmission medium, and their targeted application is LAN interconnect, interworking with IEEE 802, capable of carrying both isochronous (voice) and data traffic. Only FDDI-II, the second-generation FDDI, carries both voice and data. Differences do exist: DQDB is intended for the public network, whereas FDDI is intended for private LANs. DQDB came into existence in the mid-1980s when the IEEE 802 committee recognized the need for an access technology to be used in the public metropolitan area network (MAN). In the 802.6 subcommittee, the different shared medium topologies and access mechanism were discussed, resulting in the definition of DQDB. DQDB is based on a proposal that was initially called queued packed synchronous exchange (QPSX) received from the University of Western Australia in 1988. IEEE 802.6 adopted the concept of QPSX and renamed it DQDB. The IEEE

802.6 subcommittee was working on these standards at the same time that ITU-T was working on the ATM standards. Thus, it was possible for each standards body to influence the standards of the other, allowing DQDB standards to be included in the CCITT standards wherever possible.

3.2 Fiber Distributed Data Interface (FDDI)

In this section, concepts, standards, and architecture of FDDI are discussed.

3.2.1 FDDI concepts

Figure 3.1 illustrates the basic FDDI topology. FDDI is a token-passing LAN technology that uses a timed-token protocol. This protocol guarantees that stations can gain access to the ring within a time period negotiated between stations each time a new station joins the ring.

An FDDI is constructed of nodes that connect together via links to form the ring. The nodes can be of two types: dual-attachment stations (DAS) or single-attachment stations (SAS). A special type of node, called a *concentrator*, allows connection of stations into the ring and provides for greater fault tol-

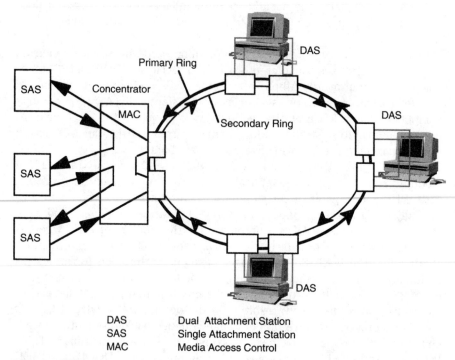

DAS	Dual Attachment Station
SAS	Single Attachment Station
MAC	Media Access Control

Figure 3.1 FDDI ring configuration.

erance than the basic ring architecture. One of the important features of FDDI is the use of dual-counter rotating rings. The dual rings are independent until a fault (such as a cable break) occurs, in which case the rings are joined together, or wrapped, to restore the ring to its operational state.

FDDI, as the name implies, is based on the use of fiber-optic technology, which allows construction of rings up to a total fiber length of 200 km. Using multimode fiber, interstation distances of 2 km are permitted, while single-mode fiber implementations enable interstation distances that exceed 20 km. Recent enhancements in the FDDI standard have allowed the use of cheaper fiber-optics for runs less than 500 m, and twisted-pair cable for runs of up to 100 m.

In addition to the dual counter rotation rings, FDDI offers other features such as the following:

- use of a token-passing, media access control (MAC) scheme based on the IEEE 802.5 token ring standard
- compatibility with IEEE 802 LANs by using 802.2 logical link control (LLC)
- ability to utilize multimode or singlemode optical fibers
- operation at a data rate of 100 Mbps
- Ability to attach any number of stations (standards assume no more than 1,000 physical attachments)
- a total fiber path of 200 km
- the ability to dynamically allocate bandwidth so that both synchronous and asynchronous data services can be provided simultaneously

3.2.2 FDDI standards

Work on FDDI standards started in the early 1980s and still continues to enhance FDDI technology to meet ever-changing customer requirements. In this section, we address some of the basic standards of FDDI; for detailed technical information, refer to the individual ANSI or ISO document listed in Table 3.1.

TABLE 3.1 ANSI/ISO FDDI Standards

Standard title	ANSI standard no.	ISO standard no.
Hybrid Ring Control (HRC)	X3.186	9314-5
Physical layer Medium Dependent (PMD)	X3.166	9314-3
Single-Mode Fiber Physical layer Medium Dependent (SMF-PMD)	X3.184	9314-4
SONET Physical layer Mapping (SPM)	T1.105	

TABLE 3.1 Continued

Standard title	ANSI standard no.	ISO standard no.
Station Management (SMT)	—	—
Token Ring Physical Layer Protocol (PHY)	X3.148	9314-1
PHY-2	—	—
Token Ring medium access control (MAC)	X3.139	9314-2
MAC-2	—	—

The FDDI standards address the physical and data link layer requirements of the OSI reference model, as illustrated in Figure 3.2. The FDDI (PMD, PHY, MAC and SMT) standards, along with the IEEE 802.2 LLC standards, provide essential networking services to devices attached to an FDDI network.

The physical layer medium dependent (PMD) is the lowest sublayer of the OSI physical layer. It includes specifications for power levels, characteristics of the optical transmitter and receiver, permissible bit-error rates, jittery requirements, acceptable media, etc. The physical layer protocol

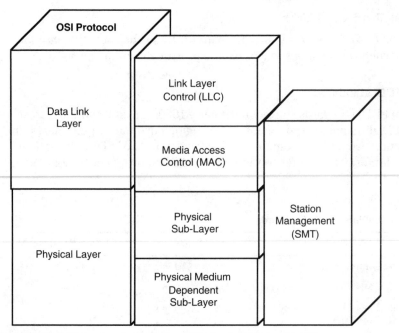

Figure 3.2 FDDI protocol standards and their relationship with OSI.

(PHY) is the upper sublayer of the OSI physical layer. It deals with such issues as the encoding scheme, clock synchronization, and data framing. The lower sublayer of the OSI data link layer is the MAC. The MAC standard defines rules for medium access, addressing, frame formats, error checking, and token management. The station management standard specifies system management applications for each of the FDDI protocol layers. In particular, it concerns itself with the control required for correct operation of a station on an FDDI network. The details of these components and their functions are covered in the next section.

The FDDI standards comprise many separate documents. Before we go further, we need to understand the background of FDDI standards development. The initial idea of developing a new high-speed data interface for computers based on optical fiber dates back to October 1982. The ANSI X3T12 subcommittee was already addressing a similar high-speed data communications network using coaxial cable called the *local distributed data interface*. The subcommittee formed an ad hoc task group to examine the fiber-based concept in detail. It proposed that the working group formally start working on FDDI physical, data link, and network layer standards.

Initial proposals for the MAC and physical layers were put forth in 1983. The MAC layer corresponds to the lower half of the OSI data link layer. The task group decided to develop the FDDI MAC to operate below the IEEE 802.2 LLC standard, also under development at that time. The LLC corresponds to the upper half of the OSI data link layer, and all IEEE 802 MAC schemes are designed to operate under it. This choice by X3T12 developed the need for data link and network layer protocols to be specific to FDDI. The FDDI MAC protocol was formally adopted as an ANSI standard in 1987.

In 1984, the subcommittee recognized that fiber technology was changing rapidly, making the physical layer standard nearly impossible to complete in a timely fashion. The committee then decided to divide the physical layer into two protocol sublayers. The PHY corresponds to the upper half of the OSI physical layer and describes those physical layer issues independent of the network medium. Development of the PHY standard could continue in parallel with the MAC standard, and the PHY was later adopted by ANSI in 1988.

The second layer, the physical layer medium dependent (PMD) sublayer, corresponds to the lower half of the OSI physical layer and deals with media-specific issues. The original PMD standard is written for multimode fiber (MMF) and was formally adopted in 1990. In 1987, meanwhile, the task group realized that the distance limitations of MMF were too confining, and it started to work on a single-mode fiber PMD (SMF-PMD).

Some FDDI users do not want to incur the expense of purchasing, installing, and managing their own optical fiber transmission facilities. Instead, they would prefer to use the high-speed transmission services available from public network providers. In 1989, recognizing that FDDI had applications using public network facilities, the task group created an interface between FDDI's PHY protocol and the emerging SONET standards. The standard defined an alternative to the PMD standards and allowed mapping of FDDI transmissions directly onto SONET-based networks. It was completed in 1992.

In 1984, the subcommittee also recognized the need for a separate standard describing station management issues. This standard had to conform to work already in progress by both the IEEE 802.1 committee and ISO on station and network management. With the development of integrated voice/data networks, and particularly integrated services digital network (ISDN), the subcommittee anticipated the requirement for a new type of network to carry this mixture of traffic. It started to develop plans for a second-generation FDDI capable of carrying voice, image, and video in addition to high-speed data. This second generation FDDI is commonly called FDDI-II. We do not address FDDI-II because of ongoing standards activity in this area.

3.2.3 FDDI protocol architecture

The previous section introduced the FDDI protocol model. Here, we look into the different components of FDDI and their functions. As mentioned in chapter 2, the lowest level of the OSI model is the physical layer. This level defines the transmission of bits on the physical medium. FDDI standards subdivide this OSI physical layer into two sublayers: PMD and PHY. These two sublayers separate the physical medium and transmission details into two distinct parts.

3.2.3.1 Physical layer medium dependent. FDDI has standardized two PMDs. As highlighted in Figure 3.3, PMD standards define how nodes (stations) physically attach to the FDDI ring and how stations are physically interconnected on the network by media type (optical fiber or copper).

The four types of PMD are the following:

- *PMD.* Uses light-emitting diodes (LEDs) and multimode fiber. It was the first PMD developed by ANSI.

- *SMF-PMD.* Uses laser diodes and single-mode fiber. It is used to interconnect stations separated by distances that exceed the 2-km limit imposed by the PMD.

- *LCF-PMD.* Currently in development by ANSI, the low-cost fiber PMD also uses LEDs and multimode fiber. It spans 500 meters between stations and is envisioned to be a low-cost alternative to PMD.

- *TP-PMD.* Also currently in development, the twisted pair PMD will operate over copper media—shielded twisted pair (STP) and some categories of unshielded twisted pair (UTP). It is anticipated that the transmission distance between stations will be limited to 100 meters.

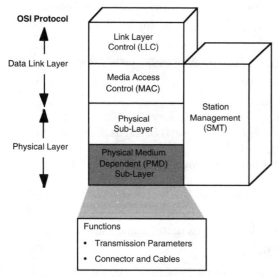

Figure 3.3 Physical medium functions.

Data is transmitted between stations by first converting the data bits into a series of signals, then transmitting these signals over the cable linking the two stations. The PMD standards deal with all areas associated with physically transmitting the data:

- optical and electrical transmitters and receivers
- fiber-optic or copper cable
- media interface connector
- optical bypass relay (optional in optical PMDs)

3.2.3.2 Physical layer protocol. The physical layer protocol (PHY) standard, shown in Figure 3.4, defines those portions of the physical layer that are media independent. Thus, new media such as twisted pair can be added without changing the PHY parameters.

The physical layer protocol provides the following functions:

- *Clock and data recovery.* recovers the clock signal from the incoming data.
- *Encode/decode process.* converts data from the MAC into a form for transmission over the FDDI ring.
- *Symbols.* Smallest signaling entities used for communication between stations. Symbols consist of 5 code bits.

- *Elasticity buffer.* accounts for clock tolerances between stations.
- *Smoothing function.* prevents frames from being lost because of shortened preambles.
- *Repeat filter.* prevents the propagation of code violations and invalid line states.

3.2.3.3 Media access control. The second level of the OSI reference model is the data link layer. As shown in Figure 3.5, the data link layer is divided into two sublayers, MAC and LLC.

The FDDI MAC standard defines the following functions:

- fair and equal access to the ring through the use of the timed-token protocol
- communication between attached devices using frames and tokens
- construction of frames and tokens
- transmitting, receiving, repeating, and stripping frames and tokens from the ring
- various error-detection mechanisms
- ring initialization
- ring fault isolation

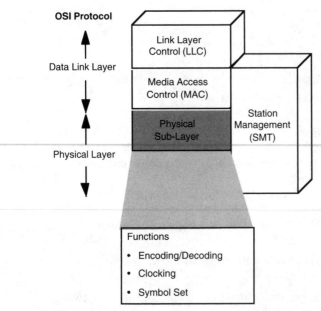

Figure 3.4 Physical sublayer functions.

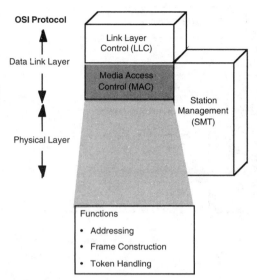

Figure 3.5 MAC functions.

3.2.3.4 Logical link control.
Although LLC is not part of the FDDI standard, FDDI requires LLC for proper ring operation and transmission of user data. Logical link control defines link-level services that allow the transmission of a frame of data between two stations. Figure 3.6 shows the relationship between LLC and FDDI. FDDI assumes implementation of the IEEE 802.2 LLC standard. Here, frames are used to transfer information between MAC layers in FDDI. FDDI defines three different types of frames:

- MAC frames that carry MAC control data
- SMT frames that carry FDDI-specific management information between stations
- LLC frames that carry LLC information

3.2.3.5 Station management.
The SMT standard provides the necessary services at the station level to monitor and control an FDDI station. SMT allows stations to work cooperatively within the ring and ensures proper station operation. FDDI stations can have multiple instances of PMD, PHY, and MAC entities, but only one SMT entity.

Station management contains three major components:

- connection management (CMT)
- ring management (RMT)
- SMT frame services

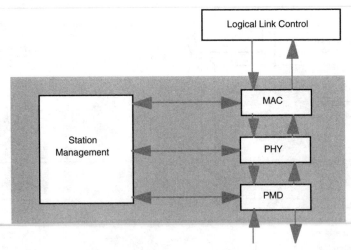

Figure 3.6 Relationship between LLC and FDDI.

Figure 3.7 shows how these three SMT components are incorporated into the FDDI architecture.

3.2.3.6 Connection management. CMT is the portion of station management that performs physical-layer insertion and removal of FDDI stations. Remember that FDDI stations can have multiple PHYs and MACs. Therefore, one of the functions of CMT is to manage the switch configuration (Figure 3.8) that connects PHYs to MACs and other PHYs within a station. Connection management functions include:

- connecting a PHY to its neighboring PHY
- connecting PHYs and MACs via the configuration switch
- using trace diagnostics to identify and isolate a faulty component

    As shown in Figure 3.8, CMT contains the following components:

- Physical connection management (PCM) provides for managing the physical connection between adjacent PHYs, including establishing the connection, testing the quality of the link before the connection is established (link confidence testing), and continuously monitoring error rates once the ring is operational (link error monitoring).
- Configuration management (CFM) provides for configuring PHY and MAC entities within a station.
- Coordination management (ECM) provides for controlling bypass relays, signaling to PCM that the medium is available, and coordinating trace functions.

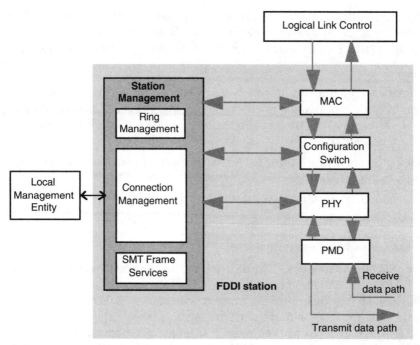

Figure 3.7   Relationship of SMT components in FDDI.

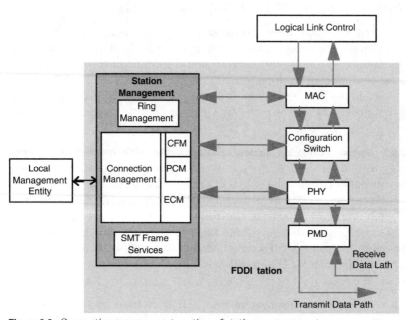

Figure 3.8   Connection management portion of station management.

3.2.3.7 Ring management. The ring management (RMT) portion of SMT receives status information from MAC and CMT. Ring management then reports this status to SMT and higher-level processes. Services provided by RMT are the following:

- Stuck-beacon detection, in which a specialized frame called a beacon is used by MAC to announce to other stations that the ring is broken. A beacon indicates that a station cannot resolve a ring error condition detected in its receive path.

- Tracing process, which provides a recovery mechanism for beacon conditions on the FDDI ring.

- Determination of a MAC's availability for transmission.

- Detection of duplicate addresses, which prevent the proper operation of the ring even if the ring becomes operational.

3.2.3.8 SMT frame services. The SMT frame provides services on portions of SMT that control and observe stations on the FDDI network. These services are implemented by different SMT frame classes and types. Frame class identifies the function that the frame performs, such as neighborhood information frame (NIF); status information frame (SIF); parameter management frame (PMF); and status report frame (SRF). Frame type designates whether the frame is an announcement, request, or response to a request. These different classes of frames are described in the following subsections.

3.2.3.8.1 Neighborhood information frames. Stations determine their upstream and downstream neighbors by exchanging NIFs as part of the neighbor notification protocol. Stations also use the protocol to determine the existence of duplicate address conditions. Once upstream neighbor addresses are known to the stations, these addresses can be used to create a logical ring map showing the order in which each station appears in the token path.

3.2.3.8.2 Status information frames. Stations use SIFs to exchange more detailed information about their characteristics and configuration. SIFs also contain information about the status of each port in a station. The information in SIFs can be used to create a physical ring map that shows the position of each station, not only in the token path (logical ring), but in the topology as well. SIFs are divided into two types—SIF configuration frames and SIF operation frames. SIF configuration frames show the configuration details of a station, while SIF operation frames show operational detail such as error rates.

3.2.3.8.3 Parameter management frames. PMFs are used by the parameter management protocol to manage an FDDI station. Management is achieved by operations on the station's management information base (MIB) attributes. Operations are performed by exchanging frames between the manager and the station. If an attribute is initiated by a PMF Get Request frame from the management station, it is followed by a PMF Get Response frame from the target station. Changing an MIB attribute requires a Get Exchange (to check the current value), followed by a PMF set request/response exchange.

3.2.3.8.4 Status report frames. SRFs are used by the status report protocol to announce station status to management stations. SRF frames report conditions and events. Conditions are declared when a station enters certain states, such as Duplicate Address Detected. Events are instantaneous occurrences, such as the generation of a trace.

3.2.4 How FDDI works

In FDDI, the standards define the functions that control ring operation and maintenance, whereas the technology attempts to provide survivability in case of failures. After receiving the token, an active station transmits a frame as a stream of symbols to the next active station on the ring. As each active station receives these symbols, it regenerates and repeats them to the next active device on the ring (its downstream neighbor). When the frame returns to the originating station, it is stripped by that station from the ring.

3.2.4.1 Dual counter-rotating ring. The dual counter-rotating ring is one of the basic concepts in the FDDI standards. It consists of two rings: a primary ring and a secondary ring. Both rings can carry data. As shown in Figure 3.9, the data flows in opposite directions on the two rings. In most cases, particularly in high-bandwidth applications, it is best to use the primary ring for data transmission and the secondary ring as backup. This guideline is especially important when the FDDI ring undergoes its self-healing process during a fault condition. The configuration complexity and therefore cost of FDDI increases if both rings are used for data transmission.

FDDI limits total fiber length to 200 km. Because the dual-ring topology effectively doubles media length in the event of a fault condition, the actual length of each ring is limited to 100 km.

The dual counter-rotating ring is designed with the ability to restore ring operation if a device fails or a cable fault occurs. The ring is restored by wrapping the primary ring to the secondary ring, thus restoring the transmission path. This redundancy in the ring design provides a degree of fault tolerance not found in other network standards. In an FDDI ring, if a cable failure occurs, the stations on either side of the failure reconfigure by themselves. They wrap the primary ring to the secondary ring, effectively isolating the fault, restoring continuity to the ring, and thus allowing normal operation to continue. Figure 3.10 shows the fault-isolation technique.

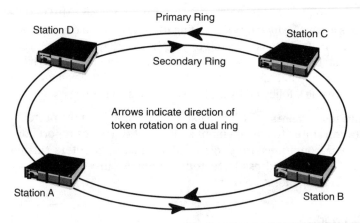

Figure 3.9 Dual-ring topology.

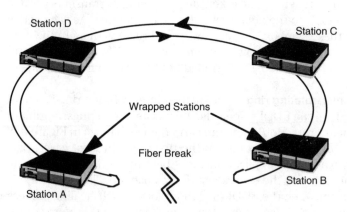

Figure 3.10 Dual-ring with fiber cut.

When a wrap occurs, the dual-ring topology changes to a single-ring topology. When the fault is corrected, the topology reverts back to dual-ring. If multiple faults occur, the ring segments into many independent rings. Therefore, when more than one fault occurs in the dual ring, communication between all stations is not possible, but communication is possible within a portion of the ring.

We mentioned earlier that if an attached station fails, the devices on either side of the failed station reconfigure to isolate the station from the ring. Figure 3.11 shows this fault-isolation technique. In the figure, station A is isolated from the ring. The ring remains operational by wrapping the primary and secondary rings at stations B and D. They remain wrapped until the fault is corrected, which is detected by transmission of a beacon frame, as shown in Figure 3.12.

3.2.4.2 Ring operation. FDDI ring operation includes the following stages:

1. connection establishment
2. ring initialization
3. steady-state operation
4. ring maintenance

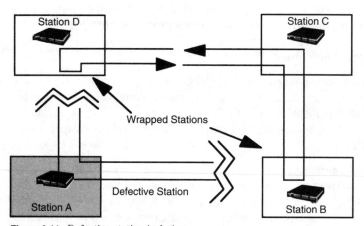

Figure 3.11 Defective station isolation.

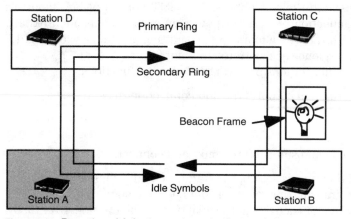

Figure 3.12 Detection of defective station using beacon frame.

Each of these stages is monitored by timers that regulate the operations. Three timers are used by each FDDI station to perform the function of regulating the ring operations. They are:

- the token-rotation timer (TRT)
- the token-holding timer (THT)
- the valid-transmission timer (TVX)

3.2.4.2.1 Token-rotation timer. The TRT times the duration of operations in a station. This timer is critical to the successful operation of the FDDI network. This timer controls ring scheduling during normal operation and fault recovery times when the ring is not operational. The token rotation timer is initialized to different values, depending on the state of the ring. During steady-state operation, the token rotation timer expires when the target token rotation time (TTRT) has been exceeded. Stations negotiate the value for TTRT via the claim process.

3.2.4.2.2 Token-holding timer. THT controls the length of time that a station can initiate asynchronous frames. A station holding the token can begin asynchronous transmission if THT has not expired. THT is initialized with the value corresponding to the difference between the arrival of the token and the TTRT.

3.2.4.2.3 Valid-transmission timer. TVX times the period between valid transmissions on the ring. TVX detects excessive ring noise, token loss, and other faults. When the station receives a valid frame or token, the valid transmission timer resets. If TVX expires, the station starts a ring initialization sequence to restore the ring to proper operation. Having seen timers used in the operation of FDDI, we can now address the different operations.

3.2.4.2.4 Connection establishment. To form the ring, stations must establish connections with their neighbors. The CMT portion of SMT controls this physical connection process. At power up or on a connection restart, stations recognize their neighbors by transmitting and acknowledging defined line state sequence as illustrated in Figure 3.13.

To establish a link, stations perform the following functions:

- exchange information on port type and connection rules
- negotiate the length of the link, which checks the quality of the links between stations
- exchange results and status on the links and connections

If the connection type is set up, the stations could complete the physical connection by transmitting another defined line-state sequence. This process is repeated for each link in the dual ring by each station. Eventually, all stations join the ring.

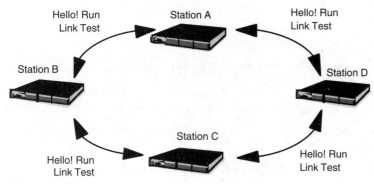

Figure 3.13 Ring power-up.

3.2.4.2.5 Ring initialization. Once the connection is established, FDDI protocol requires the stations to bid for the right to initialize the ring (i.e., generate a token) by negotiating a TTRT, ensuring that stations receive guaranteed service time. The TTRT can be set based on the:

- number of attached stations
- length of ring
- time required by each station to transmit data over the ring
- balance of low latency and adequate bandwidth

During ring initialization, stations negotiate the TTRT with other stations to determine which station can issue the token. This process of negotiation is called the *claim process*. The claim determines which station should initiate the ring. The claim process begins when the MAC entity in one or more stations enters the claim state. In this state, the MAC in each station continually transmits claim frames. A claim frame contains the station's address and bid for the TTRT. Stations in the ring compare the incoming claim frames with their own bid for target token rotation time. If the frame has a shorter time bid, the station repeats the claim frame and stops sending its own. If the frame has a longer time bid, the station removes the claim frame and continues sending frames with its own bid for TTRT.

When a station receives its own claim frame, that station wins the right to initialize the ring. If two or more stations make identical bids, the station with the longest and highest address wins the bidding.

The winning station initializes the ring by issuing a token. This token passes around the ring without being captured by any station. Instead, as each station receives the token, it sets its own TTRT to match the TTRT of the winning station. On the second token rotation, stations can send synchronous traffic. On the token's third rotation, stations can transmit asynchronous data.

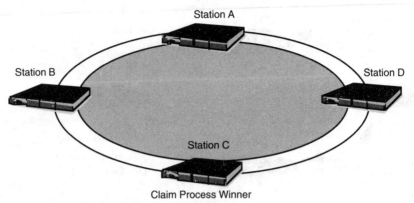

Figure 3.14 Station negotiating process.

Figure 3.14 illustrates an example of the working of a claim process. The stations negotiate (or bid) for the right to initialize the ring. In this example, stations A and C are issuing shorter bids than stations B and D. Station C is issuing a shorter bid than station A. The process of negotiation is as follows:

1. All active stations start issuing claim frames.
2. Station D receives a shorter claim from station C, stops sending its own claim, and repeats station C's shorter claim to station A. Meanwhile,
   - Station B receives a shorter claim from station A, stops sending its own claim, and repeats station A's shorter claim to station C.
   - Station C receives station A's claim frame but continues sending its own shorter claim frame.
3. Station A receives station C's shorter claim from station D, stops sending its own claim, and repeats station C's shorter claim to station B.
4. Station B receives station C's shorter claim from station A and repeats station C's shorter claim to station C.
5. Station C receives its own claim from station B. Station C sets the TTRT and issues a token to initialize the ring.

Thus, the negotiating process is achieved by supporting fair access to the ring.

3.2.4.2.6 Steady-state operation.   Once the ring is initialized, FDDI goes into steady-state operation. When in the steady state, stations exchange frames using the timed-token protocol. The ring remains in the steady state until a new claim process is initiated, such as when a new station joins the ring.

3.2.4.2.7 Asynchronous and synchronous services. As mentioned, two types of services are provided by FDDI: asynchronous and synchronous.

Asynchronous services are designed for bandwidth-insensitive applications such as datagram traffic. Asynchronous frames are designed to be transmitted even when the station does not require the bandwidth, whereas synchronous frames are sent at any time, as long as the negotiated bandwidth is available. This service is useful for frames that must have guaranteed delivery within a time period of 2 × TTRT. Such frames include compressed audio and video, among others.

3.2.4.3 Timed-token protocol. The timed-token protocol includes a number of steps by which a station acquires the right to begin transmission. Figure 3.15 shows an overview of frame transmission using the FDDI timed-

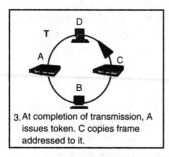

1. A has frame of data to transmit and captures token

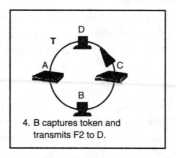

2. A begins transmitting frame 1 (F1) destined for C.

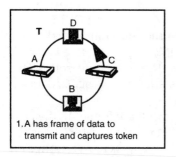

3. At completion of transmission, A issues token. C copies frame addressed to it.

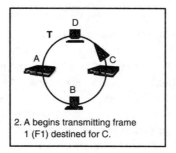

4. B captures token and transmits F2 to D.

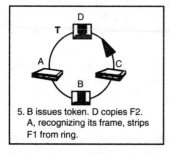

5. B issues token. D copies F2. A, recognizing its frame, strips F1 from ring.

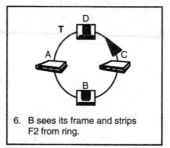

6. B sees its frame and strips F2 from ring.

Figure 3.15 Frame transmission process.

token protocol. The following are steps involved for an FDDI station to transmit a frame.

When an FDDI station wants to transmit a frame it:

1. waits until it detects the token

2. captures the token

3. stops the token repeat process (because no token is on the ring this action prevents other stations from transmitting data onto the ring)

4. begins sending frames (frames can be sent until there is no more data to send, or the token holding rules require surrender of the token)

5. releases the token onto the ring for use by another station

Once all active stations except the sending station on an FDDI network receive and repeat each frame, each station on the ring compares the destination address of each frame with its own and checks for frame errors. If addresses match, the receiving station copies the frame and sets status symbols (control indicators) to show that the station has recognized its address and copied the frame.

Repeating stations check for errors and retransmit the frames to the next station. If the station detects an error, it sets an error indicator. After the frame circles the ring, the station that sent the frame removes (strips) it from the ring. This stripping can cause partial frames, called *fragments*. These fragments are left on the ring. Each FDDI station must ensure that frame fragments do not degrade ring operation. The fragments are removed either by transmitting or operating a repeat filter at each station's PHY.

3.2.5 Scrubbing

Every time a device is attached to or removed from the ring, the ring reconfigures itself. During this process, stray frames can be introduced to the new topology. Sometimes, these stray frames have been generated by a device that is no longer part of the new topology. The frames can no longer be identified as belonging to any particular attached device. To remove these stray frames, a station sends a series of idle symbols to the ring. At the same time, MAC strips the ring of frames and tokens. Once this process is completed, the active stations enter the claim process. The time taken to scrub the ring guarantees that all frames on the ring have been created after the reconfiguration occurred, preventing old frames from continually circulating on the ring.

3.2.6 Ring maintenance

The responsibility for monitoring the ring is distributed among all stations on the ring. Each station monitors the ring for conditions that require ring initialization, such as the following:

- ring inactivity longer than the valid transmission time
- physical or logical break in the ring

3.2.7 FDDI network topologies and configurations

Like most other networks, FDDI networks can be described in terms of their physical and logical topologies. The physical topology refers to the physical arrangement of the stations and their physical interconnections. The logical topology refers to the path between MAC entities over which the information flows.

FDDI networks can be organized in many ways. A network configuration with only dual-attachment stations results in a logical and physical ring. FDDI allows for more versatility than this one configuration. For instance, when concentrators are added to the network, a branching tree topology develops.

FDDI can be implemented in three ways:

- as a high-speed backbone connecting mid-speed LANs, such as those found in IEEE 802.3 and IEEE 802.5 applications
- as a high-speed workgroup LAN that connects workstations or servers
- as a high-speed connection between host computers, or host computers-to-peripheral equipment, such as those found in a data center

The FDDI standards permit a number of topologies. The following four topologies are of particular importance:

- standalone concentrator with attached stations
- dual-ring
- tree of concentrators
- dual-ring of trees

3.2.7.1 Standalone concentrator topology. The standalone concentrator topology consists of a single concentrator and its attached stations (Figure 3.16). These stations can be either single-attachment station (SAS) or dual-attachment station (DAS) devices. For example, the concentrator can connect multiple high-performance devices in a workgroup. Figure 3.16 shows an independent workgroup topology that uses existing structured wiring, thus affording significant cost savings in prewired sites. The logical ring is formed by the stations and the concentrator, with the token path illustrated by the direction of the arrows.

3.2.7.2 Dual-ring topology. The dual-ring topology consists of dual-attachment stations connected directly to the dual ring. This topology is useful

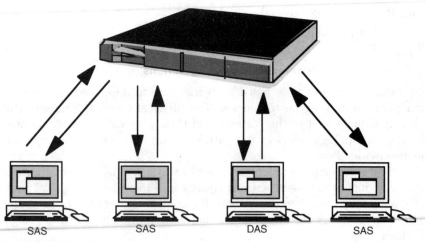

Figure 3.16 Standalone topology.

when there are a limited number of stations, as illustrated in Figure 3.17. A dual ring of DAS devices, however, does not easily lend itself to additions, moves, or changes. Since each station is a part of the backbone wiring, the behavior of each user is critical to the operation of the ring. The simple act of a user disconnecting a dual attachment workstation causes a break in the ring.

In the event of a single failure, a dual ring self-heals by wrapping the primary and secondary rings. Multiple failures result in two or more segmented rings. Each ring is fully functional, but no access exists to the other rings. For this reason, dual rings should only be implemented when little risk exists of users disturbing the network connection.

3.2.7.3 Tree-of-concentrators topology. The tree of concentrators is the preferred choice when wiring together large groups of user devices.

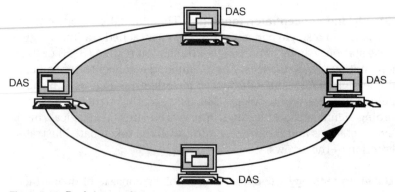

Figure 3.17 Dual ring topology.

Concentrators are wired in a hierarchical star topology with one concentrator serving as the root of the tree. In this configuration, one FDDI concentrator is designated as the root (Figure 3.18). Cables run from this concentrator to SASs, DASs, or other concentrators. This topology provides greater flexibility for adding and removing FDDI concentrators and stations, or changing their location without disrupting the FDDI LAN.

Additional concentrators can connect to the second tier of concentrators, as needed, to support new users. The tree configuration can connect all stations in a single building or a large number of stations on one floor of a building.

The tree topology is well-suited to structured cabling systems. Tree topologies also allow network managers to better control access of end-user systems to the network. Inoperative systems can be easily removed from the network by the concentrator. The network manager can also remotely access the concentrator to bypass the station.

3.2.7.4   Dual-ring-of-trees topology.   The fourth topology described in the FDDI standards is the dual ring of trees. In this topology, concentrators cascade from other concentrators connected to a dual ring, as illustrated in Figure 3.19, placing the dual ring where it is needed most—in the backbone.

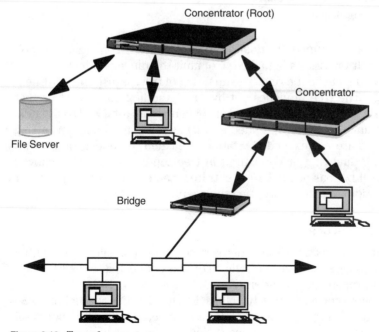

Figure 3.18   Tree of concentrators.

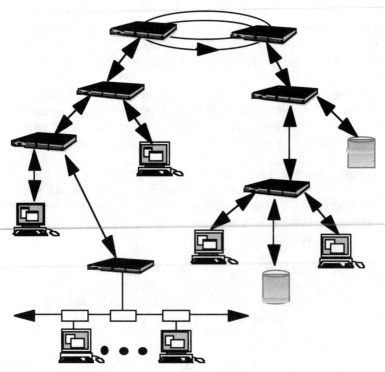

Figure 3.19 Dual ring of trees.

Of all topologies defined, the dual ring of trees is the recommended topology for FDDI. It provides a high degree of fault tolerance and increases the availability of the backbone ring. The dual ring of trees is also the most flexible topology. The tree branches out by simply adding concentrators that connect to the ring through upper-level concentrators attached to the dual ring. Tree branches can be extended as long as the station number or ring distance limits are not exceeded. Stations attached to concentrators connected to the dual ring or configured in tree topologies can be removed from the FDDI LAN as needed. Concentrators can easily bypass inactive or defective stations without disrupting the overall network.

3.2.8 Applications of FDDI

FDDI was designed as a backbone network to connect smaller LANs within a building or college campus. Typically, networks connected to the FDDI are Ethernet and token ring at different speeds, such as 4, 10, and 16 Mbps. FDDI backbones typically run at 100 Mbps. FDDI's effective throughput is 80 Mbps, depending on the number of stations connected to it. Although FDDI II was designed to carry isochronous (voice) traffic, in real-world applica-

tions, it usually carries data only because voice on a typical campus environment is carried over PBX via the conventional public telephone network.

3.3   Distributed Queue Dual Bus

In the mid 1980s, IEEE 802 started working on MAN standards. In subcommittee 802.6, the different shared-media topologies and access mechanisms
were discussed and compared, resulting in DQDB. The standards adopted
by IEEE 802 were very much in line with ATM standards because the DQDB
standardization process by IEEE and the ATM standardization process by
ITU-T were almost parallel, with each process influencing the other.

3.3.1   DQDB features

DQDB provides many features:

- use of dual-bus architecture, where operation of each bus is independent
  of the other
- compatibility with IEEE 802.X LAN standards
- utilization of many types of media for transmission systems
- looped dual-bus topology option for fault tolerance
- operability of data at rates varying from 34 Mbps to 155 Mbps
- operability independent of the number of users
- simultaneous support of both circuit-switched and packed-switched
  services

3.3.2   DQDB concept

A DQDB network is a dual-bus structure that consists of two unidirectional
buses with opposite transmission directions. Such a bus is shared by multi-
ple-access nodes, as depicted in Figure 3.20.

Each node is linked to a transmitter and receiver module through two
unidirectional buses. Therefore, full-duplex mode communication is possible between the nodes. The operations of the two buses are independent of
one another with respect to data transmission; hence, a DQDB network has
twice the capacity of a comparable system that employs only a single bus.
The transmission of data in each bus is formatted into fixed-length units
called slots. Nodes on the bus can write into the slot according to the rules
of the access protocol. All slots originate at the head of the bus and terminate at the end of the bus, as shown in the figure. This concept is the most
important one of DQDB. The other components of the DQDB architecture

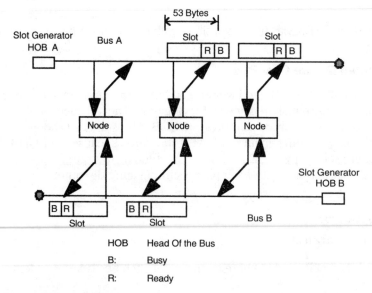

Figure 3.20 DQDB dual-bus structure.

are the nodes. Each node on a subnet consists of an access unit (AU) and the physical attachment of the AU to the two buses, as shown in Figure 3.21.

The access unit has two main responsibilities:

- It performs the node's DQDB layer functions, including access control and generation of information to place in the slots.

- It provides the physical attachment to each bus with a single read and write connection.

The AU writes on the bus using an OR-write process. It is called OR-write because writing on the bus is equivalent to the logical OR operation of the bit stream already on the bus and the bit stream that the node wants to transmit.[1] The read function occurs logically prior to the write function so that incoming data copied by a node is not affected by data written by the node. The read-and-write functions can be implemented using either active or passive techniques. An active scheme would actually read data from the incoming bus, then regenerate it on the outgoing bus. A passive system would merely read the incoming data but never remove it from the bus. A passive technique limits the number of stations on the bus because each additional station becomes a source of signal loss.

---

[1] The Boolean operator OR outputs 1 (true) if one or both inputs are 1. If both inputs are 0 (false), the output is zero.

A key feature of DQDB is that the operation of the bus is independent of the operation of the individual AUs. The node at the head of the bus has special functions that it must perform, called head-of-bus (HOB) functions. One of the functions is to generate the transmission slots mentioned earlier. The HOB must regularly create empty slots written into it by the other nodes on the bus.

3.3.3 DQDB protocol

The IEEE 802.6 standard describes two protocol layers of the OSI protocol, namely the physical layer and data link (MAC) layer, as shown in Figure 3.22. The physical layer of DQDB corresponds to the OSI physical layer and specifies how to use different underlying transmission media and speeds. The physical layer convergence protocol (PLCP) is part of the physical layer that adapts the capabilities of the transmission systems to provide the services expected by the DQDB layer. The PLCP is different for every transmission system option, but it is this part of the physical layer that allows the wide range of media and speed options supported by the network.

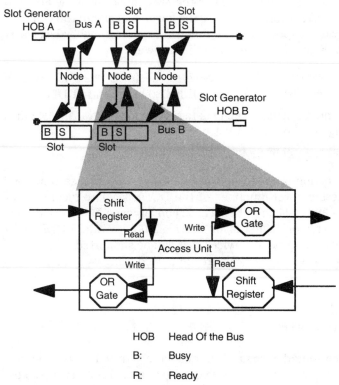

Figure 3.21 DQDB node architecture.

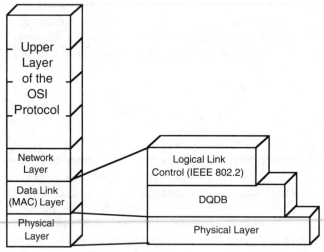

Figure 3.22 OSI relation with DQDB.

The 802.6 DQDB layer is equivalent to the MAC sublayer of the 802.3 to 802.5 LAN standards and corresponds to the lower sublayer of the OSI data link layer. It provides support for the following types of higher layer services, as shown in Figure 3.23. They are the following:

- connectionless (datagram) MAC service to the IEEE 802.2 logical link control sublayer, consistent with other IEEE 802 LAN and FDDI layers
- connection-oriented (virtual circuit) data service for the transfer of burst data, such as signaling or packet voice
- connection-oriented isochronous (circuit-switched) service

The MAC connection-oriented and isochronous convergence functions enhance the access control functions of the DQDB layer to meet the requirements of the higher-layer entity. These convergence functions are different for each type of higher-layer service and provide DQDB with enormous flexibility in terms of services that can be supported.

The higher layer services in DQDB are supported using two different access methods:

- queued arbitrated (QA) access
- pre-arbitrated (PA) access

3.3.3.1 Queued-arbitrated access. The queued-arbitrated access method supports those services not time-sensitive, such as asynchronous or packet-switched data transfer. A distributed queued-access method allows users to request access to the medium as needed and is used in conjunction with the connectionless MAC service and the connection-oriented data service such as file transfer, database access, etc.

3.3.3.2 Pre-arbitrated access. The pre-arbitrated access method assigns specific octet positions within a transmission slot for use by different stations with time-sensitive applications, such as circuit-switched data transfer. This access method supports isochronous connection-oriented services such as voice and video.

3.3.4 How does DQDB work?

In a DQDB network, as shown in Figure 3.24, the first node of each bus performs the HOB function, which entails continuously producing 53-byte empty slots or management information slots and releasing them into the bus. This node operates like a traffic cop, controlling and releasing the traf-

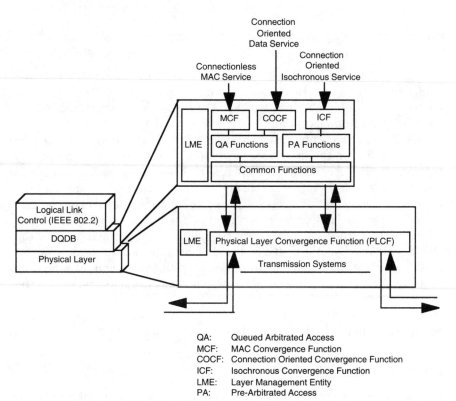

QA: Queued Arbitrated Access
MCF: MAC Convergence Function
COCF: Connection Oriented Convergence Function
ICF: Isochronous Convergence Function
LME: Layer Management Entity
PA: Pre-Arbitrated Access

Figure 3.23 Functional structure of DQDB.

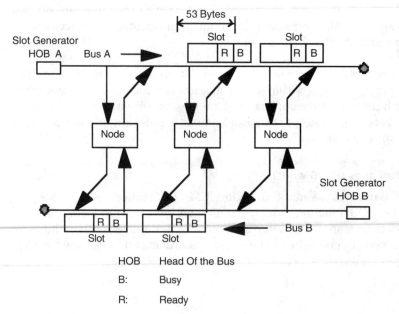

HOB   Head Of the Bus

B: Busy

R: Ready

Figure 3.24 DQDB structure.

fic onto the bus. A node with data to transmit does so by loading it onto the empty slots sent by the HOB with its own information using the distributed queuing method. Slots are divided into QA for nonisochronous traffic (data) and PA for isochronous traffic (voice or video).

DQDB is basically a bus structure, so a malfunctioning node can be easily disconnected from the bus so as not to affect the operation of the overall network. In the case of DQDB, if the first node in the bus acting as a traffic cop fails, this node is disconnected, and the rest of the nodes in the bus take the responsibility of traffic cop and generate slots and put them on the bus. If a node needs to put some information on the bus, depending on the information type, the node reads corresponding slots with the R bit from the bus. The node writes the traffic on the slots with the help of the access unit in the node. The node sets the bit in the slots to B (Busy) and puts the slots in the bus. The destination node on the bus reads the slots and copies the information from the slots and resets the Busy bit. This procedure enables the slots to be used by other nodes that need to present information. Thus, the information is transferred in a DQDB environment.

3.4 Summary

In this chapter, we explained the similarities and differences of two of the most popular access technologies, FDDI and DQDB. Both are based on fiber-optics as the transmission medium. Their targeted application is LAN inter-connect, they interwork with IEEE 802, and both are capable of carrying voice and data traffic. In the case of FDDI, only FDDI II is capable of carrying both voice and data traffic. The main difference between FDDI and DQDB is in that DQDB is designed for access to the public network, whereas FDDI is designed for private intercampus or intracampus applications. Numerous books are available that describe these technologies in detail, some of which are mentioned in the reference section at the end of this book.

Frame Relay

4.1 Overview

Frame relay is an emerging data communication concept sanctioned by ITU-T and ANSI. It is a fast packet technology developed to improve upon the X.25 packet technology. The development of fast packet was possible after the emergence of the various technologies mentioned earlier in chapter 2. Frame relay gets its name from the frame designation of the OSI data link layer. It was essentially developed through two technologies:

- widespread development of optical fiber-based transmission
- intelligent customer premises equipment (CPE)

The frame relay protocol supports data transmission over connection-oriented paths. The interface from the user to the frame relay network is called frame relay interface (FRI). FRI currently supports the following access speeds:

- 56 kbps
- n × 64 kbps
- 1.544 Mbps (T1)
- 2.048 Mbps (E1 for Europe)

In addition, there are some intermediate speeds, but the highest is T1 or E1.

FRl is an enhanced version of X.25 without error correction at the inter-mediate nodes. Error corrections in a frame relay network occur in the in-telligent CPE at either end of the connection, thus making frame relay much faster than X.25. Frame relay relies on the equipment to perform the functions previously done by the network. Advances in transmission tech-nology and, of course, the terminal equipment has allowed this change to occur. These advances have made frame relay a standard in data communi-cations from 56 kbps to T1- or E1-based applications and is slowly replac-ing the X.25-based packet-switched network. For speeds below 56 kbps, the X.25 packet-switched network is still ideal.

4.2 Frame Relay Standards

The ITU-T has divided the functions and features of frame relay into three major standards, illustrated in Table 4.1:

- service description
- core aspects
- signaling

4.2.1 Service description
standard (ANSI T1.606, ITU-T I.233)

This standard outlines the overall frame relay service description and spec-ifications. In addition, recent documents examine connection management and ISDN multiplexing and rate adaptation. Connection management de-fines the speed at which the user is assigned to transmit data over the net-work as well as the burstiness of user data. Connection management also describes how the network and end-user devices handle an overabundance of data traffic if every user transmits more than the committed data rate. The details of the service description are addressed in chapter 5.

4.2.2 Core aspects (ANSI T1.618, ITU-T Q.922)

This standard describes the core aspects of frame-relay specifications. The basics of frame relay, such as frame format, are discussed, along with the

TABLE 4.1 Frame Relay Standards Documents (ITU-T/ANSI)

Standards categories	ITU-T standards	ANSI standards
Service description	I.233	T1.606
Core aspects	Q.922	T1.618
Signaling	Q.933	T1.617

functions of different fields within the frame format, such as the DLCI field. In addition, the congestion control mechanisms and methods of managing congestion are covered. The details of the core aspects are addressed in chapter 5.

4.2.3 Signaling (ANSI T1.617, ITU-T Q.933)

This standard specifies a protocol for establishing and releasing switched frame relay virtual connections and provides a means to inform users of permanent virtual circuits of failures and restorations. It sets out the procedure for the user-to-network signaling to support the frame relay calls. The details of signaling are addressed in chapter 5.

In addition to these standards, others such as data link control (ANSI: no standard planned, ITU-T Q.922) provides an optional end-to-end mechanism for ensuring the correct delivery of information across the network. This protocol is designed for implementation in end-user devices. It is not designated to be implemented in a frame relay network to carry user traffic because end-user devices usually rely on other protocols to ensure data accuracy.

4.3 Frame Relay Field Format

This section provides an overview of basic frame information. The details are described in chapter 5. Figure 4.1 shows the basic format of a frame-relay field. This format roughly conforms to the high-level data link control (HDLC) frame format, which is common to other protocols, such as IBM's Systems Network Architecture (SNA), X.25, and ISDN.

As shown in Figure 4.1, the frame relay format has five different fields, each of varying lengths. These fields are the flag field, frame relay header, information field, frame check sequence field, and trailing flag field. Each of these fields is responsible for a specific function. Of these, the most important field is the frame relay header field.

Figure 4.2 illustrates the field that forms the frame relay header format. On each end of the frame relay format, flags delimit where the data frame starts and ends.

4.3.1 Data link connection identifier

The central part of the frame relay format is the data link connection identifier (DLCI). The DLCI distinguishes separate virtual circuits across each access connection. Thus, data coming into a frame relay node are transmitted by specifying a DLCI rather than a destination address. At the network node, this connection specification is confirmed. If the specification is in error, the frame is discarded. If not, the frame is relayed to its destination. The details of this PLCI file and its usage are discussed in chapter 5.

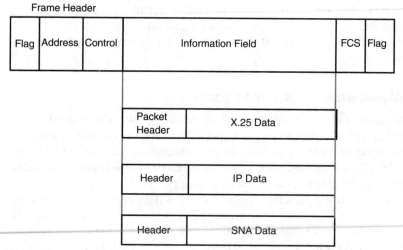

Figure 4.1 Frame relay field format.

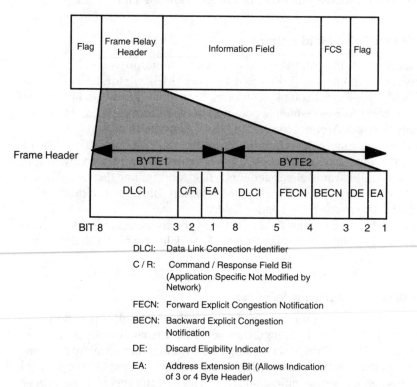

Figure 4.2 Frame relay header field format.

4.3.2   Discard eligibility

Discard eligibility (DE) is a one-bit field that can be set in low-priority frames so they are discarded first in the event of frame loss. The DE bit is set by the frame relay network in any frame that exceeds a user's subscribed rate. The network assumes anything that exceeds the user's subscribed rate is low priority. The DE bit can also be set by the end-user equipment if it knows that certain frames are more important than others.

4.3.3   Forward explicit congestion notification

Forward explicit congestion notification (FECN) is similar to the congestion information field used in protocols such as Digital Equipment Corporation's Network (DECnet). In this network, the network sends a signal to the receiving or destination end point, advising it to slow down the receipt of information. At the end point, the destination device checks the FECN bits and if the field states that the user has exceeded the threshold, it alerts the sender accordingly.

4.3.4   Backward explicit congestion notification

Backward explicit congestion notification (BECN) is usually used with an SNA-type network, where the network informs the source or transmitting end point and advises it to slow down the sending of information immediately. FECN and BECN perform congestion management in the frame relay network. The implementation of FECN and BECN is called *explicit congestion management*. Some end-to-end protocols use FECN, while others use BECN. Both work well, but they are usually mutually exclusive options in end-user equipment.

4.3.5   Information field

In Figure 4.2, the information field is shown as a part of the frame relay format that contains the actual information transmitted. The maximum allowed length of this information field can vary, depending on the design requirements of the network, from 262 to 8,000 or more octets.

4.3.6   Frame check sequence

Frame check sequence (FCS) performs error checking for the frame. Frame relay uses the error-checking technique known as cyclic redundancy check, or CRC. The frame relay CRC generates two bytes that are added at the end of the frame to detect bad data. The CRC algorithm uses an algebraic method to generate a unique bit pattern, which is recalculated at the far end. If the FCS at the source matches the FCS at the destination, the frame's integrity has (in most cases) been preserved.

When frame relay does discover a frame error, it merely drops the frame. As we have seen, frame relay does not request a retransmission of data if it finds errors in the transmission, but leaves this task up to the LAN station, X.25 switch, or other intelligent devices that are connected to the network.

4.4 Frame Relay Architecture

Figure 4.3 shows the relationship between frame-relay architecture and the corresponding protocol stacks. For example, the router performs the function of the bottom three layers of the OSI protocol: network, data link, and physical. The protocol on the left shows the protocol stack for the originating terminal with all seven layers of the protocol stack.

The data is generated from this terminal and then is passed on to the originating router to which the terminal is connected. The router uses these three layers of the OSI protocol to route the data. From the router, the data goes to the frame relay network, which uses the second layer of the OSI protocol and forwards the frames. At the terminating router, the data goes through the three layers and the terminating terminal for which the data was destined.

In the frame relay network, the bottom two layers of the OSI protocol are used to forward the forms. The last, or physical, layer actively transports bits from one point to another. The next, or data, link layer, performs the error check on the frames (or data) at each node in the network through which the data is routed. This error checking is performed by the FCS field in the frame relay format. If this field reveals that the frame was corrupted during transmission across the communication channel, the frame is discarded by the node. The field does not, however, send any acknowledgment back stating that the frame has been discarded. It goes ahead and performs error checking with the next frame.

4.5 How Does Frame Relay Work?

Figure 4.4 shows a typical frame-relay network. Two routers are connected to the network via a standard FRI protocol. One router acts as an originator, and the other acts as a receiver. The data is originated and terminated at the end terminals.

Figure 4.5 shows the source side of the frame relay network, where the originating router adds the header to the information payload with the upper-layer protocol. The header contains a 10-bit number, called data link connection identifier (DLCI), which corresponds to a particular destination node. The router maps the actual destination address to the corresponding

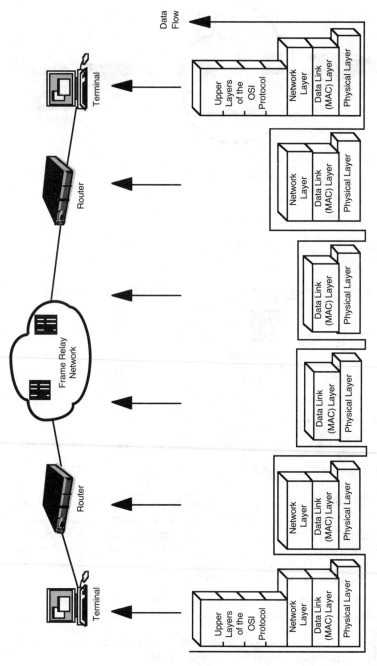

Figure 4.3 Frame relay protocol stack.

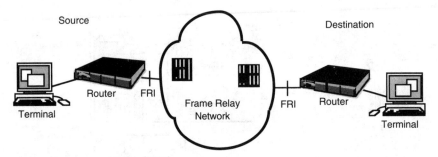

Figure 4.4 Sample frame relay network.

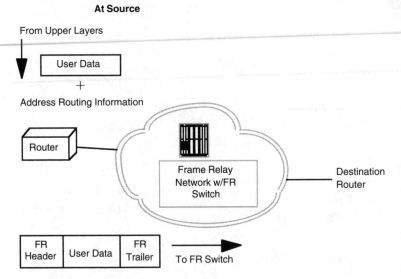

Figure 4.5 Sample frame relay network, Phase I.

DLCI. This DLCI number is later used in the frame relay network to route the frames. In the case of LAN interconnect, as in this example, the DLCI would correspond to a port to which the destination router is attached. With the appropriate DLCI in the address field, the router ships the frame to the frame relay switch in the network.

Figure 4.6 shows the second phase of the frame relay service. The packet from the router is received at the first switch in the network with the DLCI number in the header. The data at the nodes follows three simple processes:

1. The frame relay switch checks the integrity of the frame with the FCS field. If the frame has a flaw, the switch discards it.

2. The lookup table in the switch verifies the validity of the DLCI. If the DLCI is not defined in the table, the switch discards the frame.

3. If the DLCI is valid, the switch routes the frame to the appropriate port, in this case, to the destination router. This process is illustrated as a flow chart in Figure 4.7.

The frame relay switch works on the simple rule that if it finds an error in the frame, it discards that frame.

Once the frame reaches the router destination, the reverse function of the originating router is done, i.e., the header field is stripped from the frame and the frame is forwarded based on the destination in the address field of the frame relay header. This process is illustrated in Figure 4.8. So far, we have seen how the data is transmitted in the frame relay network, but we have not addressed what happened to the discarded frames and why they were discarded. We address those issues now.

A frame can be discarded due to one of the following reasons:

- A bit error is found by the FCS field in the frame.

- Network congestion occurs when the buffer at any node is full. When the node is full, the node begins to discard the frames so that the buffer can be made empty.

As mentioned earlier, the frame relay network does not perform any error recovery. Error recovery is performed by the CPE at the other end of the connection.

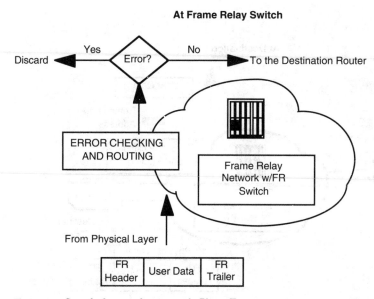

Figure 4.6 Sample frame relay network, Phase II.

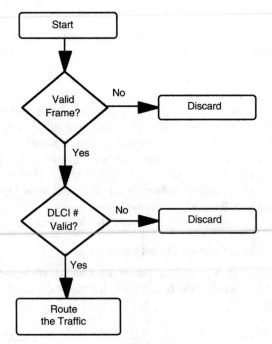

Figure 4.7 Flow chart for the frame relay switch functions.

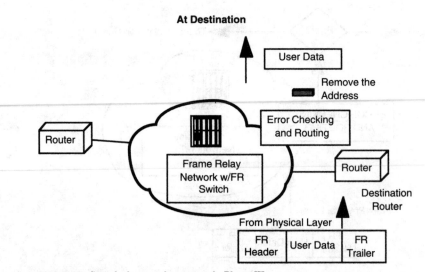

Figure 4.8 Sample frame relay network, Phase III.

4.6 Frame Relay Services and Features

The frame relay services are implemented using virtual circuits. The two types of virtual circuits are permanent virtual circuits based on ANSI T1.617 and ITU-T 933 and switched virtual circuits (currently not available).

Permanent virtual circuit (PVC) is a connection set by the service provider in a public network or by network operations in a private network. The PVC that is established on request can exist for weeks, months, or even years; use nailed-up circuits; or forgo call establishment, hence eliminating call setup delay.

In the case of a switched virtual circuit (SVC), the connection is set up like a regular telephone call with a call setup delay. The connection exists only for the duration of the call, as in an ordinary phone call. The standards for switched virtual circuit were completed in 1993. Currently, no vendors have added this service to their list of products. Vendors are planning to implement SVC sometime in 1994. Thus, this service is not currently offered by the service providers. A few vendors are offering PVC services at national and international levels, including:

- BT (British Telecom)
- CompuServe
- Sprint
- AT&T
- Wiltel
- MCI

Frame relay provides certain features along with the basic service, including:

- interface signaling
- committed information rate (CIR)
- priority levels
- multicast and global addressing

Each are discussed in the following subsections.

4.6.1 Interface signaling

Interface signaling is the signaling across the FRI between the user and network. It is accomplished by sending a frame with a DLCI field of 1023, called a *consolidated link layer management* (CLLM) message. Figure 4.9 illustrates this feature.

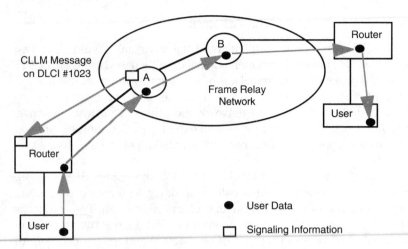

Figure 4.9 Use of CLLM in signaling congestion.

Interface signaling was included in the standard as an option. Service providers now are implementing it as a service differentiator against their competitors. Signaling:

- signals the CPE that there is congestion in the network
- communicates the status of the PVC
- keeps the user informed of the access rate

Signaling is accomplished by using certain bits called *explicit congestion notification* (ECN) bits in the header and by designating certain DLCI for specific purposes. The function of these bits is to inform the network that congestion has occurred. The two types of ECNs are FECN and BECN, which were discussed previously. As you recall, their function is to inform the forward and backward switches in the path about congestion. This concept is depicted in Figure 4.10.

Another type of FRI signaling is keeping the user informed of the status of the virtual connections, for example, sending "keep alive" messages to keep the user informed that the connection is up and alive. This feature is useful when the users have not used the connection for a long period of time. It also informs the user about the congestion status of the nodes.

4.6.2 Committed information rate

A parameter called *committed information rate* (CIR) allows the user to stipulate his average or normal traffic during a busy period. The network then measures the information flow on a dynamic basis. As long as the CIR is not exceeded, relay traffic (frames) is unaltered. If the user exceeds the

CIR, the DE (Discard Eligibility) bit is set on those frames to signal the frame relay switch to discard these frames first if congestion occurs. The CIR feature is explained further in chapter 5.

4.6.3 Priority levels

Frame relay networks can assign priority levels to users so that lower priority traffic can be discarded first during congestion. Priority level is assigned by setting the DE bit. For example, the transaction traffic that is delay-sensitive would be given higher priority, and bulk file transfers would typically be given lower priority.

4.6.4 Multicasting and global addressing

Multicasting allows the broadcast of communications by assigning a certain DLCI to indicate multiple destinations. Global addressing allows DLCI not to be originating port-specific. In other words, a certain DLCI can route to a given terminal, no matter which port it is sent from.

4.7 Frame Relay Applications

We discussed some of the applications of broadband communications in chapter 2. Most of its applications are applicable to the technologies (including frame relay) addressed in this book. As we know, the driving application of frame relay is LAN interconnect. Other applications can also benefit from frame relay:

- IS (information systems) applications
- client-server computing
- CAD/CAM applications
- graphics applications
- other applications that generate bursty (variable bit rate) traffic

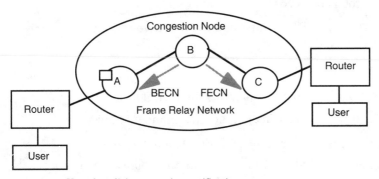

Figure 4.10 Use of explicit congestion notification.

Let's look at an example of the economics of frame relay service for the LAN interconnection application. Figure 4.11 shows a typical LAN interconnect solution. If there are five locations with routers at each place, these can be connected via a network to provide LAN interconnect service across the United States. Each router is connected to many LANs within a region, such as different buildings across a city. Let's assume that the network traffic is intermittent but has high peak rates. Users thus have point-to-point leased lines of different capacities across the nation, depending on the traffic pattern.

Based on this network, we have the following:

Components of the network	Units	Cost per unit*	Total cost* (No. of units × cost per unit)
Number of CSU/DSU	14		
Number of router ports	14		

* To be filled by the reader based on the prevailing cost.

The other cost components are the link lease from the interexchange carrier (IEC) and the local exchange carrier (LEC). They are not separated as two different components because the IEC often provides direct access to the customer.

Number of links	Link capacity	Distance in miles	Cost/month (Include the LEC and IEC)*
4	56 kbps		
1	256 kbps		
2	T1 or 1.544 Mbps		

* To be filled by the reader based on the prevailing cost.

To find the total cost, use the following formula:

TOTAL COST = {one-time capital investments}

cost per CSU/DSU × number of CSU/DSU +

cost per router port × number of router ports + **{monthly recurring cost}**

cost per local loop per month × number of local loops +

cost per 56-kbps link per month × number of links +

cost per 256-kbps link per month × number of links

cost per T1 (1.544 Mbps) link per month × number of links +

usage cost based on number of packets or frames transmitted.

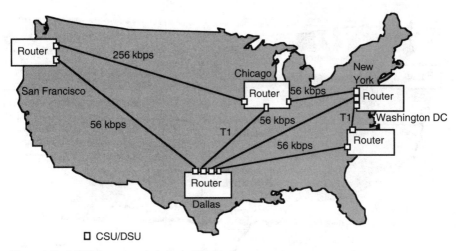

Figure 4.11 LAN interconnect via leased lines.

Note: Cost of different speed varies with distance of the link.

Adding these components gives the exact prevailing cost.

Now, having seen the traditional method of calculating the cost for LAN interconnect, let's compare this cost with the cost when frame relay is used. Figure 4.12 shows the same network with a frame-relay backbone. Based on this network, we have the following:

Components of the network	Units	Cost per unit*	Total cost*
Number of CSU/DSU	5		
Number of router ports	5		

* To be filled by the reader based on the prevailing cost.

The other cost components are the frame relay interface costs for the different speeds of access to the network and the usage cost (usually in number of packets or frames transmitted). Based on current rates from the IEC providing the service, the savings obtained by using frame relay is approximately 32 to 50 percent.

Number of links	Link capacity	Cost/month (Include the LEC and IEC)*
0	56 kbps	
1	256 kbps	
4	T1 or 1.544 Mbps	

* To be filled by the reader based on the prevailing cost.

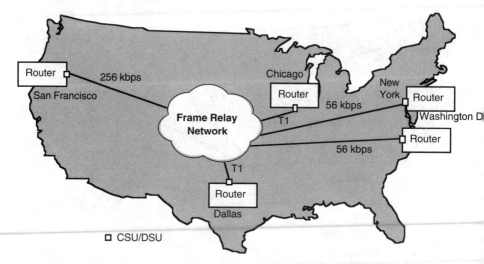

Figure 4.12 LAN interconnect via frame relay network.

This example clearly demonstrates the cost savings obtained by migrating to frame relay. Of course, frame relay might not be a good solution for other applications. The only way to find out is to perform the calculation based on the usage pattern of that particular traffic.

4.8 Summary

As the name implies, frame relay relays the frames one behind the other in a predetermined path. Frame relay is a switched service positioned to improve communications performance through reduced delays, more efficient bandwidth utilization and decreased equipment cost. Advantages of frame relay include the following:

- port and link sharing
- bandwidth on demand
- high throughput and low delays
- ease of network expansion
- ease of transition from existing X.25 network
- simplified network administration
- technology based on global standards
- cost advantages

One of the obvious limitations of frame relay is the maximum access rate of T1 (1.544 Mbps) in the U.S. and E1 (2.048 Mbps) in Europe. The user cannot request more than T1 of bandwidth on a single connection. LAN inter-connect is the driving application for the frame relay, but one must realize that frame relay is a connection-oriented service whereas LAN interconnect is an inherently connectionless application with traffic patterns of varying burstiness.

5

Advanced Frame Relay

5.1 Overview

In chapter 4, we introduced the frame relay standards, which are based on service description, core aspects, and access signaling. This chapter looks at these in detail. It also addresses the new standards for a network-to-network interface (NNI) between frame-relay nodes.

5.2 Service Description

This section addresses the frame relay service description specification published in the ANSI T1.606 and ITU-T I.233 documents. We focus on service attributes such as information transfer, access, and other attributes of a frame-relay session. In addition, we look into the performance parameters used to measure frame relay. Examples of performance parameters are throughput, delay, and committed information rate.

5.2.1 Service attributes

In frame relay, the service attributes are categorized into information transfer attributes, access attributes, and general attributes. Each is discussed in the following subsections.

5.2.1.1 Information transfer attributes. Information transfer attributes are related to type of transfer mode; information transfer rate, i.e., the speed in bits per second; type of communication establishment, such as switched virtual circuit (SVC) or permanent virtual circuit (PVC); method of transfer; and how traffic is sent across the network, such as point-to-point, point-to-multipoint or broadcast. Table 5.1 summarizes the different information transfer attributes.

5.2.1.2 Access attributes. Both ANSI and ITU-T have defined the access attributes that determine the type of channels used and the type of protocols required at each layer for information and signaling (control access). Table 5.2 summarizes the access attributes.

5.2.1.3 General attributes. ITU-T and ANSI differ in their definition of general attributes. Currently, ITU-T has a list of provisional supplementary services, whereas ANSI has supplementary services defined for further study. Both ITU-T and ANSI have not listed much information to date. Table 5.3 summarizes the ANSI and ITU-T general attributes.

5.2.1.3.1 Performance. ANSI T1.606 has several definitions of frame relay performance parameters, such as throughput and committed information rate (CIR). These parameters are used by the network operator when establishing service contracts with customers. In addition to these parameters, several others are described.

TABLE 5.1   Information Transfer Attributes

Information transfer mode	Frame (packet in T1.606)
Information transfer rate	Less than or equal to maximum user channel bit rate (max. access line capacity)
Transfer capability	Unlimited
Structure	SDU integrity
Communication establishment	SVC, PVC
Configuration	Point-to-point, point-to-multipoint

TABLE 5.2   Access Attributes

Access channel	D, B, and H channels
Signaling access protocol layer 1	I.430 or I.431 (ITU-T)
Signaling access protocol layer 2	Q.921 (ITU-T)
Signaling access protocol layer 3	Q.930 series (ITU-T)
Information access protocol layer 2 (core function)	Core function of Q.922
Information access protocol layer 2 (data link control)	User specified

TABLE 5.3 General Attributes

Supplementary services	For further study (ANSI), provisional services (ITU-T)
Quality of service	For further study
Interworking possibilities	ITU-T I.500 services, ANSI T1.606
Operational and commercial	For further study

5.2.1.3.2 Throughput. Throughput for frame relay is defined as the number of protocol data units (PDUs) successfully transferred in one direction per unit of time over a virtual connection. Q.933 defines the interface in bits per second. A virtual connection can include any number of intermediate nodes between two end users. Figure 5.1 shows a virtual connection via two frame-relay nodes. A PDU includes all bits between the flags of the frame-relay frame. It includes the bits between the address field and the frame check sequence (FCS) field, i.e., the information field (see Figure 4.1). The term *successful transfer* means that the FCS check has acknowledged that the transfer has been completed successfully.

5.2.1.3.3 End-to-end delay. End-to-end is defined as the delay measured between two end points or end users. Typically, it includes the multiple transit delay, which is the delay measured between a pair of boundaries. ITU-T in its X.13 defines *boundary* as follows:

"A boundary separates a network section for the adjacent circuit section, or it separates an access circuit section from the adjacent DTE (end user)."

Transit delay could define a boundary between international networks, national networks, or local-access and long-distance networks. Figure 5.2

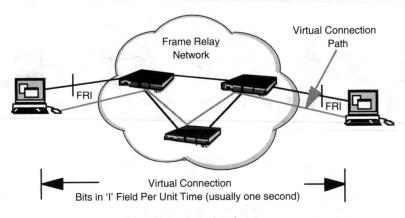

Figure 5.1 Throughput.

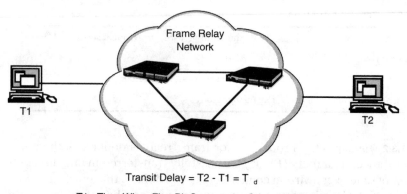

Transit Delay = T2 - T1 = T $_d$

T1 - Time When First Bit Crosses the Originating Point
T2 - Time When Last Bit Crosses the Destination Point

Figure 5.2 Transit delay.

shows the definition of transit delay, and Figure 5.3 shows the definition of end-to-end delay.

Figure 5.2 shows the transit delay where it starts at time T_1 when the first bit of the PDU crosses the boundary at the originating point. It ends at time T_2, when the last bit of the PDU crosses the second boundary. Thus, the transit delay time is the time difference of T_2 and T_1, i.e., transit delay = $T_2 - T_1 = T_d$. When this delay is added across multiple boundaries, it is called *end-to-end delay*, which is illustrated in Figure 5.3.

5.2.1.3.4 Committed information rate (CIR). The committed information rate (CIR) describes the information transfer rate that the network must com-

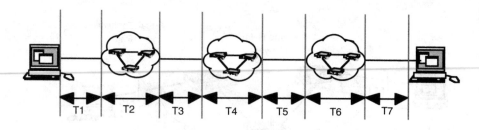

$$\text{End-to-End Delay} = \sum_{n=1}^{m} T_n$$

T_n = Transit Delay Between Two Points
m = Number of Transit Delays

Figure 5.3 End-to-end delay.

mit to support a user during normal operation. In other words, the CIR is the throughput that the user asks for and the network guarantees for a particular SVC or PVC connection. For SVC, the CIR is negotiated during call setup. Figure 5.4 shows the CIR definition.

Other measurement parameters work in conjunction with CIR, including measurement interval (T_c), committed amount of data (B_c), and excess amount of data (B_e). Figure 5.5 shows these parameters.

The measurement interval (T_c) is defined as the time interval in which the user can send a committed amount of data, B_c, and an excess amount of data, B_e. B_c is defined as the data rate committed by the user and guaranteed to be delivered error-free by network operators. B_e is defined as the amount of data that exceeded the committed amount of data B_c during the measurement interval.

The relationship between these parameters is as follows:

If   CIR > 0, B_c > 0 and B_e = 0,
      then T_c = B_c/CIR

If   CIR = 0, B_c = 0 and B_e > 0,
      then T_c = B_c/access rate (maximum access speed of T_1)

The other parameter used for measurement is residual error rate (RER),

$$R = \frac{1 - \text{total correct SDUs (service data units) delivered}}{\text{total offered SDUs}}$$

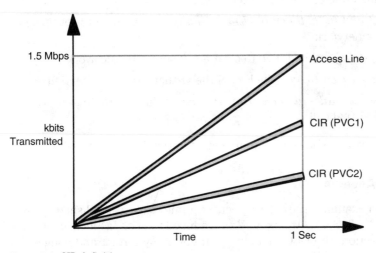

Figure 5.4 CIR definition.

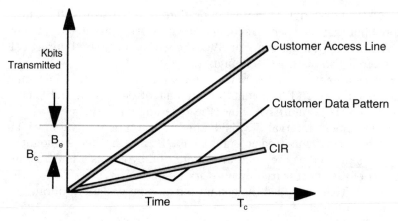

Figure 5.5 CIR parameters.

An incorrect frame is one in which one or more bits are in error, such as:

- *Delivered error frame.* a frame delivered when the values of one or more of the bits in the frame are discovered to be in error.

- *Delivered out-of-sequence frame.* the arrival of a frame that is not in sequence relative to previously delivered frames.

- *Lost frame.* frame is declared to be lost when the frame is not delivered correctly within a specified time.

- *Misdelivered frame.* one delivered to the wrong destination. In this situation, data link control identifier (DLCI) interpretation can be in error, the routing table can be out of date, etc.

- *Switched virtual call establishment delay.* time taken to set up a call across the network.

- *Clearing delay.* time taken to clear a call across the network.

- *Premature disconnect.* the loss of the virtual circuit connection.

- *Switched virtual call clearing failure.* a failure to tear down the switched virtual call.

These are some of the parameters used to provide frame-relay service.

5.3 Core Aspects

This section examines the core aspects of frame relay, as published in ANSI T.618 and ITU-T Q.922/Annex A. We introduced the core aspects in chapter 4. In this section, the DLCI field in the frame-relay format and congestion control management are explained thoroughly.

5.3.1 Frame-relay format

In chapter 4, we introduced the DLCI field and its usage. The DLCI field in the frame-relay frame can vary in size from 2 to 4 bytes allowing the use of more DLCI numbers, and thus enabling more virtual circuit connections. Figure 5.6 illustrates the three formats.

In the figure, the extended address (EA) bits are set to 0 to indicate that the header contains additional bytes. If the header and the EA are set to 1, it indicates the end of the header. The D/C field (1 bit) is called DLCI or DL-core control indication. This bit is set to 0 if the last DLCI byte contains DLCI bits. If the bit is set to 1, it contains DL-core information. The fields forward explicit congestion notification (FECN), backward explicit congestion notification (BECN), and discard eligibility (DE) were explained in chapter 4.

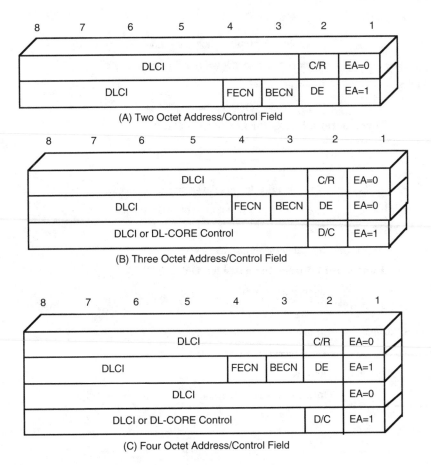

Figure 5.6 DLCI frame formats.

5.3.1.1 DLCI values

We learned in chapter 4 that the DLCI field identifies a logical connection multiplexed across a physical channel. The DLCIs with the same value always identify the same logical channel across a particular physical circuit. The DLCI values are explained in the ITU-T and ANSI core aspects document. The values and ranges depend on whether frame relay is being transmitted across the D or B channel. DLCI values vary depending on the use of 2, 3, or 4 bytes. The values are listed in Table 5.4.

TABLE 5.4 DLCI Values Using Different DLCI Formats

DLCI value	Function
Two-octet address format	
0	In-channel signaling
1–15	Reserved
16–991	Assigned using frame-relay connection procedures
992–1001	Layer two management of frame-relay service
1002–1008	Reserved
1023	In-channel layer management
Three-octet address format with D/C = 0	
0	In-channel signaling
1–1023	Reserved
1024–63,487	Assigned using frame-relay connection procedures
63,488–64,511	Layer two management of frame-relay service
64,512–65,534	Reserved
65,535	In-channel layer management
Four-octet address format with D/C = 0	
0	In-channel signaling
1–131,071	Reserved
131,072–8,126,463	Assigned using frame-relay connection procedures
8,126,464–8,257,535	Layer two management of frame-relay service
8,257,536–8,388,606	Reserved
8,388,607	In-channel layer management

5.3.2 Congestion control management

As in any network, a frame-relay network must deal with congestion; it must have some form of congestion control management. The objective of such management is to meet the user's quality of service (QOS) request for each connection. To meet this requirement, an enhancement was added to frame relay. ITU-T and ANSI have developed the consolidated link layer management (CLLM) message to provide additional function to the frame-relay service, which was described in chapter 4. In the frame-relay network, prior to transferring information, certain procedures must be accomplished. This process is similar to a telephone call. In frame relay, to initiate a data transfer, the user must signal the network (by sending certain connection control messages) that it would like to transfer data.

5.4 Access Signaling

This section examines the frame-relay specification published in ANSI T1.617 and ITU-T Q.933 for setting up a switched virtual connection or call. Current frame-relay services are PVC-based.

A rather extensive modification to Q.933 is underway. The frame-relay forum has proposed a set of specifications for SVC. This proposal is different from the X.933/T1.617 specification in that it is simpler and appropriate for both UNI and NNI. We discuss in this section the high-level description of SVC proposed by the Frame-Relay Forum (FRF). FRF is an interest group formed by service providers, users and equipment manufacturers with respect to frame relay.

5.4.1 Frame-relay connection process

The messages used for connection control can be grouped into three categories: call establishment, call clearing, and miscellaneous. Figure 5.7 illustrates the different messages in a network.

SETUP: This message is sent by the frame-relay originator to the frame-relay network to establish a call. The setup message contains a number of fields that describe the type of message, the type of capabilities established along with the call, the appropriate DLCIs, the recommended end-to-end transit delay, and other parameters such as calling party address.

CONNECT: If the setup message is accepted by the called user, this user responds with a connect message sent to the local node and relayed via the network to the caller. It contains parameters such as transit delay, DLCI, etc.

CONNECT ACKNOWLEDGE: This message is sent by the network to the called frame-relay user to notify the user that information transfer can occur. The destination user is informed with an ALERTIN message.

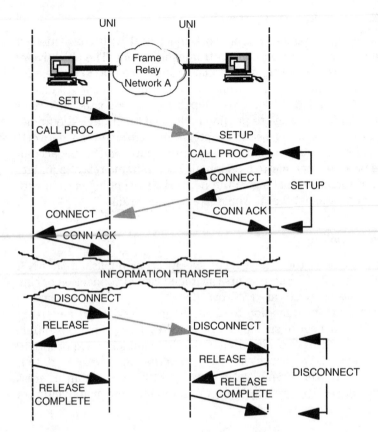

Figure 5.7 Connection establishment and release process.

CALL PROCEEDING: This message is sent by the called user to the network and relayed to the calling user to indicate that a call-establishment procedure has begun.

PROGRESS: This message is sent by the network or the user to provide status of the call. The above messages are for call establishment. After this message, the data transfer begins.

On completion of the call, the following messages are used for call disconnect.

DISCONNECT: This message requests the network to clear the frame-relay call.

RELEASE: This message is sent by the user or the network indicating that the connection occurred, and if a DLCI has been used, it is released for further use.

RELEASE COMPLETE: This message clears the call and connection, freeing the channel for reuse. In addition to these messages, others such as status and status enquiry are sent across the network to obtain information about the network (status information).

5.5 Network-to-Network Interface

The network-to-network (NNI) was developed by the Frame-Relay Forum, and is currently published as a draft in the ANSI T1S1.2 working group. We examine some of the NNI functions and its operations. The ANSI specification is important because it enables equipment from different vendors to interoperate.

Initially, work on frame relay was focused on user-network interface (UNI). Only recently has NNI been addressed. NNI defines the procedures for different networks to interconnect with each other to support the frame-relay operations. Figure 5.8 shows the NNI in a typical frame-relay network. Some of the principal operations of NNI are the following:

- notification of adding a PVC
- detection of deleting a PVC
- providing notification of UNI or NNI failure
- notification of PVC segment availability or unavailability
- verification of links between frame-relay nodes
- verification of frame-relay nodes

These operations occur through the exchange of status (S) and status enquire (SE) messages, which contain information about the status of the PVCs.

In Section 5.4, we described the SVC at UNI. The proposal also includes a specification for SVC capability at the NNI. Figure 5.9 shows the operations for a connection setup and connection release.

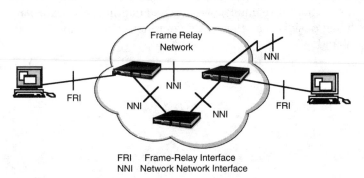

FRI   Frame-Relay Interface
NNI   Network Network Interface

Figure 5.8 Typical FR network with NNI interface.

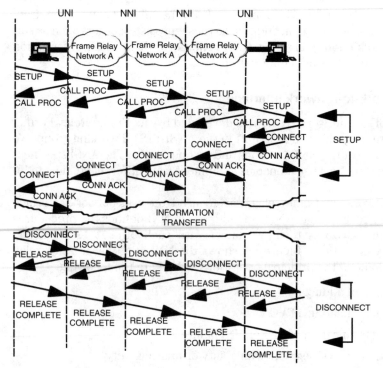

Figure 5.9 Connection setup and release in a multiframe relay network.

5.6 Summary

The advanced features in frame relay of service, description, core aspects, and signaling have been described. In addition, NNI was discussed in general. Currently, frame relay has all the important specifications required to provide SVC-based services, enabling frame relay to become even more popular for lower speed traffic requirements (less than 1.544 Mbps). In the next chapter, we see another technology—switched multimegabit data service—which is being designed as the first comprehensive technology and service for broadband.

6

Switched Multimegabit Data Service

6.1 SMDS Overview

Switched multimegabit data service (SMDS) is similar to frame relay in many aspects. For example, each has:

- an emerging data communication technique based on fast packet-switching technology
- a type of protocol for transferring data traffic
- a technology driven by LAN-interconnect traffic

SMDS is the first protocol designed for broadband communications to provide connectionless service. It is fixed cell-based and designed to be a broadband public network service and provide fully switched "any-to-any" connections.

SMDS was the creation of Bellcore, which adopted the connectionless data networking capability found in the IEEE 802.6 specification for its clients—the regional Bell operating companies (RBOCs). SMDS is designed for local, intra-LATA and wide area network (WAN) services. The connectionless nature of SMDS is the most important distinction when compared to other technologies. Bellcore designed SMDS as a public broadband service to be provided by the RBOCs. Bellcore devised a set of specifications that spelled out the subscriber network interface (SNI) specification, commonly known as a *user network interface*, or in the case of frame relay, frame-relay interface (FRI). Bellcore then defined the interswitching system interface (ISSI), which is the interface between switches within a network. Next came the intercarrier interface (ICI), which defines the interface between switches of two networks, such as local exchange and interexchange carriers. Finally came the operations systems (OSS) interface, used for billing and network administration.

In this chapter we tour SMDS, its working principles, and protocols. In chapter 7, we cover in detail some of the protocols defined within each of the interfaces described in the SMDS standard documents.

6.2 SMDS Standards

Bellcore developed SMDS standards using its knowledge and expertise in telecommunications. Thus, it was able to cover every aspect of the network, such as the subscriber network interface, interswitching system interface, and intercarrier interface, allowing local exchange network to be connected using a standard interface to any other network, such as an interexchange carrier's network (or long-distance network). In addition to these interfaces, Bellcore defined services and features not available in any new technology. Such comprehensive services would take a long time to be developed using a standards board. Bellcore developed many SMDS features that meet the needs of end users and the network operators. In fact, the international broadband committee is using Bellcore's document as the starting point in defining global BISDN standards. Table 6.1 lists the documents in which Bellcore defined in detail every aspect of the SMDS service definition and implementation.

TABLE 6.1 List of SMDS-Related Documents

Document	Title
TR-TSV-000772	Generic System requirements in support of SMDS service (May 1991)
TR-TSV-000773	Local access systems generic requirements, objectives, and interface in support of the SMDS (June 1991)
TR-TSV-000774	SMDS operations technology network element generic requirements (March 1992, Issue 3)
TR-TSV-000775	Usage measurement generic requirements in support of billing for SMDS (June 1991)
TR-TSV-001059	Interswitching system interface (ISSI) generic requirements (Dec 1990)
TR-TSV-001060	Exchange access SMDS (XA-SMDS) service generic requirements (Dec 1991)
TR-TSV-001061	Operations technology network element generic requirements in support of ISSI and XA SMDS (May 1991)
TR-TSV-001062	Generic requirements for SMDS customer network management service (Feb 1992, Issue 2)
TR-TSV-001063	Update functional requirements for BOC switching systems, in support of XA-SMDS (March 1992)

6.3 SMDS Architecture

The reference network architecture of SMDS with various interfaces as defined by Bellcore in its standard document is shown in Figure 6.1. Each interface has relevant Bellcore technical reference (TR) documents, which describe in detail the generic requirements that support SMDS.

The Bellcore specification defines how the data is to be converted to cells to pass through the subscriber network interface (SNI) via an SMDS interface protocol (SIP), which indicates the end of the public network and the beginning of the customer premises. The SIP contains three protocol layers that give the frame structure: addressing, error control, and transport of the data at the SNI. These specifications provide a definition of the basic requirements for SMDS service. One can compare SIP to the

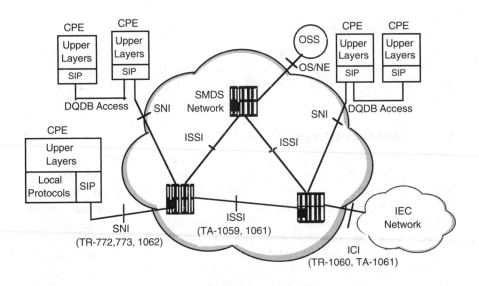

DQDB	Distributed Queue Dual Bus
SIP	SMDS Interface Protocol
SNI	Subscriber Network Interface
SS	Switching System
ISSI	Interswitching System Interface
OS/NE	Operations System/Network Element
DCN	Data Communications Network
OSS	Operations Systems
ICI	Intercarrier Interface
IEC	Interexchange Carrier

Figure 6.1 SMDS reference architecture.

X.25 interface protocol in a packet-switched network today because the SIP defines the network services and how they are accessed by the user, similar to the X.25 specification.

The other interfaces defined in SMDS are ISSI and intercarrier interface (ICI). The ISSI interface defines the interface between switching systems, such as between two SMDS switching systems. The interface can be between two SMDS switches or between an SMDS switch and an ATM/SONET-based BISDN switch. The ICI is an interface or boundary between the networks of two carriers providing SMDS service or one network providing SMDS service and another providing ATM-based broadband service. In the United States, the networks typically involved are a local exchange carrier (LEC) and an interexchange carrier (IEC), as shown in Figure 6.1. The details of the protocol involved in these interfaces are described in chapter 7.

6.4 The SMDS Protocol

The SMDS protocol is based on the three SIP layers. These three layers do not correspond to the layers of the OSI model, but the basic functionality of the bottom three OSI layers is used. For the customer to connect to the SMDS network, the customer interface must follow these three layers.

6.4.1 Layer 3 of SMDS interface protocol

Figure 6.2 shows the three-layer SIP protocol stack. The SIP provides the equivalent of the MAC and physical layers as described in the IEEE 802 standards. SIP minimizes the processing overhead performed at the SNI by the CPE. Minimal processing is particularly important because one of the goals of SMDS is to allow for simple CPE. Higher layers are not defined; instead, the protocol used in the existing bridges, routers, and gateways can be used here.

Level 3 of the SIP accepts data from the higher-layer protocols. The variable-sized SMDS SDU (SMDS data unit) has a maximum size of 9,188 octets, big enough to accommodate IEEE 802.5 (4-Mbps version) and FDDI frames. Larger frames require an internetworking protocol to fragment the frame. Figure 6.3 shows the level-3 protocol data unit (PDU). Because SMDS used the IEEE 802.6 standards, the structure is similar in many ways to the DQDB protocol structure addressed in chapter 7.

The SMDS PDU consists of the following main fields:

- header field
- information field
- packet assembler and disassembler (PAD) field
- cyclic redundancy check (CRC) field
- trailer field

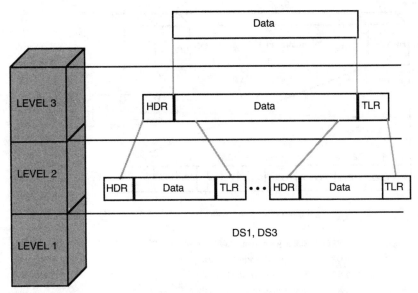

Figure 6.2 SMDS protocol layers.

In Figure 6.3, the field X+ denotes those fields added to provide alignment with the cells produced in layer 2 of the SIP. These fields are not processed by the network. Each field is described in the following subsections.

6.4.1.1 Header field. The header field is 36 bytes in length. The header is where the critical information related to the sender and destination resides. The fields that constitute the header field are the following:

- reserved (r) field
- beginning-end tag (BEtag) field
- buffer allocation size (BAsize) field
- destination address (DA) field
- source address (SA) field
- higher layer protocol identifier (HLPI) field
- PAD length (PL) field
- quality of service (QOS) field
- CRC32 indication bit (CIB) field
- header extension length (HEL) field
- bridging field

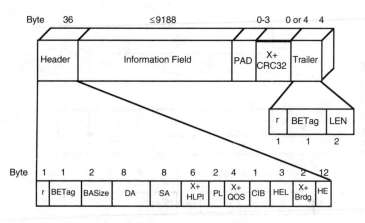

r   Reserved
BETag Beginning-End Tag
BASize Buffer Allocation Size
DA   Destination Address
SA   Source Address
X+   Unchanged by the Network
HLPI   Higher Layer Protocol
        Identifier
PL   PAD Length
QOS   Quality of Service
CIB   CRC32 Indication Bit
HEL   Header Extension Length
Brdg   Bridging
LEN   Length
HE   Header Extension

Figure 6.3 Layer 3 PDU frame format.

6.4.1.1.1 Reserved (r) field. A one-byte field reserved by the standard and set to zeros. Its use is not currently defined.

6.4.1.1.2 Beginning-end tag (BEtag) field. A one-byte sequence number that appears in both the header and trailer and ranges from 0 to 255. The number is used to associate the header and trailer of the level-3 entity at the receiving interface. A counter is maintained by the level-3 entity that is incremented after transmission of a frame so that the values 0 through 255 are cycled through as frames are sent.

6.4.1.1.3 Buffer allocation size (BASize) field. This field is two bytes and indicates the size of the frame. It lies between the destination address and CRC32 fields. It is used to allow the receiving node to determine how many level-2 cells it can expect to receive.

6.4.1.1.4 Destination address (DA) field. This field is the eight-byte address used in SMDS to allow for individual or group addressing. The first four bits of the DA field are set to 1100 for individual addresses and 1110 for group addresses. The remaining 60 bits are used for the destination address, which is the address verified by the SMDS network before forwarding the cell through the network. The SMDS numbering plan has the same structure as the numbering scheme of the ITU-T E.164 ISDN numbers (telephone numbers).

6.4.1.1.5 Source address (SA) field. This field corresponds to the individual sender's address. It is the field verified by the SMDS network with its database of subscribers. If a match exists, the network allows the source to transmit via the SMDS network. In other words, this field checks whether the sender has subscribed to the SMDS services.

6.4.1.1.6 Higher layer protocol identifier (HLPI) field. A 6-bit field that is not processed by the network; it is included to provide alignment of the SIP level-3 protocol with the cells of the SIP level-2 protocol.

6.4.1.1.7 PAD length (PL) field. A two-bit field that indicates the number of bytes in the PAD field and ensures that the level-3 entity is aligned on 32-bit boundaries. This 32-bit alignment is important for efficient implementation of this protocol on reduced instruction set computing (RISC) processors.

6.4.1.1.8 Quality of service (QOS) field. This 4-bit field is currently ignored by SMDS.

6.4.1.1.9 CRC-32 indication bit (CIB) field. This bit indicates the presence of the CRC-32 field. If the bit is set to 1, CRC is present.

6.4.1.1.10 Header extension length (HEL) field. A 3-bit field indicating the number of 32-bit words that populate the header extension (HE). At present, this field is set to 3 (binary 011) to indicate the fixed extension of 12 bytes.

6.4.1.1.11 Bridging field. This 2-byte field provides 32-bit alignment.

6.4.1.2 Information field. The next field of the layer-3 protocol is the information field, where the actual payload or data is carried. This field carries data from the higher layers. The payload can be up to a maximum of 9,188 bytes.

6.4.1.3 Packet assembler and disassembler. The PAD field ensures that the information field is aligned on 32-bit boundaries. This field varies from 0 to 3 bytes in length.

6.4.1.4 Cyclic redundancy check CRC-32. If present, this field provides 32-bit error detection for the fields, covering the fields from the destination address up to and including CRC-32. This field does error detection in the following way: on the transmitting side, a calculation is performed on the bits of the PDU to be transmitted. The result, called an *error-detecting code*, is inserted as an additional field in the packet or frame. A calculation is performed on the same CRC-32 field on the received bits, and the calculated result is compared to the value stored in the incoming frame. If a discrepancy exists, the receiver assumes that an error has occurred and discards the PDU. Thus, CRC-32 performs error detection on the receiving side.

6.4.1.5 Trailer field. This field is at the end of the frame-relay frame and consists of four bytes. The trailer field consists of three fields: reserved field, BEtag field, and length field (LEN). The first two fields are the same as in the header.

The BASize field value is placed in the length field (LEN) as an additional check to ensure the correct assembly of the SIP level-2 cells. These fields provide for the delivery of information and allow a number of checks to be conducted to ensure correct delivery. Errors that cause frames to be discarded include the following:

- if the header and the trailer BEtag fields do not match

- if the BASize field in the header and length field in the trailer do not match

- if destination and source address formats are incorrect

The SMDS standards do not define how the CPE should react when delivery does not occur. This function is assumed to be part of the higher layers and is beyond the scope of these standards.

6.4.2 Layer 2 of SMDS interface protocol

The SIP level-2 protocol provides access control to the MAN. The level-2 SIP format is shown in Figure 6.4.

Level 2 of SMDS is compatible with level 2 of IEEE 802.6; thus, equipment conforming to IEEE 802.6 (DQDB) protocol can work in an SMDS network. The level-3 frame is segmented into 44-byte data units for transmission in the level-2 cells. As with fragmented DQDB PDUs, the first cell is beginning-of-message (BOM) segment type, intermediate cells are continuation-of-message (COM) segment type, and the final cell is end-of-message (EOM) segment type, providing the receiving station with sufficient information to reassemble the frame.

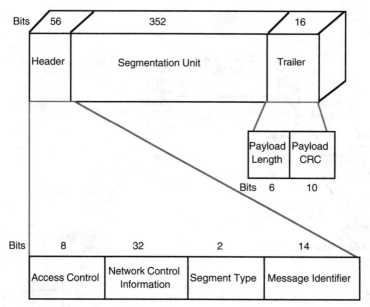

Figure 6.4 SMDS layer 2 frame format.

6.4.3 Layer 1 of SMDS interface protocol

The SMDS access path is described in SIP level 1. This specification provides for the transmission of level-2 cells across the SNI. The operation is divided into the physical layer convergence protocol (PLCP) and transmission system sublayers. The transmission system sublayers define the digital carrier systems that can be used for the SNI. Currently, the specifications support DS-1 and DS-3. The PLCP defines how the level-2 cells are mapped onto the transmission systems. Thus to maintain compatibility with IEEE 802.6 systems, the PLCP is extracted directly from the IEEE standards.

6.5 How Does SMDS Work?

The basic network of SMDS is shown in Figure 6.5. In the network, there is an originating and destination router, where the traffic is originated and terminated, respectively. The routers are connected via a high-speed serial interface (HSSI) running at 34 Mbps to an SMDS CSU/DSU, which in turn is connected to the public SMDS network at 45 Mbps, or at DS3 speed using the SIP specification. It is not necessary to have an HSSI interface; it can be bypassed and connected directly to the network.

The router generates the datagram traffic of variable frame length. On re-

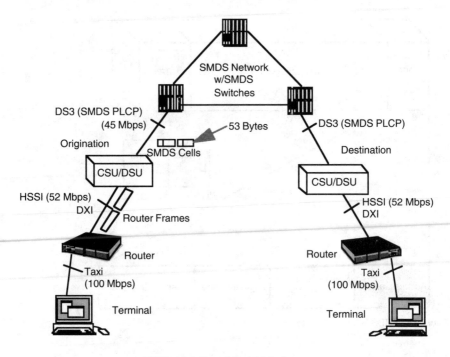

HSSI : High Speed Serial Interface
PLCP : Physical Layer Convergence Protocol

Figure 6.5 Example SMDS network.

ceiving the frame the CSU/DSU segments the frame into 48 bytes and adds the 5-byte header, to make cells totaling 53 bytes. The cells are generated at a speed of 45 Mbps. Figure 6.6 shows the SMDS cells at the SMDS switch. Once the SMDS cells reach the SMDS switch, the switch performs the following functions:

- buffers the 5-byte header and the 48-byte payload
- performs CRC check on cell payload
- reads the segment type as BOM, COM, or EOM

Once the SMDS switch receives one of these messages, it processes each datagram according to the segment type. In our case, assume that the first datagram received by the switch is of type BOM, with valid source and destination addresses.

Once the segment type is identified as BOM by the SMDS switch, the switch performs the following operations:

- records the message identified (MID) in a new temporary location called *call record* entry

- records the sequence number of this MID
- performs source address validation
- performs destination address screening
- associates destination address with MID in the call record
- if no routing table entry exists with this address, performs the shortest-path first routing calculation and stores the routing information in the routing table, which is located in cache memory for quick retrieval
- enters the routing information for the MID into call record
- increments to the next expected sequence number
- routes the cell to destination ISSI or ICI

On processing the datagram with BOM segment type, the switch starts receiving the cells. The other cells following the first arrive with COM as their segment type from the level-2 PDU of the router. On receiving the new segment type, the switch performs the following functions:

- looks up the temporary call record by the message ID for this particular SNI
- verifies sequence number

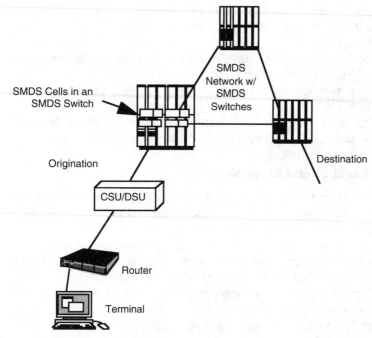

Figure 6.6 Cells at SMDS switch.

- increments to next expected sequence number
- routes PDU based on the routing information in the call record

Upon processing the cell (or cells) with COM segment type, the last cell arrives at the switch with an EOM segment type, indicating that this is the last of the series of cells and the end of the message. Once the EOM is received by the SMDS switch, the following functions are performed by the switch:

- looks up the temporary call record by the message ID for this particular SNI
- verifies sequence number
- routes PDU based on the routing information in the call record
- clears the temporary call record
- completes billing data collection at the ingress (destination) switch

If the cells must traverse more than one SMDS switch, the above-mentioned functions are performed at each switch until the destination address is matched by the SMDS switch. This process is called a *connectionless service*. In other words, there is no predefined connection between the source and the destination. Once the switch identifies the destination address and knows that it terminates at its location, it then forwards the cells to the SMDS CSU/DSU, which is usually located at the customer premises (Figure 6.7).

On receiving the cells, the CSU/DSU translates the cells into frames, maps the header address to the router address, and forwards it to the router. Thus, the SMDS switch routes each of the cells to the destination router. This router then sends the frames to the appropriate terminal, which is usually a part of the router address. One of the special features of SMDS is that customers can manage the network like their own private networks.

6.6 SMDS Features and Services

SMDS provides a number of features and services beyond the simple transmission of data through the network. In fact, SMDS differs from other protocols in the number of features provided to its users. For example, the security feature allows network security to screen the frames to verify if they can be legally transmitted to the destination or received from the source. This security feature allows a subscriber to build a virtual private network within a public network. Corporations can use the screening facilities to limit access to their resources. This feature could provide a single point of entry for external systems, allowing for the control of information flow into and out of the corporation. This feature is also used by a number

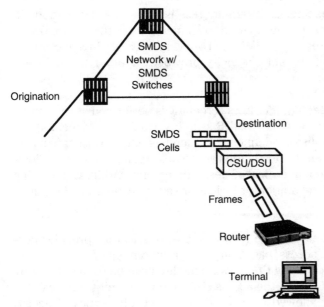

Figure 6.7 Cells at SMDS destination.

of companies to connect to the public backbone network. SMDS offers a variety of services both to the end user and the network operators. These services can be segmented into end-user features and network operator features.

6.6.1 End-user features

SMDS, as a service, provides numerous features to the user:

- address validation
- group addressing
- address screening
- access classes
- congestion control

Each is described in the following subsections.

6.6.1.1 Address validation. The network ensures the validity of the source address using address validation. The network verifies the source address at the SNI interface. If the address is invalid, the network prevents the user from using the SMDS network services.

6.6.1.2 Group addressing. A group address is used as a destination, allowing multicasting of the information to all members of the group. Group addressing is performed by the SMDS network. This feature in SMDS is similar to the point-to-multipoint capability provided by other technologies, such as ATM.

6.6.1.3 Address screening. Address screening is performed at the source on the destination address. The SMDS switch verifies the source address and allows delivery from a predefined list of senders only. If verification performed at the source on the destination addresses is valid, the switch allows transmission to a predefined list of destinations only. Address screening thus allows the implementation of logical private networks on the public network.

6.6.1.4 Access classes. As the name implies, this feature provides customers with different access, based on traffic requirements or characteristics. Although the capacity of the physical medium can be used entirely by the CPE for bursts of traffic, the access class defines a limit to the average rate of data transfer allowed over a longer period. Five types of access classes have been identified for DS3:

- 4 Mbps
- 10 Mbps
- 16 Mbps
- 25 Mbps
- 34 Mbps

For each access class, the maximum DS3 speed can be used for the first 9,188 bytes transferred in a burst. For subsequent bytes, the network operator can drop the data in excess of the subscriber's committed rate.

Access classes are useful for service provisioning. The operator can dimension a network based on the subscribed bandwidth rather than on the throughput of the access media. Access classes also provide flexibility in charging. More precisely, users can subscribe based on the power of their CPEs (i.e., the maximum bandwidth their CPE can use) rather than on the bandwidth of the access medium. In a regulated environment, access classes are mandatory for the RBOCs to be allowed to charge less than full DS3 access rates to subscribers who need only a fraction of the 45-Mbps bandwidth.

Access classes are used in performing congestion control. SMDS relies on the access classes, which correspond to the rate of traffic and the burstiness subscribed to by the user. In case of congestion, the user is neither notified explicitly nor is expected to reduce the traffic. The traffic measurement specified in SMDS simply allows the acquisition of information on traffic patterns.

6.6.2 Network operators features

Bellcore developed SMDS operations, administration, and maintenance (OAM), and billing features for network operators.

6.6.2.1 OAM. Many features are defined for OAM. The objective is to standardize the interface with service providers' OSS, which can provide a unified view, independent of vendors and switch equipment. The main functions covered are memory management, maintenance, traffic management, network data collection, customer network management, and other status and usage information.

6.6.2.2 Billing. One of the most important features that any service requires is a good billing mechanism. SMDS is a very good example of such a service. The billing system developed for SMDS consists of the description of usage measurement, performance, and operations-related information. Under performance objectives, the billing system addresses several areas. Examples are the maximum transit delay for a packet, probability of loss, error, missed delivery of packets, and service availability status. In fact, this billing concept is used as an example by the International Standards Organization (ISO) for its work on BISDN standards.

6.7 Summary

SMDS is a high-speed protocol designed by Bellcore and backed by its clients, the RBOCs. From a standards perspective, the strength of SMDS is that it is defined by a coherent organization. The objective was to design a broadband protocol to provide public broadband data services. SMDS is a good candidate for customers who require high throughput or low latency delay for data transfer between disparate locations. Currently, customers can access the SMDS network via existing DS1 and DS3 facilities using the IEEE 802.6 DQDB protocol used for layer 2. In the future, DS1 and DS3 will be replaced by SONET STS3c.

Advanced SMDS

7.1 Overview

In chapter 6, an overview of switched multimegabit data service (SMDS) and its functions was provided. This chapter investigates some of the advanced portions of SMDS, such as the interface protocols. The interswitching system interface (ISSI) and ICI (intercarrier interface) are described. The ISSI interface defines the specifications required to connect two switching systems (either SMDS switches or an SMDS switch and another switch), and the ICI specifies the interface between switches in two networks.

7.2 ISSI Specification

The ISSI is an interface between two switches in an SMDS switching system. Its specifications are based on the ISSI protocol. The protocol architecture has three levels that do not necessarily correspond to layers or the levels of any other model, such as SIP or OSI. The lower two levels are based on the Distributed Queue Dual Bus (DQDB) Metropolitan Area Network (MAN) protocol described in the IEEE 802.6 standard. Figure 7.1 shows the three levels of ISSI architecture. Each level of this architecture performs certain functions to map the information onto the transmission system according to SMDS interface specifications, so equipment from different vendors can interoperate.

ISSI level 1 of the protocol provides ISSI level 2 with a physical interface to the digital transmission network. ISSI level 2 provides for the transport of variable-length ISSI level-3 data transport protocol data units (L3-DTPDUs)

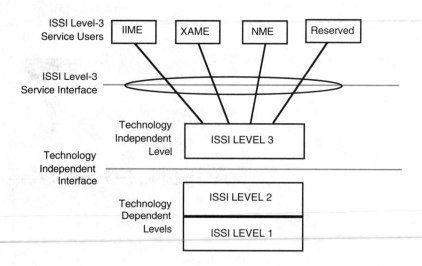

IIME: Intra-LATA Intra-network SMDS Mapping Entity
NME: Network Management Entity
SAP: Service Access Point
XAME: Exchange Access SMDS Mapping Entity

Figure 7.1 ISSI protocol architecture.

across ISSI links using fixed-length (53 bytes) ISSI cells, which are also called SMDA cells. ISSI level 3 provides routing, relaying, forwarding, congestion management, and maintenance functions to transport ISSI level-3 service user PDUs. ISSI level 3 provides a basic packet switching service to the ISSI level-3 service users. Currently, three ISSI level-3 service users have been defined:

- intra-LATA, intranetwork SMDS mapping entity (IIME)
- exchange access SMDS mapping entity (XAME)
- network management entity (NME)

These are a logical collection of SMDS switching functions that support their respective SMDS services. Thus, ISSI not only provides an interface between SMDS switching systems, but also between other switching systems, such as BISDN.

7.2.1 ISSI level-1 specification

The basic service of level 1 is to provide ISSI level 2 with an interface to the physical path. This physical path is usually the same, even for other technologies such as frame relay or ATM. ISSI level 1 is divided into two parts, as shown in Figure 7.2, the physical layer convergence procedure (PLCP), and the transmission system sublayer.

The transmission system sublayer defines the characteristics and the method of attachment to the transmission link. Currently, two types of transmission system sublayers are defined for ISSI: DS3 and SONET STS-3c. This layer is common, regardless of the protocol being carried by the transmission system.

The upper layer of level 1 is the PLCP. Its function is to adapt the service of the transmission system sublayer into the generic physical layer service. The PLCP defines a method of mapping the ISSI level-1 control information and the ISSI cells into a format suitable for the transmission system sublayer. The DS3 PLCP is based on IEEE 802.6, and SONET PLCP is based on IEEE 802.6. This PLCP is the one that varies from protocol to protocol because it must perform protocol mapping from another protocol to the transmission system protocol. Figure 7.3 shows how the SMDS cells are mapped into a SONET STS-3c frame. In the figure, the cells are carried on the SONET payload (described in chapter 9) with a pointer that indicates the beginning of sequenced cells.

7.2.2 ISSI level-2 specification

ISSI level-2 function is to transport the variable-length ISSI level-3 data transport protocol data units (L3-DTPDUs) between SMDS switching systems. The level-2 data transport protocol (L2-DTP) is based on the IEEE 802.6 standards, which are similar to the DQDB protocol. The level-2 protocol provides functions such as bit error detection and framing for the variable-length level-3 data units. In addition to providing error detection, this layer provides segmentation of the variable length data into 44-byte SMDS cells and reassembly of the data.

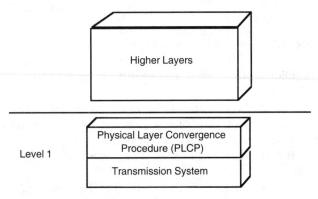

Figure 7.2 Level-1 layers of ISSI protocol.

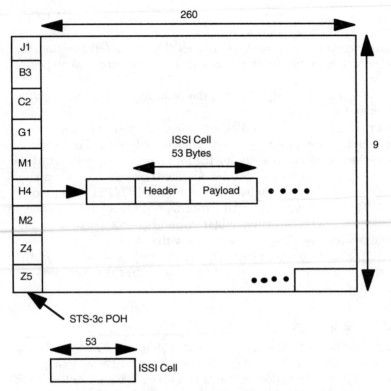

Figure 7.3 Mapping SMDS cells to SONET STS-3c frame.

In level 2, data units are divided into three functional blocks:

- convergence function block
- segmentation and reassembly function block
- cell function block

These functional blocks are shown in Figure 7.4.

7.2.2.1 Convergence function block.   On the transmitting side, the convergence block performs the function of creating the variable-length convergence PDU (CVG-PDU). A data header and trailer are then added to the level-2 service data unit. This CVG-PDU, its priority, and the virtual channel value are forwarded to the next functional block in level 2.

On the receiving side, the convergence block validates data units, including checking for bit errors, etc. It extracts the ISSI L2-PDU from the validated C-PDU. The CVG-PDU is then forwarded to level 3 of ISSI protocol.

7.2.2.2 Segmentation and reassembly function block. On the transmission side, the SAR block provides segmentation of the variable-length CVG-PDU into a fixed 44-byte segmented unit. A two-byte header and a two-byte trailer are appended to this unit, thus making the cell length 48 bytes. This segment unit is called SAR-PDU. It, along with its priority and the virtual channel value, is forwarded to next functional block.

On the receiving side, the SAR block validates the SAR-PDU and re-assembles the CVG-PDU to be forwarded to the next functional block in level 2.

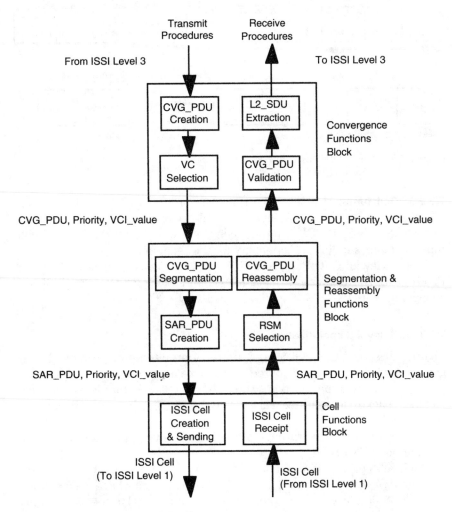

Figure 7.4 ISSI level-2 functional architecture.

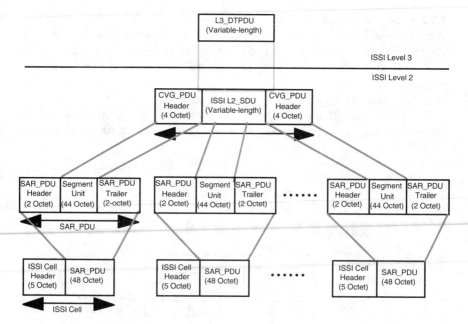

Figure 7.5 Overview of ISSI layers interworking.

7.2.2.3 Cell function block. For transmission of an L3-DTPDU, the cell function block creates the 53-byte ISSI cells by adding a 5-byte cell header to each 48-byte SAR-PDU. On the receiving side of the L3-DTPDU, the cell function block supports functionality for the reception of the ISSI cell.

An overview of CVG-PDU, SAR-PDU, and ISSI cells formats is shown in Figure 7.5. The detailed formats for CVG-PDU, SAR-PDU, and ISSI are shown in Figures 7.6, 7.7, and 7.8, respectively.

7.2.3 ISSI level 3 specification

Figure 7.9 shows the ISSI level-3 protocol architecture, positioned as a generic packet-switching resource. A *resource* is an entity capable of transporting a variety of packet types through the network. ISSI level-3 service includes packet forwarding, relaying and receiving, route management, congestion management, and other layer management functions related to the transport of packets through a multi-SMDS switching system network that supports SMDS.

ISSI level-3 protocols and their respective procedures collaborate to provide the ISSI level-3 service. Three ISSI level-3 protocols have been defined so far. They are level-3 data transport protocol (L3-DTP), the ISSI routing management protocol (RMP) and the level-3 layer management protocol (L3-LMP).

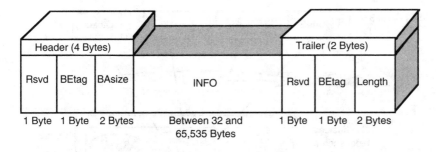

Rsvd   Reserved

BEtag   Beginning-End Tag

BASize   Buffer Allocation Size

Figure 7.6 Convergence protocol data unit (CVG-PDU) format.

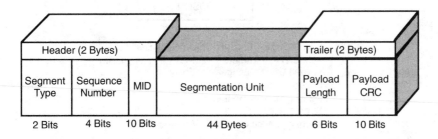

MID: Message Identifier
CRC: Cyclic Redundancy Check

Figure 7.7 Segmentation and reassembly (SAR) SAR-PDU format.

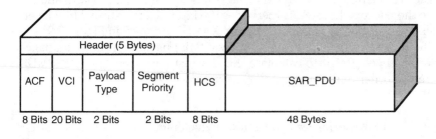

ACF   Access Control Field
VCI   Virtual Channel Identifier
HCS   Header Check Sequence

Figure 7.8 ISSI cell format.

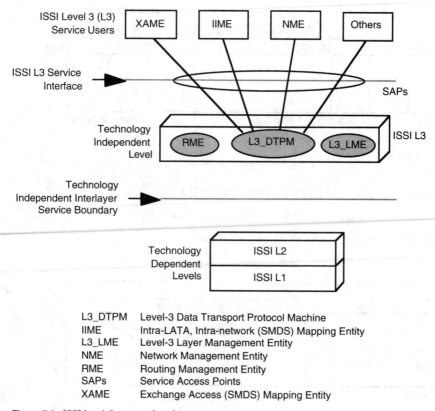

Figure 7.9 ISSI level-3 protocol architecture.

L3_DTPM	Level-3 Data Transport Protocol Machine
IIME	Intra-LATA, Intra-network (SMDS) Mapping Entity
L3_LME	Level-3 Layer Management Entity
NME	Network Management Entity
RME	Routing Management Entity
SAPs	Service Access Points
XAME	Exchange Access (SMDS) Mapping Entity

Figure 7.9 shows the different functions relating to the level-3 protocol.

The service function provided by layer 3 is to assemble, forward, and receive the packets. Level 3 also serves as the interface to external ISSI level-3 service users, such as IIME, NME and XAME, as shown in Figure 7.9.

The ISSI routing management entity (RME) uses the routing management protocol (RMP) to calculate the shortest path from the host SMDS switch to other SMDS switches in the network. This shortest path information is used by L3-DTPM to forward and relay the packets in the network.

The L3-LME performs congestion management functions and other layer management functions that ensure the correct operation of ISSI level 3. This layer ensures that no congestion exists on the route through which the packet is transmitted. It also checks the status of any congestion in the link.

Table 7.1 maps the functionality defined in level 3 to the level-3 protocols.

TABLE 7.1 Protocol-to-Entity Mapping

Protocols	Functional entity
Level-3 data transport protocol (L3-DTP)	L3-DTPM (machine)
Level-3 routing management protocol (RMP)	Routing management entity (RME)
Level-3 layer management protocol (L3-LMP)	Layer management entity (L3-LME)

7.2.3.1 Level-3 data transport protocol machine specification (L3-DTPM).
The service provided by ISSI level 3, specifically L3-DTPM, allows a level-3 service user to exchange information with its peer users. ISSI level 3 provides point-to-point and point-to-multipoint connectionless service modes. This service has the following characteristics:

- No acknowledgment is sent if the data transfer must be expedited.

- The order of information is preserved under normal conditions. In case of congestion or any such abnormal condition, the order of information is not preserved.

- Both syntactical and semantic errors are detected. Data units containing such errors are discarded.

Figure 7.10 shows the L3-DTPM format, which is where the actual data is carried. The data transport protocol machine format carries the necessary information to perform its functions of forwarding, relaying, and receiving. Most of the header fields are self-explanatory, such as source address, destination address, etc.

In addition to these fields, some fields identify the priority of the information, the length of the information carried, and the type of service level protocol, such as RMP, L3-LMP or L3-DTP. The alignment field ensures the alignment of the L3-DTPDU format with that of IEEE 802.6 protocol formats.

7.3 Intercarrier Specification

The ICI is an interface or boundary between the networks of two carriers that provide SMDS service. Typical networks involved are an IEC and an LEC that exchange access SMDS service, and two LECs that serve in some LATA or between IECs of two different countries.

The ICI is a part of an ICI transmission path that designates the boundary between the carrier networks. The ICI typically begins at the distribution frame or other equipment where the LEC's access ends and where cross-connection, testing, and service verification occur. An ICI transmission path is a telecommunications facility that connects the switching systems of two carriers. The protocol for communication across an ICI transmission path is the ICI protocol. This transmission path can be DS3 or SONET/SDH.

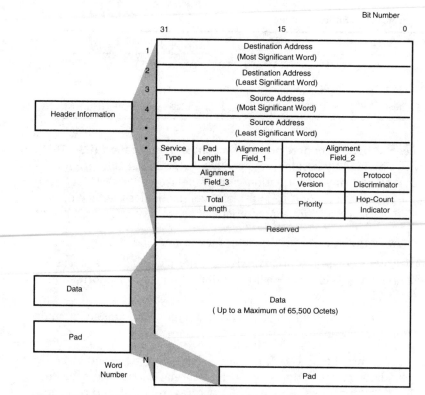

Figure 7.10 Level-3 data transport protocol machine (L3-DTPM) format.

The intercarrier interface protocol (ICIP) is a connectionless protocol for transporting variable-length data units. The protocol operates on a point-to-point basis and provides functions such as transport of variable-length data units, addressing, error detection, framing, and access to the transmission path. The protocol meets the requirements of service features, performance, and usage measurement. It also supports growth for new capabilities. Figure 7.11 shows an example of the physical arrangement of ICI. Figure 7.12 illustrates the functional organization of the ICIP. The ICIP provides a collection of functions to its service users.

The services defined for the ICI protocol are the following:

- originating exchange access SMDS service
- terminating exchange access SMDS service
- network functions for inter-LEC serving arrangement
- system management

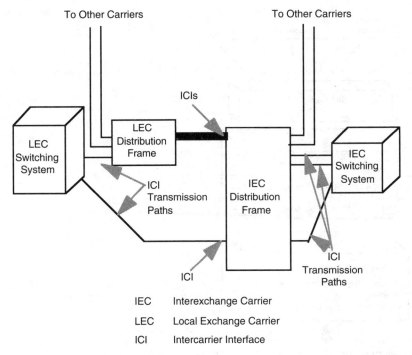

Figure 7.11 Example of physical arrangement.

Each of the above services provide certain functions that support SMDS. They use variable-length data transfer layer functions to transfer information. The information transferred consists of service-specific information needed to provide SMDS service. The service-specific functions also provide special padding functions to simplify interworking across the networks of the end user, IEC, and LEC.

The functions of the variable-length data transfer layer are needed for the connectionless transfer of variable-length data units. This layer segments and reassembles the variable-length data unit into fixed-length units. It also detects bit errors and lost or misordered segments.

The function of the physical layer provides access to the transmission path between adjacent switching systems.

7.3.1 ICI protocol

Figure 7.13 illustrates the organization of the protocol and its relation to the functions. Level 3 of the ICIP implements functions for the connectionless transfer of variable-length data units to the adjacent switching system. Level 3 provides addressing, transferring of the necessary service-specific information, transferring of user data, and detecting the loss of ICIP L2-PDUs.

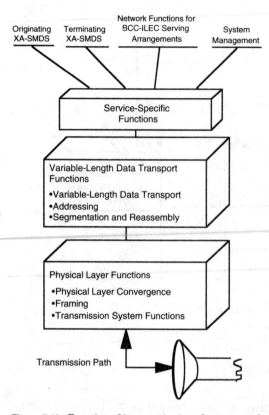

Figure 7.12 Function of intercarrier interface protocol.

Level 2 of the ICIP segments and reassembles the variable-length data units from level 3 into fixed-length data units. It also provides bit error detection. Level 1 provides access to the transmission path between the adjacent switching systems. It can be either DS3 using PLCP or SONET/SDH.

Figure 7.14 shows how the data from each level are encapsulated to the next level in the ICI protocol. Figures 7.15 and 7.16 show the level 3 and level 2 PDU frame formats, respectively. Level 1 is standard DS3 or SONET.

7.4 Summary

The interfaces in the SMDS reference network architecture, such as ISSI and ICI, were described. In addition, the function, protocol, and frame format for each level of the protocol are explained. The ISSI interface protocol describes the interface between the two SMDS switching systems of the same network. The ICI interface protocol describes the interface between two different networks: the local access network and backbone network.

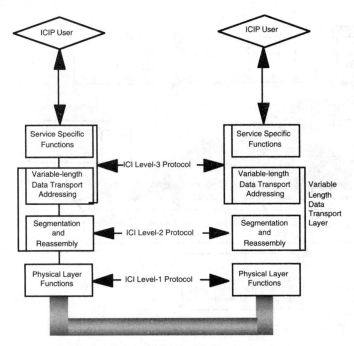

Figure 7.13 Organization of the ICI protocol.

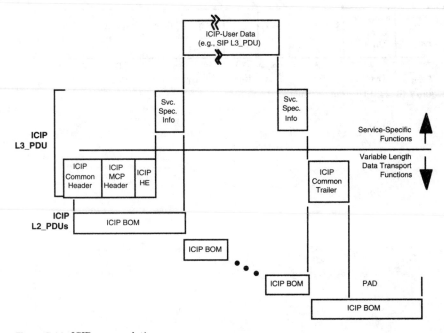

Figure 7.14 ICIP encapsulation.

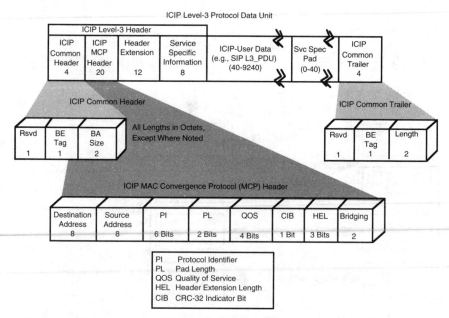

PI   Protocol Identifier
PL   Pad Length
QOS   Quality of Service
HEL   Header Extension Length
CIB   CRC-32 Indicator Bit

Figure 7.15 L3-PDU frame format.

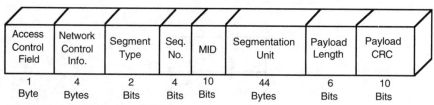

Figure 7.16 L2-PDU frame format.

8

ATM

8.1 ATM Overview

ATM does not stand for automatic teller machine. In the telecommunications world, it stands for *asynchronous transfer mode*, whereby information packets are transferred asynchronously. This mode is another fast-packet switching mode. The first research on ATM and its related techniques was published in 1983 by two research centers, CNET and AT&T Bell Labs. In 1984, the research center of Alcatel Bell in Antwerp started to develop the ATM concept.

ATM has the same basic characteristics of packet switching, but also the delay characteristics of circuit-switching technology. This combination is obtained by reducing the network functionality to a minimum. Initially, different names were proposed, and the standards organization settled on the famous acronym *ATM*. ITU selected ATM technology as the switching technology, or the transfer mode, for BISDN, the foundation for all broadband communications. ATM is regarded as the technology of the 21st century, and its impact is expected to be similar to PCM (pulse code modulation), which is used widely around the world in telecommunications today.

The word *asynchronous* is used because ATM allows asynchronous operation between the sender clock and receiver clock. The difference between both clocks can be easily solved by inserting and removing empty or unassigned cells (packets that do not contain any information). We look into this in detail when we study ATM in the BISDN protocol. One of the special features that ATM provides is that it guarantees successful transport of any service (CBR or VBR) regardless of the charac-

teristics of the originating service in terms of bit rate, quality requirements, or the bursty nature of traffic. ATM is applicable in all network environments, which is why most of this book is dedicated to the
implementation of ATM. A network with such a service-independent
transfer technique cannot suffer from disadvantages when compared to
other transfer modes in terms of service dependence, inefficiency in the
use of available resources, etc.

Before we leave this section, an analogy is provided to help better understand ATM. We have all seen railway trains and railway lines (tracks). Have
you ever wondered why all coaches are the same size, regardless of the type
of cargo or passenger they carry? Why can't the railway department build
differently sized coaches—one size for passengers, one for cargo? The reason is that it is easier to build only one size. The railroad then has the flexibility to add or drop coaches at intermediate junctions so that coaches can
be put onto different tracks to reach different destinations. The same is
true with ATM cells. By having a uniform ATM cell size, routing, adding,
dropping, and multiplexing of ATM cells can be done faster without worry
about the information carried within the ATM cells. In sum, the advantages
ATM offers are:

- flexibility—ATM can be easily evolved for future services
- efficiency in the use of its available resources
- it is one simple universal network
- reduction of operation, administrative, and maintenance costs
- reduction of transport costs (by statistical multiplexing)
- provision of dynamic bandwidth allocation

8.2 The ATM principle

This section addresses the basic principle behind ATM: divide and conquer.
Sometimes, the best way to manage large chunks of information is to split
the information into the smallest possible units, thus making the units easy
to handle. For example, most of us have traveled in our lifetime with children. Sometimes the children pack all their stuff in one bundle and expect
you to pack it into your suitcase. But you don't have space in your suitcase
to keep the whole bundle as one. So what do you do? You unpack the sack
and distribute the contents to different parts of the suitcase, wherever
space is available.

Keeping in mind the strength of segmentation, let's come back to the
principle behind ATM. ATM does not care what the information is or its
form. It simply cuts the information into equal-sized packets or cells and at-

taches a header so the packet can be routed to its destination. The headers in ATM have very little functionality, so they can be processed by the network without delay. Figure 8.1 shows how traffic of different speeds, namely 64 kbps, 2 Mbps, and 34 Mbps, is chopped into equal-sized packets or cells by a "chopper" or "cell slicer." The different cells are put into a huge transmission pipe that mixes all the cells from different sources in such a way that the transmission pipe is optimized. The optimization is done via a technique called *statistical multiplexing*.

In an ATM network, several sources are combined or multiplexed on a single link. In a conventional time division multiplexing (TDM) network, the effective bandwidth is simply a sum of the individual sources' bandwidth. If two sources of bandwidth are x bps and y bps, their effective bandwidth is $(x + y)$ bps. The effective bandwidth in an ATM network, however, is z bps, where $z < (x + y)$ because all the information in bits is packed into ATM cells. The ATM switch then multiplexes the cells that carry valid information, thus discarding cells with zero or invalid information. Thus, effective bandwidth is reduced. This bandwidth can also carry other users' traffic. Figure 8.2 shows a comparison of conventional TDM with statistical multiplexing. From the figure, you can see that bandwidth is wasted in TDM due to fixed allocation of the bandwidth. In the ATM environment no bandwidth is wasted, because of statistical multiplexing. To achieve statistical multiplexing, all traffic, including voice, must be packetized, thus creating VBR traffic.

The ATM switch does not differentiate between the type of traffic carried within a cell. All it knows is that it has an input port, where cells come in, and a destination port, where cells go.

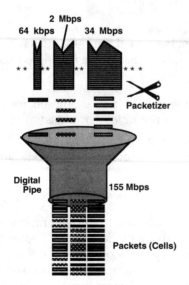

Figure 8.1   ATM principle.

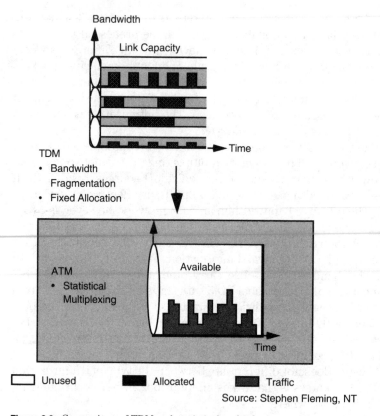

Figure 8.2 Comparison of TDM and statistical multiplexing.

A decision must also be made on the size of the cells or packets. Many issues come into play when making this decision. The most important among them are:

- *Transmission efficiency*—the larger the packet, the higher the delay. The smaller the packets, the higher the ratio of overhead to information.

- *Delay*—different delays are encountered by a packet, such as basic packet transitive delay, queuing delay at each switching node, jitter, packetization, and depacketization.

- *Implementation complexity*

Many other contradicting factors in addition to the ones mentioned above influence the choice of cell sizes. After long debate and argument by the ITU-T committee, the final decision was to choose between 32 and 64 bytes. This choice was mainly based on delay characteristics, transmission efficiency, and implementation complexity. Europe was more in favor of 32 bytes (because of echo cancelers for voice), whereas the United States and Japan were more in favor of 64 bytes for transmission efficiency. Thus, in June of 1989 in Geneva, ITU-T in its SGVIII meeting reached a compromise of 48 bytes. It then added 5 more bytes for header information. Thus, the peculiar 53-byte packet size came into existence for ATM.

8.3 ATM Protocol

The 53-byte ATM packets are routed through the network. As mentioned earlier, these cells carry an information payload of 48 bytes and a 5-byte header, as shown in Figure 8.3. Table 8.1 compares the packet switching

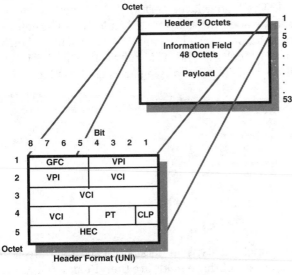

Figure 8.3 ATM cell format.

TABLE 8.1 Comparison of Technology Functionalities

Functionality	X.25 packet switching	Frame relay	ATM switching
Packet retransmission	√	—	—
Frame delimitation	√	√	—
Error checking	√	√	—

technologies X.25, frame relay, and ATM in terms of functionality. The existing packet-switched network (X.25) does all three functions mentioned, such as packet retransmission, frame delimitation, and error checking.

As mentioned earlier, one of the advantages of ATM is the reduced functionality of the ATM network, which is caused by using a smaller ATM header. As can be inferred, the reason for the reduced header is to simplify the switching and processing functionality in the network.

Figure 8.4 shows the protocol stack for ATM. The functions not performed by the ATM layer are addressed by the upper layers in the protocol. ATM basically forms layer 2 of the BISDN protocol (further explained in chapter 10).

The higher layers convert the information into chunks of 48 bytes that can be processed by ATM. The higher layer performs the functions mentioned earlier. The ATM layer adds the 5 bytes of header to each cell as shown in Figure 8.3. The header attached at the ATM layer carries sufficient information to route the cells in the ATM network. The ATM cell header consists of six different fields with varying sizes (in bits) based on their functions:

- generic flow control (GFC)
- virtual channel identifier (VCI)
- virtual path identifier (VPI)
- payload type (PT)
- cell loss priority (CLP)
- header error control (HEC)

Each field has a certain functionality. The details of each field are addressed later. The most important of these fields are the VPI and VCI, used for routing information in the ATM network. Before we explain routing in

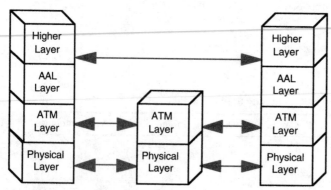

Figure 8.4 ATM protocol stack.

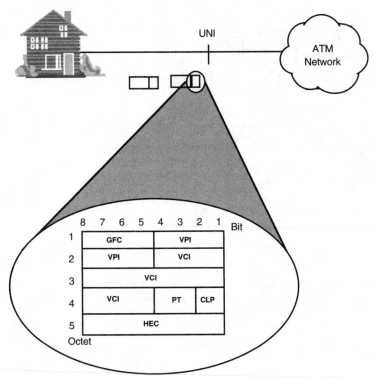

Figure 8.5 ATM cell at user-network interface (UNI).

the ATM network, let's compare the VPI and VCI to something used in to-day's telecommunications world:

VCI-based switching is similar to TDM in circuit switching.

VPI-based switching is similar to digital cross-connect (or slow switch-ing).

Within the ATM cell header, there are two formats: the user-to-network interface (UNI format), which is the header format for the cells between the user and the network as shown in Figure 8.5, and the network-to-node in-terface (NNI format), which is the header format when the cell is between the switching nodes, as shown in Figure 8.6.

8.3.1 Generic flow control

The GFC is envisaged to provide contention resolution and simple flow con-trol for shared medium-access arrangements at the customer premises equipment (CPE). Thus, the GFC field is present at the cells between the user and the network.

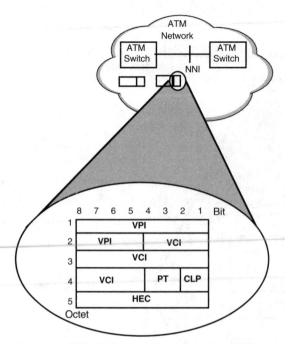

Figure 8.6 ATM cell at network-node interface (NNI).

8.3.2 Virtual channel identifier

The VCI is used to establish connections using translation tables at switching nodes that map an incoming VCI to an outgoing VCI. Circuits established using VCI's connections are referred to as virtual circuits, and VCI's end-to-end connection is called a *virtual connection*. In this sense, that bandwidth is not utilized unless user information is actually transmitted. The VCI field in the header of the ATM has 16 bits. VCI is further discussed later in this chapter.

8.3.3 Virtual path identifier

The VPI is used like VCI to establish a virtual path connection for one or more logically equivalent VCIs in terms of route and service characteristics. The VPI allows simplified network routing functionality and management. The VPI field has 8 or 12 bits, depending on the location of the ATM cell. The VPI is used in setting up the end-to-end virtual path connection of multiple virtual path segments. A virtual path contains multiple virtual channels.

8.3.4 Payload type

The PT, a three-bit field, is used to differentiate the cells traversing the same virtual circuit. Cells can contain operation, administration, and maintenance (OAM) information or user information.

8.3.5 Cell loss priority

The CLP is used to explicitly indicate cells of lower priority, by setting this one-bit field to 1. The lower-priority cells might be discarded by the network, depending on network conditions. This bit is the only priority provided by the ATM network. Any other priorities are at a higher layer of BISDN protocol.

8.3.6 Header error control

The HEC performs a CRC calculation on the first four bytes of the header field for error detection and correction. The HEC sequence is utilized for reducing cell loss and misrouting because of cell header errors. The function of the HEC is described later. HEC performs error control only on the header of the ATM cell. No error control is performed by the ATM network for the actual payload.

The details of these two formats, their differences, and the use of each of the fields are explained in later chapters when we address the ATM layer in the BISDN reference model.

8.4 How Does ATM Work?

So far, we have seen ATM as a technology and a set of principles. Now we see how this technology can be applied as a service. A user can get ATM service in two ways—by setting up either a permanent virtual circuit (PVC) or a switched virtual circuit (SVC). Currently, completed standards are not available for SVCs. In this section, we address both types of methods. Table 8.2 shows the relationship of these two methods to today's environment.

TABLE 8.2 Comparison of PVC and SVC to Today's Environment

ATM service	Today's equivalent
Permanent virtual circuit	Private line service (PL)
Switched virtual circuit	Switching concept similar to telephone network for voice from a user perspective

Let's first address how a PVC service is provided in an ATM environment. PVCs, as mentioned earlier, are similar to private lines. In today's private line environment, the user calls the service provider requesting a private line from point A to point B. The service provider, based on circuit path and availability, "nails up" a circuit based on the capacity requested by the user. It usually takes from 10 days to two months to get the circuit nailed up. Usually, the contract by the user with the service provider is for a couple of years. The user is committed to pay for the circuit even if it is not used for the duration of the contract.

8.4.1 PVC

Figure 8.7 shows the ATM PVC in an ATM network. In setting up a PVC, the following procedures, similar to requesting a private line service, are performed:

1. User calls the service provider with a request for PVC.
2. User provides the destination address, average bandwidth requirement or committed information rate, and duration of the PVC circuit.
3. Operator enters the information on the control terminal to set up circuit path. This step is done almost in real-time, as the user is on the telephone requesting a connection.
4. Circuit is established as requested.
5. User pays a monthly fee for a circuit and pays only for usage of that circuit. If that circuit is not used, the user pays only the monthly circuit fee. This contact is just like a basic monthly telephone bill, where the customer pays a fixed amount regardless of the usage of the telephone service.

PVC is similar to a private line but has numerous advantages, including the following:

- negligible provisioning time (almost real-time)
- almost real-time availability of the circuits
- bandwidth on demand
- no call-establishment procedures
- nailed-up connection, which means circuit always exists between points. The service provider is simply connecting the circuit via a remote terminal with the click of a button.
- easy extension of the circuit or disconnection of the circuit if not used. If the user needs more time for the circuit, it can be extended by a simple request to the service provider.

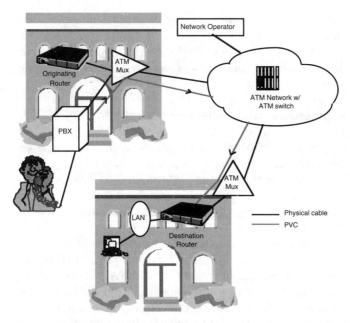

Figure 8.7 ATM PVC in an ATM network.

8.4.2 SVC

As mentioned earlier, SVC operation is similar to making a direct-dialed telephone call. When the call is set up, the default, or the only capacity or bandwidth assigned, is exactly 64 kbps, and once the call is set up, the circuit is assigned to the user and dedicated for the user's use, whether the user actually transmits information or not. Of course, the user pays for the duration of the call regardless of usage because the circuit is dedicated to the user as long as the circuit is up and cannot be used by others.[1] You can see how the resource is wasted.

Having recollected the basics of a direct-dialed telephone call, let's see how the same call works on an ATM environment. The following procedures exist in either environment:

- call setup procedure
- call establishment procedure
- data transfer procedure
- disconnect procedure
- billing procedure

---

[1] Many techniques such as TADI, TASI have been developed to use those circuits for other purposes.

These procedures occur at different stages of a call. Figures 8.7 and 8.8 show a portion of the call process. The major differences between a telephone call in a regular network and one in an ATM network are in the call setup and call establishment parameters. We examine this difference next.

8.4.2.1 Call setup procedure. Figure 8.8 shows a call setup from a telephone over an ATM network. In an ATM network, one can set up a connection for video and data similarly. For a telephone connection, the call originator dials the destination number. The call is routed via the local PBX to the ATM hub, which adapts the signaling information to ATM cells. The ATM hub verifies the bandwidth requested, using the ATM payload information. The ATM hub can also identify the default bandwidth, depending on the CPE terminal connected to it. In this case, it is 64 kbps.

Default values are used because it is expected that existing, nonintelligent CPE will be used in the ATM environment for a long period of time, and this equipment cannot request variable bandwidth. The ATM hub thus sets the default values if it receives no specific bandwidth requests for connection. If the user attempts to set up a video connection, the user must specify the bandwidth required, and the ATM hub fills in that information. In the

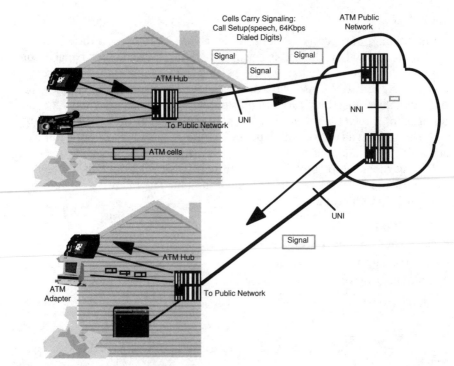

Figure 8.8 ATM call setup: sample voice call.

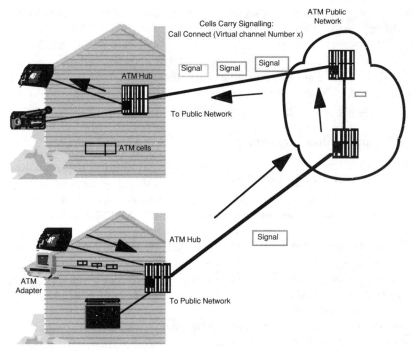

Figure 8.9 ATM call establishment: sample voice call.

case of a telephone call, the ATM hub fills in 64 kbps because it knows the CPE's capability. The ATM converts the information into cells using the signaling information, and these cells are carried into the network to the destination telephone for the circuit to be set up.

Once the cells reach their destination, the ATM hub at the destination address sends cells back with information for the virtual channel on which the originator needs to be connected establish the connection. This connection is accomplished by sending the ATM cells to the originator with the VPI information carried on the payload of the cell, as illustrated in Figure 8.9. Once the cells are received at the originating end, the ATM hub connects the call by assigning the cells to the appropriate VCI value. The network now knows where to route the cells so they can reach their destination.

Once the connection is set up, the information is carried in the ATM cells with the VCI number identified. This VCI, along with the VPI value in the ATM cell header, is used in routing the cells across the ATM network. In the network, each VPI and VCI value on the incoming cell is mapped to an outgoing VPI and VCI. These outgoing VPI and VCI values need not be the same as the ones defined by the end user during call setup; they have only local significance. Therefore, when the cells traverse the network, they can be mapped onto different VCI and VPI values in any ATM switch. All the network needs to know is that when leaving the network it must map the

VPI and VCI values to the ones negotiated at call setup. (The details of how this mapping occurs within an ATM switch is covered in Part 4.)

One of the users hangs up the telephone on completion of the call, so the call is disconnected and the destination switch stores the billing information. The billing in ATM can be done in various ways, such as by number of cells transmitted, PVCs bandwidth used, etc. The billing system for SVC in ATM environment is still in the early stages of definition.

8.5 ATM Application Environment

ATM technology has an upper hand when compared to other technologies, such as SMDS, frame relay, FDDI, etc., because ATM technology cuts across all spectrums of networks, LAN, WAN, public, and GAN. ATM is a technology for both switching and transmission. It can function as a switch, as a multiplexer, and as a cross-connect.

ATM can be used in many different ways in a LAN environment. The most obvious way is by having an ATM LAN hub, where terminals with ATM adapters are connected directly to the hub, as shown in Figure 8.10. This configuration is a typical business environment where the ATM hub is the interface to the outside world.

We call the hub environment a *pure ATM environment*. This scenario is far from reality because in today's environment, with different existing LAN technologies, there will be a migration strategy to ATM that requires very little investment to connect existing LANs to ATM. Thus, our ATM hub is in a slightly different environment from a regular ATM switching system. The differences and similarities between two application environments of ATM technology are given below:

- both switch ATM cells along their backplane
- both look similar in functionality
- both directly connect workstations, support a high-speed backplane, and provide access to public WAN
- an ATM hub supports Ethernet, Token ring, FDDI, TAXI and ATM adaptation interface, whereas a conventional ATM switch does not
- an ATM switch supports ATM interfaces operating at speeds from DS3, OC3, OC12, or higher as it becomes available, whereas an ATM hub has a maximum of OC3 interfaces
- an ATM hub focuses on performing the adaptation function for the ATM switch, whereas an ATM switch focuses more on switching, routing, and call management features

Thus, an ATM hub in a LAN environment can be used as a backbone switching mechanism, as shown in Figure 8.11. The details of this application of ATM are addressed in chapter 15.

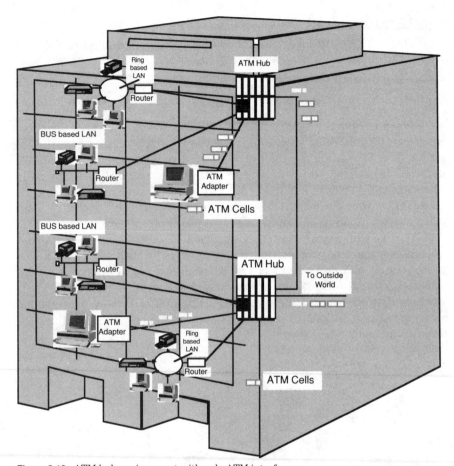

Figure 8.10 ATM hub environment with only ATM interfaces.

Having seen ATM in a LAN, let's address ATM in a WAN or GAN. Both environments can exist in private and public networks but are found mostly in public ones. ATM in both environments has the following characteristics:

- Initially, ATM will support some subset of BISDN services.
- Its interface speeds of up to 150 Mbps (OC3) UNI are already defined, DS3 or ATM forum's 45-Mbps public UNI are used initially, and OC3 UNI and OC3 NNI will be used later.
- It will support frame relay and switched multimegabit data service (SMDS) access.
- It will support SVCs through proprietary solution until standards are available in 1994.

- Initial applications are likely to be:

  ~PVC for VPN (virtual private network) to connect multiple ATM LAN for LAN interconnect applications (data applications)

  ~SVCs for dialup service

  ~backbone connection for frame relay (FR) and SMDS services.

Figure 8.12 shows a typical ATM network in WAN/GAN. In this section, we referred to an ATM environment where we deploy the standard ATM switch. The ATM switch we refer to here uses two fields, called VPI and VCI, in the ATM header to do switching of ATM cells to reach their destination. The same ATM switch, if it uses only VPI to switch and route the cells, is called *ATM cross-connect*. ATM cross-connect will be used as a traffic concentrator where an ATM switch cannot be justified.

ATM cross-connects can be used in place of existing digital cross-connects. Their multiplexing feature enables them to be used as an ATM multiplexer (mux) and cross-connect (XC) simultaneously in public networks. The details of an ATM switch are discussed in Part 4. A typical ATM network configuration uses ATM XC at the access interface where multiplexing and adaptation of traffic from different sources is performed; switching is done in the backbone network.

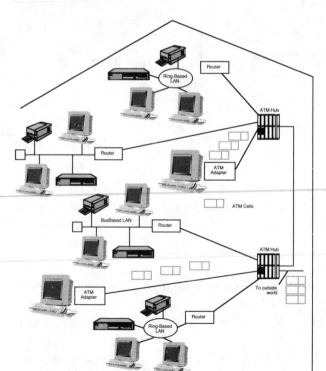

Figure 8.11 ATM hub environment.

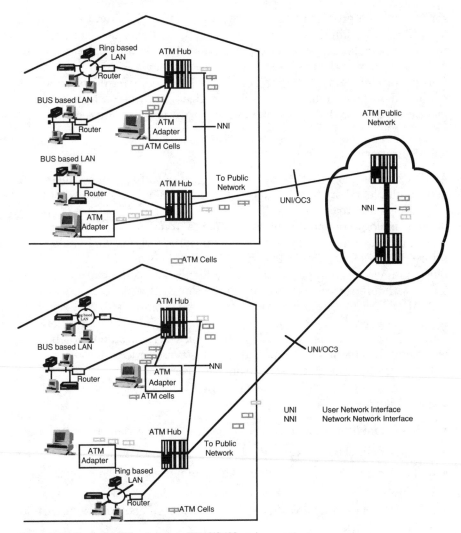

Figure 8.12 Typical ATM network in WAN/GAN environment.

The initial ATM PVC services became available from large IEC carriers like:

- AT&T in 1994
- MCI by the end of 1994
- Sprint by the end of 1993
- Wiltel by the end of 1993
- MFS by the end of 1993

Even though carriers are planning to provide a wide range of services, their initial plans are for VPN-based LAN interconnect services. The other services will be provided based on customer demand. Later, as standards for SVCs and the implementation of SVCs on the switches become available, the demand might increase. There is still much work to be done related to ATM, especially in the areas of VBR and the switching of such VBR traffic. Current products on ATM are all hardware. Once hardware is finalized or mature, more work must be done in intelligent software, where management functions for the ATM switches and network are performed.

8.6 Summary

ATM is a technology that can handle all types of traffic (voice, video, and data) multiplexed on the same network. In an ATM network, bandwidth can be reassigned in real-time to different traffic based on demand. ATM is the only technology common to all environments from LAN to GAN. The main reason for ATM's popularity is its selection as the switching technology for future BISDN services and can handle future unknown services. ATM combines both circuit- and packet-switching modes and is thus able to handle traffic with different characteristics on the same network.

ATM as a switching technology is different from ATM-based services. ATM technology is the converting of information into cells and then the routing of them to their destination. ATM as a service is an end-to-end service using ATM technology to provide higher-layer services. This ATM technology is transparent to the service and the user.

Some of the advantages provided by ATM in an ATM service environment are the following:

- statistical multiplexing of all types of traffic (voice, video, and data)
- flexible channel bandwidth allocation
- reduced number of overlay networks (data network, voice network, and video network)
- protection of existing investment for users by connecting existing system to ATM networks
- savings of administration costs (because of uniformity in technology)
- support of multimedia applications
- high-speed access to the network (begins at DS1, or 1.544 Mbps)

9

SONET/SDH

9.1 Overview

So far, we have discussed technologies for switching systems. This chapter describes the standard transmission technology to be used for the next generation of broadband communications based on the BISDN protocol. Fiber-optics has been selected as the medium of transmission. To have a uniform optical transmission interface, standard bodies around the world have worked on developing a set of specifications that enable any vendors' transmission systems to be interconnected. Thus, synchronous optical network (SONET) and synchronous digital hierarchy (SDH) standards came into existence. The SONET specification was designed for the United States and Canada, while the SDH specification was designed for the European Community and other countries. As not much difference exists between SONET and SDH, this chapter uses SONET as an example and describes transmission systems with a reference to SDH whenever applicable.

9.1.1 SONET

SONET was first conceived by R.J. Bohm and Y.C. Ching of Bellcore. It was proposed as an optical communications interface standard to the ANSI T1 committee at the end of 1984. The objective was to produce a common standard for fiber-optic transmission systems that would provide the operating companies a common, simple, economical, and flexible transmission network to operate. Just five years later, a stable base set of standards had emerged for SONET. SONET has been chosen as the transmission technology for the next-generation protocol, the BISDN, in the United States because it is capable of providing the infrastructure for the next generation of communication into the next century.

9.1.2 Synchronous digital hierarchy

In July 1986, CCITT, with SG XVIII playing the central role, began the process of standardizing SDH. Like SONET, SDH is an optical transmission standard that operates by appropriately managing the payloads and transporting them through a synchronous transmission network. Before the advent of SDH, the most common digital hierarchy used was plesiochronous digital hierarchy (PDH), which is still widely used in Europe as E1, E2, and E3, and in North America as DS1, DS2, DS3, and DS4 (these are asynchronous transmission hierarchies). These PDH signals are multiplexed into an (STM-n) signal. Compared to PDH, SDH appears to be extremely simple in operation. For example, mapping PDH tributaries into STM-n signals via synchronous multiplexing is not a trivial matter.

The term *synchronous* in SDH comes from the fact that multiplexing plesiochronous tributaries into STM-n adopts a synchronous multiplexing structure. Advantages of using synchronous multiplexing structure are:

- simplified multiplexing/demultiplexing technique
- direct access to low-rate tributaries without demultiplexing/multiplexing all the intermediate signals
- enhanced operations and OAM capabilities
- easy transition to higher bit rates of the future in step with the evolution of transmission technology

Hence, one can conclude that synchronous multiplexing structure is the very essence of SDH.

9.1.3 Advantages of SONET/SDH

Although slight differences exist between SONET and SDH, the advantages provided by both are similar because the same motivation is behind the development of both—develop a standard for a fiber-based synchronous network. In this section, we address the advantages that SONET/SDH provide:

- Both SONET and SDH standards are based on the principle of direct synchronous multiplexing, which is the key to cost-effective and flexible telecommunications networking around the world. In essence, individual tributary signals can be multiplexed directly into a higher rate of SONET/SDH signals without intermediate stages of multiplexing. SONET/SDH network elements can then be interconnected directly with obvious cost and equipment savings over the existing network.

- Both SONET and SDH provide advanced network management and maintenance capabilities required in a flexible network to manage and maintain that flexibility effectively. Nearly five percent of the SONET/SDH bandwidth is allocated to support advanced network management and maintenance procedure and practices.

- Both SONET and SDH signals can transport all the tributary signals defined for the networks in existence today. Thus, SONET/SDH can be deployed as an overlay network to the existing network, and where appropriate, provide enhanced network flexibility by transporting existing signal types. In addition, SONET/SDH has the flexibility to readily accommodate new types of customer service signals that network operators will want to support in the future.

- Both SONET and SDH can be used in all three traditional telecommunications application areas: long-haul networks (backbone networks), local networks (access networks), and loop carriers. They can also be used in a CATV network to carry video traffic.

9.2 Standards

The following subsections address the different stages of SONET and SDH standards.

9.2.1 SONET standards

In 1988, work by standards committees resulted in the publication of a national standard for SONET. The SONET standards allow vendors to build equipment to transport information point-to-point, but they do not spell out the nature of the messages or commands that conduct performance monitoring or control, or allow equipment from different vendors to function together. SONET standards have been introduced in three phases. Each phase presents additional levels of control and operations, administration, maintenance, and provisioning (OAMP).

9.2.1.1 Phase I. This phase, approved by ANSI in 1988, defines transmission rates and characteristics, formats, and optical interfaces. This phase primarily defines the hardware specifications for point-to-point data transport. Phase I supports the initial requirement of an optical carrier-n (OCn) midspan meet at payload level only. It also defines the standard data communications channels (DCC) with basic functions, as well as the basics of framing and interfaces.

9.2.1.2 Phase II. Phase II was built upon the midspan meet defined in Phase I for multiple vendor connectivity and management. Phase II defines:

- OAMP procedures
- synchronization
- SONET-to-BISDN interconnectivity
- pointer adjustments for wander and jitter
- central office electrical interfaces and network advantages
- imbedded operation channels
- common management information service elements (CMISE)
- point-to-point, add/drop multiplexer capabilities

In addition, it defines an intraoffice optical interface (IAO), which allows equipment to be interconnected at the central office.

9.2.1.3 Phase III. This phase is built upon Phase II by providing all of the OAM&P required for a midspan meet. Additional network management, performance monitoring, and control functions are added, as are DCC standard message sets and addressing schemes for identifying and interconnecting SONET network elements, allowing the passing of DCC information between various vendor implementations of SONET. Phase III also provides for ring and nested protection switching using automatic protection switching (APS) mechanism.

SONET standards set 51.84 Mbps as the base signal for the new multiplexing hierarchy, called the synchronous transport signal level 1 (STS-1). Its mirror signal for transmission over fiber-optical lines is optical carrier level 1 (OC1), which is a direct conversion from electrical to optical signaling.[1] Upper-level signals are multiples of OC1. Thus, OC3 carries three times more capacity than OC1, or 155.52 Mbps. Table 9.1 shows the SONET multiplexing hierarchy and line rates.

TABLE 9.1 SONET Multiplexing Hierarchy

Optical No.	Electrical No.	Speed	Multiple of DS3	Multiple of DS1	Multiple of DS0
OC1	STS-1	51.84 Mbps	1	28	672
OC3	STS-3	155.52 Mbps	3	84	2,016
OC12	STS-12	622.08 Mbps	12	336	8,064
OC24	STS-24	1.244 Gbps	24	672	16,128
OC48	STS-48	2.488 Gbps	48	1,344	32,256
OC192*	STS-192	9.6 Gbps	192	5,376	129,024

* Recently defined standard interface.

[1] There is a slight difference in speed between STS-1 and OC-1.

TABLE 9.2   ITU-T's SDH Recommendations

G.702	Digital hierarchy bit rates
G.703	Physical/electrical characteristics of hierarchical digital interfaces
G.707	Synchronous digital hierarchy bit rates
G.708	Network node interface for the synchronous digital hierarchy
G.709	Synchronous multiplexing structure
G.773	Protocol suites for Q interfaces for management of transmission systems
G.781	Structure of recommendations on multiplexing equipment for the synchronous digital hierarchy
G.782	Types and general characteristics of synchronous digital hierarchy multiplexing equipment
G.783	Characteristics of synchronous digital hierarchy multiplexing equipment
G.784	Synchronous digital hierarchy management
G.955	Digital line systems based on the 1.544-Mbps hierarchy on optical-fiber cables
G.956	Digital line systems based on the 2.048-Mbps hierarchy on optical-fiber cables
G.957	Optical interfaces for equipment and systems relating to the synchronous digital hierarchy
G.958	Digital line systems based on the synchronous digital hierarchy for use on optical-fiber cables
G.652	Characteristics of a single-mode optical-fiber cable
G.653	Characteristics of a dispersion-shifted single-mode optical-fiber cable
G.654	Characteristics of a 1,500-nm wavelength loss minimized single-mode optical-fiber cable
M.30	Telecommunications management network

9.2.2   SDH standards

ITU-T recommendations G.707, G.708, and G.709 were the result of efforts to create a worldwide standard for SDH. The recommendations were based on ANSI's North American standard for SONET. After much discussion and compromise, American and European parties arrived at a unified standard that accounted for both the European hierarchy, with its 2.048 Mbps (E1) basic bit rate, and the North American hierarchy, with its 1.544 Mbps (T1 or DS1) bit rate. The new SDH standard has a common bit rate of 155.52 Mbps.

In addition to these three recommendations, the ITU-T specified a number of supplemental recommendations for SDH. Table 9.2 shows the full range of recommendations related to SDH.

9.3 SONET/SDH Protocol

The current asynchronous broadband network has grown on an ad hoc basis, where problems in construction, operation, or maintenance must be resolved individually, resulting in an excessively complex network structure. Consequently, these networks are difficult to operate, maintain, or expand. The SONET standards groups have sought to resolve this problem by defining a hierarchical layered structure. Also, to manage information better in SONET, information is accessed at the byte level instead of the bit level, as in asynchronous systems. Each layer can handle intralayer communications independently and is responsible for a portion of the overall link management. Although in some respects analogous to the layering of the OSI communications model, SONET layering as described in this section is concerned with the frame itself, which only applies to the OSI data link layer. The exception to SONET/SDH is the communication overhead bytes, used for operations and maintenance, which transmit information generated by lower layers to higher OSI layers as required.

Figure 9.1 shows the different layers in SONET and how they interact. The SONET layers have a hierarchical relationship—each layer, starting with the path layer, requires the services of all lower-level layers to perform its own functions.

From a bottom-up approach, each layer builds on the services provided by the lower layers. The four layers are:

- photonic
- section
- line
- path

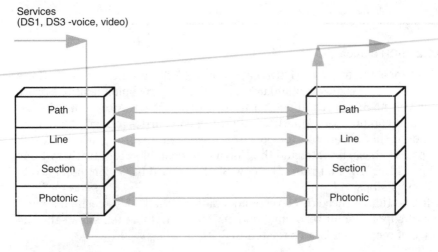

Figure 9.1 SONET protocol stack.

9.3.1 Photonic layer

This layer provides optical transmission at a very high bit rate. Issues dealt with at this layer include optical pulse shape, receiver and transmitter power levels, and operating wavelength. Electro-optical equipment communicates at this level. The main function of the photonic layer is to convert the electrical signal to optical signals and map the electrical STS-n frame into an optical OC3-n frame. The reason for this mapping is that the higher layers perform their functions in an electrical domain, whereas the physical transmission system is in an optical domain.

9.3.2 Section layer

The section layer deals with the transport of STS-n frames across the physical medium. Functions include framing, scrambling, section-error monitoring and communicating, and adding the section-level overhead. A section of the transmission facility includes termination points, between either a terminal network element and a repeater or two repeaters.

9.3.3 Line layer

The line layer deals with the reliable transport of the path layer payload and its overhead across the physical medium. The line layer provides synchronization and multiplexing for the path layer. A *line* is the transmission medium required to transport information between two consecutive network elements (e.g., an OC-n/OC-m multiplexer), one of which originates the line signal and another that terminates it. The network elements are also called terminating equipment because the signals terminate in them.

9.3.4 Path layer

The path layer deals with the transport of services (e.g., DS1 or DS3) between path terminating equipment (PTE). The main function of the path layer is to map the services of the path overhead (POH) into an STS SPE, which is the format required by the line layer. The path overhead uses pointers to identify the beginning of DS1 or DS3 signals.

9.4 Basic SONET Frame

Figure 9.2 shows the basic SONET frame format, with a bit rate of 50.688 Mbps. The concept of the synchronous transport system goes beyond the basic needs of a point-to-point transmission system to include the requirements for telecommunications networking. SONET can therefore be used in all three of the traditional network applications (voice, video, and data) where the growth in bandwidth and the provisioning of new customer services are expected to happen in the near future.

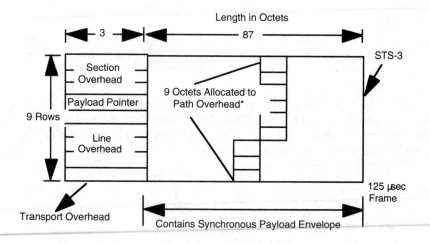

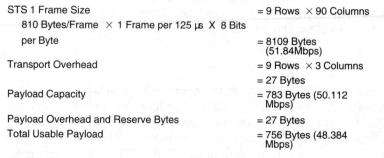

STS 1 Frame Size	= 9 Rows × 90 Columns
810 Bytes/Frame × 1 Frame per 125 μs X 8 Bits	
per Byte	= 8109 Bytes (51.84Mbps)
Transport Overhead	= 9 Rows × 3 Columns
	= 27 Bytes
Payload Capacity	= 783 Bytes (50.112 Mbps)
Payload Overhead and Reserve Bytes	= 27 Bytes
Total Usable Payload	= 756 Bytes (48.384 Mbps)

Figure 9.2 STS-1 SONET frame.

This basic signal is STS-1. Before we describe the different components of the SONET frame, we need to understand the frame size in bytes. The lowest SONET frame is STS-1. This frame consists of 9 rows and 90 columns, where each row and column is 1 byte, thus making the total frame 810 bytes. The time taken to transmit this frame is 125 microseconds (μs) with 8 bits per byte. This frame has the capacity of 51.84 Mbps. It is divided into different components to perform certain functions. The components are:

- transport overhead (TOH)
- section overhead (SOH)
- line overhead (LOH)
- path overhead (POH)
- STS synchronous payload envelope (SPE)
- envelope payload

Each is discussed in the following subsections.

9.4.1   Transport overhead

In Figure 9.2, the first three columns of the STS are transport overhead (TOH), which are reserved for transport information. The TOH has been assigned 27 bytes, of which 9 bytes are reserved for the section overhead, and 18 bytes are reserved for the line overhead.

9.4.2   Section overhead

The section overhead contains the information required for the section elements only (i.e., repeaters). This section overhead information is processed at each section terminating point, which is usually between any two pieces of SONET transmission equipment. Section overhead provides the following functions:

- detection of STS-1 frame alignment
- section performance monitoring and fault isolation
- data communications channel for OAM&P
- channel for voice communications for maintenance personnel

9.4.3   Line overhead

Line overhead contains the information required between the line termination equipment (such as an add-drop terminal). It provides the following functions:

- STS payload pointer
- line performance monitoring and signal failure detection
- automatic protection switch (APS) signaling channel (bidirectional)
- data communications channel for alarm gathering, remote provisioning, and other OAMP
- channel for voice communication for maintenance personnel

9.4.4   Path overhead

The path overhead is assigned to and transported with the payload until the payload is demultiplexed. The POH is carried with the payload envelope and supports the transport of the payload from the point it enters the SONET network to the point it leaves. For every payload, there is a corresponding POH. The POH performs the following functions necessary to transport the payload between path-terminating equipment:

- monitors end-to-end transport of the payload and its performance when transferring the network
- identifies that a correct connection was made

- identifies payload type
- provides user channel for carrying service provider's information

9.4.5 Envelope payload

Envelope payload is the bandwidth within the STS frame and is aligned to the STS frame that carries the STS SPE. This bandwidth can be combined with several STS-1s to carry a higher bandwidth payload. The STS SPE consists of 87 columns by 9 rows of bytes (the remaining bytes of TOH), which is the capacity reserved for the payload and path overhead, as shown in Figure 9.2. Column 1 contains the STS path overhead (9 bytes), the remaining 774 bytes are available for the payload to carry the actual information. The STS SPE can begin anywhere in the STS envelope capacity (it can start in one frame and end in the next), as illustrated later in this chapter.

9.5 SONET Overhead Capabilities

As mentioned earlier, the SONET overhead consists of three components, section, line, and path. Before we address each one of the overheads, we need to understand the relationship of these overheads to the network elements. Figure 9.3 shows the originating and terminating element for the section, line, and path overheads.

For network management and maintenance purposes, the SONET network can be described in terms of three different network spans:

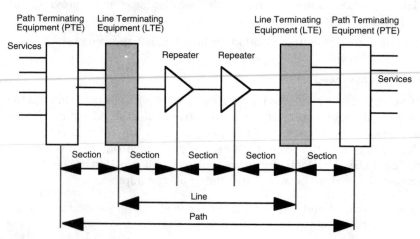

Figure 9.3 SONET network spans.

- The path span allows network performance to be maintained from a service end-to-end perspective.

- The line span allows network performance to be maintained between transport nodes (or between two active terminals).

- The section span allows network performance to be maintained between the SONET network elements. Section spans can be regenerators or repeaters between any SONET network elements.[2]

Let's look into the functions provided by each of the SONET overheads and the details of bytes within each.

9.5.1   Path overhead

Figure 9.4 shows the POH within a SONET frame. The functions provided by the path overhead are:

- end-to-end transport of services
- sequencing of cells
- path-terminating element status
- continuity
- error detection
- user-defined functions

The POH of the SONET frame consists of 9 bytes of the STS. It is carried in the STS SPE and comprises the following bytes:

J1	The J1 byte is used to repetitively transmit 64-byte information. It consists of a fixed-length string, so that continued connection to the source of the path signal can be verified at any receiving terminal along the path.
B3	The B3 byte provides BIP-8 (bit interleaved parity) path error for monitoring. The path BIP-8 is calculated over all bits of the previous SPE, and the computed value is placed in the B3 byte before scrambling.
C2	The C2 byte indicates the construction of the STS SPE by means of a label value assigned from a list of 256 possible values (8 bits).

---

[2] Repeaters and regenerators are actually different. A repeater amplifies the incoming signal and transmits it. A regenerator or recreates the original signal and transmits it. In the former, the noise in the signal is also amplified, whereas in the latter the noise is removed.

G1   The G1 byte is used to convey back to the originating STS PTE the path termination status and performance. This feature allows the status and performance of a two-way path to be monitored at either end or at any point along the path.

F2   This byte is allocated for the user's purpose between the path terminations.

H4   This byte provides a multiframe phase indication for VT payloads.

Z3 to Z5   These three bytes are reserved for future use.

9.5.2 Line overhead

Figure 9.5 shows the LOH bytes in an STS-1 frame. The functions provided by the line overhead are:

- communication between the line-terminating equipment (LTEs)
- synchronization between LTEs
- payload location or identification within the payload
- multiplexing
- error detection
- automatic protection switching

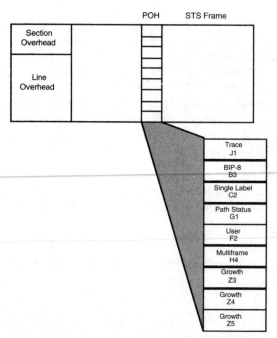

Figure 9.4 Path overhead (POH) in STS-1 frame.

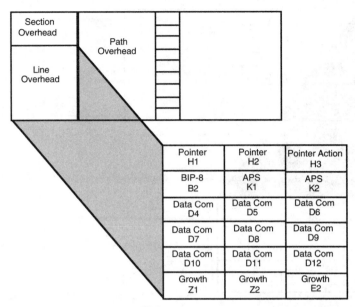

Figure 9.5 Line overhead of STS-1.

The 18 bytes of the STS-1 line overhead comprised as follows:

H1 to H3   These three facilitate the operation of the STS-1 payload pointer and are provided for all STS-1s in an STS-n.

B2   This byte provides BIP-8 line error monitoring. The line BIP-8 is calculated over all bits of the line overhead and payload envelope capacity of the previous STS-1 frame before scrambling, and the computed value is placed in the B2 byte. This byte is provided for STS-1 in an STS-n signal.

K1 to K2   These two bytes provide APS signaling between line-terminating equipment and are defined only for STS-1 number 1 in an STS-n signal.

D4 to D12   These nine bytes provide a data communications channel at 576 kbps for messages of administration, monitoring, maintenance, alarms, and other communications needs between line termination equipments. These bytes are defined only for STS11 of an STS-n signal.

Z1 to Z2   These two bytes are reserved for functions not yet defined.

E2   This byte provides an express orderwire channel for voice communications between line-terminating equipment and is only defined for STS-1 of an STS-n signal.

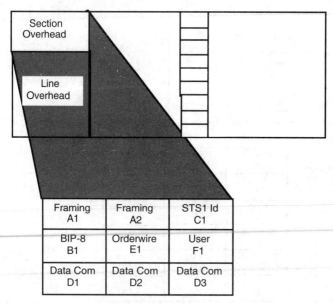

Framing A1	Framing A2	STS1 Id C1
BIP-8 B1	Orderwire E1	User F1
Data Com D1	Data Com D2	Data Com D3

Figure 9.6 Section overhead (SOH).

9.5.3 Section overhead

The last of the three overheads in the transport portion of the SONET header is the section overhead. Figure 9.6 shows the bytes of the section overhead.

The functions provided by the section overhead are:

- frame alignment pattern
- STS-1 identification
- parity check
- data communication channel
- voice communications (orderwire)
- user channel

The nine bytes of the STS-1 section overhead are made up as follows:

A1 to A2   These two bytes provide a frame-alignment pattern (11110110 00101000). These bytes are provided in all STS-1s within an STS-n. These two bytes identify the beginning of the SONET STS-1 frame.

C1   This byte is set to a binary number corresponding to its order of appearance on the byte-interleaved STS-n frame and can be used in the framing and de-interleaving process to determine the position of other signals. This byte is provided in all STS-1s within an STS-n, with the first STS-1 being given the number 1 (0000 0001).

B1	This byte provides section error monitoring by means of a bit-interleaved parity (BIP 8) code using even parity. In an STS-n, the section BIP-8 is calculated over all bytes of the previous STS-n frame after scrambling, and the computed value is placed in the B1 of the STS-1 number before scrambling.
E1	This byte provides a local orderwire channel for voice communications between the regenerators and network elements.
F1	This byte is allocated for the user's purpose and is terminated at all section-level equipment.
D1 to D3	These three bytes provide a data communications channel for messages of administration, monitoring, alarm, maintenance, and other communications needs at 192 kbps between section-termination equipment.

9.6   How Does SONET/SDH Work?

To explain the operation and functions of SONET/SDH, we take SONET as an example. By now we know that SONET works on the principle of synchronization. So first let's look into a basic synchronous signal structure, shown in Figure 9.7. A synchronous signal comprises a set of bytes (8 bits each) organized into a frame structure. Within this frame structure, the identity of each byte is known and preserved with respect to a framing or marked byte. In Figure 9.7, a single frame in the serial signal stream is shown in a two-dimensional format. It consists of N rows and M columns of boxes, each representing a byte (8 bits) of the synchronous signal. A framing byte, or F-byte, appears on the top left corner of the box to provide the frame reference-byte location. This signal is transmitted in a sequence starting from the top lefthand corner byte (the F-byte), followed by those in the second byte in row 1 and so on, until the bits in the Mth byte, i.e., the last byte in row 1, are transmitted. Then, the bits in the first byte of row 2 are transmitted, followed by the bits in the second byte of row 2, and so on, until the Mth byte of the Nth row is transmitted. This whole sequence is repeated for the next STS frame.

Figure 9.8 shows the next stage of the synchronous signal structure, called the synchronous transport frame. A synchronous transport frame comprises two distinct and readily accessible parts within the frame:

- synchronous payload envelope (SPE)
- transport overhead (TOH)

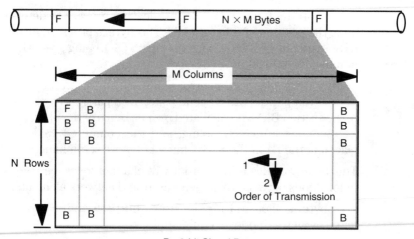

B - 8-bit Signal Byte
F - 8-bit Frame Byte

Figure 9.7 Synchronous signal structure.

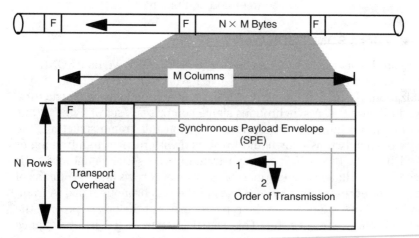

Figure 9.8 Synchronous transport frame structure.

SPE is where the actual information is carried by the signal. In SPE, individual tributary signals, such as DS1 and DS3, for instance, are mapped onto the payload. These signals are assembled and disassembled only once, even though they might be transferred from one transport system to another many times on their route through the network to its destination.

A portion of the signal capacity is reserved for transporting what is called *signal overhead*. We explained the function of the overhead in Section 9.4. The overhead, in general, provides information such as alarms, maintenance, bit error monitoring, etc., which is carried across the network to support and maintain the transportation of the SPE between the nodes in a synchronous network.

Within each STS-1 frame, there is a pointer to the payload envelope, known as the STS-1 payload pointer. This pointer provides a method for flexible and dynamic alignment of the STS SPE within the STS envelope capacity. The alignment is independent of the actual contents of the envelope. This dynamic alignment means that the pointer can accommodate differences in the phases and frame rate of the STS SPE and the transport overhead (i.e., when network elements are running at slightly different clock rates). This difference occurs when the frame is transported from one network to another, when each derives its master clock from different sources. Synchronization and timing are covered in Part 4.

Figure 9.9 shows the link between the transport overhead and the SPE. To facilitate efficient multiplexing and cross-connection of signals in the synchronous network, the SPE is allowed to float within the payload capacity provided by the STS-1 frames. Thus, the STS-1 SPE can begin anywhere in the STS-1 payload capacity and is unlikely to be wholly contained in one frame. Usually, the STS-1 SPE begins in one frame and ends in the next. When an SPE is assembled into the transport frame, additional bytes, referred to as payload pointers, are made available in the transport overhead. These bytes contain a pointer value that indicates the location of the first byte (J1 is part of the POH byte) of the STS-1 SPE. The SPE is allowed to float freely within the space made available for it in the transport frame, so that timing phase adjustments can be made as required between the SPE and transport frame. The payload pointer maintains the accessibility of the SPE by identifying the location of the first byte of the SPE.

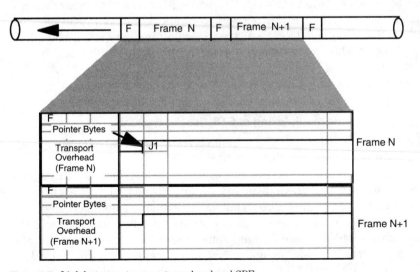

Figure 9.9 Link between transport overhead and SPE.

We have discussed the pointer in the link between the transport overhead and SPE. Now let's address the functions of payload pointers. Basically, payload pointers allow asynchronous operation in a synchronous network. Because SONET works only on a synchronous network, mapping asynchronous network traffic can be achieved using pointers. Ideally, this means that all synchronous network nodes should derive their timing signals from a single master network clock. Current synchronous network timing scenarios, however, do allow for the existence of more than one master network clock.

For example, networks owned by different network operators or service providers must have their own independent master timing references. These clocks would operate independently and, therefore, at slightly different rates. Also, situations exist in which a network node loses its timing reference and operates on a standby clock, which might not be as stable as the master clock. Therefore, synchronous transport must be able to operate effectively between network nodes operating asynchronously within certain limits. To accommodate clock offsets, the SPE can be moved (justified) positively or negatively one byte at a time with respect to the transport frame. Moving the SPE is achieved by simply recalculating or updating the payload at each SONET network node. In addition to clock offsets, updating the payload pointer also accommodates any other timing phase adjustments required between the input SONET signals and the timing reference of the SONET node.

Of course, payload processing does introduce a new signal impairment known as *payload adjustment jitters*. This jitter impairment appears on a received tributary signal after recovery from an SPE that has been subjected to payload pointer changes. Excessive jitter on a tributary signal influences the operation of the network equipment that is processing the tributary signal immediately downstream. Therefore, care should be taken in the design of the timing distribution of the synchronous network to minimize the number of payload pointer adjustments and level of tributary jitter that could be accumulated through synchronous transport.

9.6.1 Synchronous multiplexing of STS

So far, we have seen how an STS-1 frame is transported carrying a payload. Now we study an example. Figure 9.10 shows how synchronous multiplexing is done. Synchronous multiplexing of STS is defined in SONET as a procedure for multiplexing the individual bytes of a signal such that each signal is visible within the multiplexed signal, i.e., it eliminates complete demultiplexing of an STS-n to access one STS-1. An STS-n signal is formed by byte interleaving of n STS- 1 signals. For example, an STS-12 comprises 12 STS-1s, each separately visible (meaning each can be accessed). The individual STS-1s could be carrying different payloads, each with a different destina-

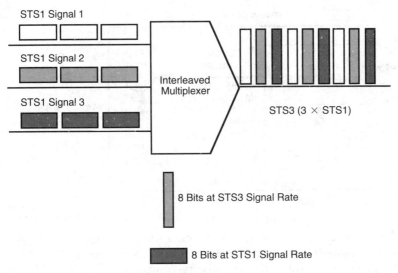

STS1 Signal 1

STS1 Signal 2

Interleaved
Multiplexer

STS1 Signal 3

STS3 (3 × STS1)

8 Bits at STS3 Signal Rate

8 Bits at STS1 Signal Rate

Figure 9.10 Example of synchronous multiplexing.

tion. Figure 9.10 shows an example of three STS-1 signals multiplexed onto an STS-3 signal. Here, the three STS-1s are multiplexed to form an STS-3 signal by making the output of the equipment operate at a rate three times the input rate. For example, at the input each STS-1 takes 125 μs to transmit, whereas on the output, the STS-3 frame takes only 125 μs to transmit. Thus, three STS-1 signals can be multiplexed onto one STS-3 frame.

Multiplexing of SONET signals provides benefits such as transporting narrowband signals and broadband signals together in the most economical way. This feature reduces the number of network elements required and allows the transport elements to be dynamically provisioned.

9.6.2 Payload mapping

So far, we have addressed some of the basics of the SONET frame and its multiplexing. Now let's look into mapping information into the payload. As mentioned earlier, SPE is the location in the SONET frame where the actual information is stored and carried in the network. The process of carrying the information consists of the SPE assembly process and the SPE disassembly process.

9.6.2.1 SPE assembly process. Figure 9.11 shows the assembly process of the STS payload. In this example, the process of mapping a tributary signal such as DS3 into a synchronous payload envelope is explained. Once assembled, this payload is transported across a synchronous network. This process is part of the fundamental principle defined in the SONET stan-

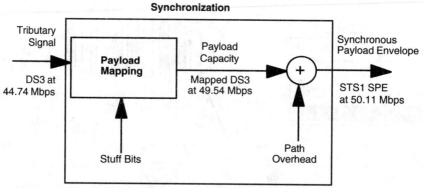

Figure 9.11 SPE assembly process.

dards. The process of assembling the tributary signal into an SPE is referred to as *payload mapping*. To provide uniformity across all SONET transport capabilities, the payload capacity provided for each individual tributary signal is always slightly greater than that of the required tributary signal. Thus, the essence of the mapping process is to synchronize the tributary signal with the payload capacity provided for transport by adding extra stuffing bits to the signal stream as part of the process as shown in Figure 9.11. For example, in the figure, a DS3 tributary signal at a nominal rate of 44.74 Mbps needs to be desynchronized with a payload capacity of 49.54 Mbps provided by the STS-1 SPE. The addition of path overhead completes the assembly of the STS-1 SPE and increases the bit rate of the composite signal to 50.11 Mbps.

9.6.2.2 SPE disassembly process. Figure 9.12 shows the disassembling process of the STS payload. At the disassembling point, the tributary signal that has been transported over the network must be recovered from the SPE that provided transportation facility for the original signal. The process of disassembling the original tributary signal from the SPE is referred to as *payload demapping*. The SPE consists of path overhead, which was added during the payload mapping process, and the tributary signal to carry the payload. Thus, the essence of the demapping process is to desynchronize the tributary signal from the composite SPE signal and reproduce this tributary signal to its original form as precisely as possible. So, in our example, an STS-1 SPE carrying a mapped DS3 payload arrives at the disassembly location with a signal rate of 50.11 Mbps. Stripping the path overhead and stuffing bits from the SPE results in a discontinuous signal, representing the transported DS3 signal with an average signal rate of 44.74 Mbps. These timing discontinuities are reduced by means of a desynchronizing phase locked loop (PPL) to produce a continuous DS3 signal at the same average signal rate.

9.6.3 Virtual tributary

In a SONET environment, mapping DS3 onto an STS-1 payload is relatively straightforward because of the compatibility between the speed of the two signals. In a real-world environment, however, very few users have DS3 amount of traffic. Usually, this amount of traffic is available only in the backbone. So SONET standards-making bodies defined mapping of T1 or DS1 onto the SONET STS-1 payload in the form of virtual tributaries (VTs), enabling access to each of the VTs without the need to demultiplex the whole payload to get one VT. We know a DS3 can accommodate 28 DSIs. Thus, to access each one of the VTs (DSI) within an STS payload, different types and modes of VT are present. The STS SPE can be divided into VTs for transporting and switching payloads smaller than the STS-1 rate (e.g., DS1, DS2, PCM30). Four types of VTs are defined, which are shown in Table 9.3.

Let's go through an example of a VT mapping onto an STS-1 SPE. For simplicity, we take VT1.5 (DS1), which is popular among VTs because it maps the DS1 signal. Figure 9.13 shows how VT1.5 is packaged onto an STS-1 payload.

In an SPE payload, 28 VT1.5s can be packaged for transportation. The 3-column by 9-row structure of the VT1.5 fits neatly into the same 9-row

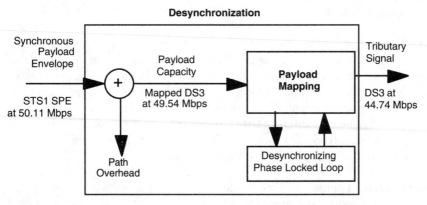

Figure 9.12 SPE disassembly process.

TABLE 9.3 Different types of VT

Optical levels		Payload example	
VT size (type)	Capacity (Mbps)	Name of payload	Rate (Mbps)
VT1.5	1.728	DS-1	1.544
VT2	2.304	E-1	2.048
VT3	3.456	DS-1C	3.152
VT6	6.912	DS-2	6.312

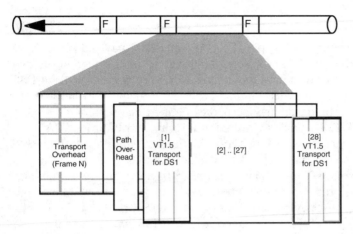

Figure 9.13 VT1.5 packaged in STS-1 payload.

structure of the STS-1 SPE. Thus, 28 VT1.5s can be packaged into the 86 columns of the STS-1 SPE payload capacity, still leaving two spare columns in the STS-1 SPE payload capacity. These spare columns are filled with fixed-stuff bytes that allow the STS-1 SPE signal structure to be maintained in case of nonsymmetric timing. These VTs can operate in two modes—floating and locked.

9.6.3.1 Floating mode. This mode has been designed to minimize the network delay and provide efficient cross-connection of transport signals at the VT level within the synchronous network. This goal is achieved by allowing VT SPE to float with respect to the STS-1 SPE to avoid the use of unwanted slip buffers at each VT cross-connect between different transport systems without unwanted network delay. This mode allows a DS1 to be transported effectively across a SONET network. The problem in this mode is to identify the beginning of the VT because the source of VT is not fixed.

9.6.3.2 Locked mode. Locked mode has been designed to minimize the interface complexity and support bulk transport of DS1 signals for digital switching applications. This goal is achieved by locking individual VT SPEs in fixed positions with respect to the STS-1 SPE. Each VT1.5 SPE is not provided with its own payload pointer. It is not possible with this mode to route a selected VT1.5 through a SONET network without unwanted network delay and extra cost caused by having to provide slip buffers to accommodate the timing synchronization issues. This mode has the advantage of mapping VTs in a predefined location within the payload. The disadvantage is the delay and inefficient use of the SPE payload.

9.6.3.3 SONET synchronization. In a synchronous system such as SONET, all clocks are locked onto a reference frequency. Every clock can be traced back to a highly stable reference clock source. The STS-1 rate remains at a nominal 51.84 Mbps. A synchronous system allows simpler multiplexing and direct payload visibility, as compared to an asynchronous environment, where the bit stuffing does not allow direct payload visibility. Because of the use of payload pointers, the STS-1s and their contents (the individual bytes) are easily accessed at higher STS-n rates. SONET can thus transport between smaller networks with different clocks using a pointer-adjustment facility or when part of the network loses its timing reference because of fault conditions.

9.7 Summary

In this chapter, we addressed one of the most important fiber-optics transmission technologies for broadband communication, SONET. The reason for its importance is that SONET has been selected as the transmission technology for fiber-based transmission systems for BISDN. We also addressed SDH, which is the European counterpart of the SONET as defined by ITU-T for the European Community and others who adopt the E1-based standard. Although some differences exist in the basic frame format, SONET and SDH are the same beyond the STS-3 signal level, as shown in Table 9.4.

To generalize the terminology, one could use STM-n for SDH and STS3-n for SONET. Standards have been developed in such a way that both are compatible from the second level (i.e., STS-3 or STM-1) and their lower tributaries can be mapped interchangeably between the two formats from there onwards. We also discussed the overheads, the working of SONET payload mapping, and tributary mapping onto the SONET payload.

TABLE 9.4 Comparison of SONET and SDH Frames

Speed	SONET	SDH	Optical equivalent
50 Mbps	STS-1	—	OC1
150 Mbps	STS-3	STM-1	OC3
622 Mbps	STS-12	STM-4	OC12
2.4 Gbps	STS-48	STM-16	OC48

Broadband Architecture

In this part, the BISDN protocol reference model is described as defined by ITU-T. Here, each layer of the BISDN is covered in detail. In addition to defining each of the BISDN protocol layers, the other aspects of BISDN, such as service, network, and traffic management, are covered. This part gives the reader a background on broadband protocol, which is the basis of the broadband communications network architecture described later in this book.

10

BISDN Lower Layers

10.1 Overview

In telecommunications, broadband services are referred to as BISDN, for broadband integrated services digital network; it is an extension of ISDN as the name implies, at least in terms of naming terminology. The intent of the name was to state that BISDN is an extension of ISDN in terms of capabilities, i.e., it not only has the narrowband capability of ISDN, but also broadband capability. Thus, since the introduction of BISDN, ISDN has been renamed N-ISDN, or narrowband- ISDN. N-ISDN, according to ITU-T, is defined as any service requiring bandwidth less than 64 kbps or up to a regular voice channel. Anything above 64 kbps and below 1.544 Mbps is defined as wideband. As mentioned earlier, any service inquiry with a speed greater than 1.544 Mbps is defined as broadband. Thus, any communications based on this speed are called *broadband communications.*

BISDN is an extension of ISDN only in terms of the name. Everything else is different, including its protocol, architecture, transmission and switching technology, and platforms. If someone is an expert in ISDN, he is not necessarily knowledgeable about BISDN.

BISDN is not a totally new concept. Many of the ideas are extracted and enhanced from ISDN and other telecommunications and data communication protocols. The reason for such extraction is that the fundamental objective of BISDN is to achieve complete integration of services, ranging from low-bit-rate bursty signals to high-bit-rate continuous real-time signals. Services include voiceband services, such as telemetry, low-speed data, telephone, and facsimile and broadband services, such as high-quality video conferencing, high-definition television (HDTV) video transmission, and high-speed data transmission.

Thus, to meet its objective, BISDN must adapt the characteristics of each of the different services and integrate them into a common transmission and switching platform. For instance, the packet switching concept is used for data transmission, and the circuit switching concept is used for voice transmission. In BISDN, both these concepts are used so that both types of traffic can be handled. Of course, some technical difficulties exist in achieving such an objective. But, in this case, the benefits outnumber the difficulties.

Why do we need such a protocol that is completely different from existing ones? The reason is the increasing demand for various types of broadband services. Although some broadband services are available in a disintegrated form, these services currently available in private networks are expensive because of the inefficient utilization of such network resources. The BISDN protocol is the right solution for such inefficiencies. From the user's perspective, BISDN will offer a single interface that provides for all required communication needs: voice, data, video, and carrying both signal and user information. The interface between the user and the network will be identical for all users of BISDN. This network provides a high level of transparency to the user, which means that the user will not be concerned with the mechanics of how the service is provided, but only with the services received from network. Figure 10.1 shows the basic concept of BISDN from the user's perspective.

Having looked at BISDN from a user's perspective, let's see how a network operator or service provider views it. A typical network operator might be a telephone operating company or a large company. The following are some of the objectives of a network operator for any network:

- maximum efficiency or utilization of the network, considering grade of service to maintain performance levels demanded by the users of the network

- single-operation interface of all network elements and terminals attached to the network

- ability of network operators to rapidly identify, isolate, and thus minimize the impact of faults occurring in the network

- ability to manage the traffic transiting through the network in terms of automatic rerouting in case of congestion

Although these are objectives of any network operator, it is not that easy to achieve them. One has to keep in mind considerations such as whether all equipment can talk with each other in the same language or protocol, which is far from the reality of today's world. Figure 10.2 shows a view of the network from the operator's perspective.

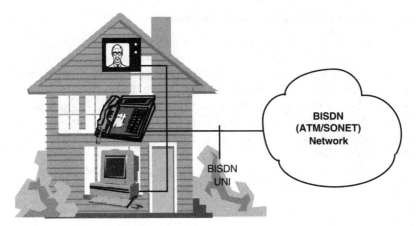

Figure 10.1 BISDN user network interface.

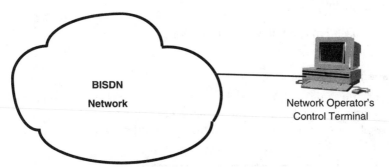

Figure 10.2 Network operator's view of BISDN network.

To come anywhere close to the above-mentioned objectives, one has to perform a lot of protocol conversions that degrade the system's efficiency. BISDN is the first technology or network that addresses these problems and achieves most of the network operators' objectives, at least in the long term. BISDN can achieve these goals because of its capability to carry different types of traffic under a single network and still maintain the application characteristics independent of another application's traffic. Currently, no complete list of applications or services is defined that could be provided by BISDN. But ITU-T has listed some of its potential current and future applications, which are mentioned in chapter 12.

Of course, *Rome cannot be built in a day*. To achieve a complete BISDN network, one must go through different phases. Some of the transitions for different environments where BISDN/ATM is applicable are addressed in Part 5 of this book.

In this chapter, we address the lower layers of the BISDN protocol for the proposed target network.

10.2 Broadband Protocol Reference Model

BISDN protocol architecture is a vertical layered architecture covering the transport, switching, signaling and control, user protocols, and applications or services. The architecture model covers the complete set, including management. For an individual function such as switching or transmission, a subset of the protocol model applies, along with its respective upper-layer functions, such as the management functions. The protocol architecture divides the functions of each layer so that appropriate functions are used to support a given application.

The BISDN reference protocol model, or simply BISDN PRM, consists of three planes as shown in Figure 10.3. The three planes are:

- management
- user
- control or signaling

Each is discussed in the following subsections.

10.2.1 Management plane

Two types of functions exist in this plane: layer management functions and plane management functions. All the management functions that relate to the whole system (end-to-end) are located in the plane management. Its

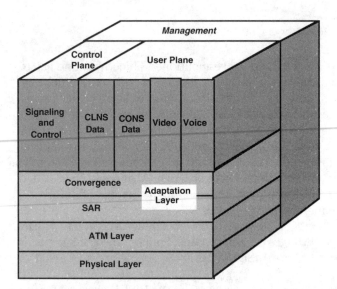

Figure 10.3 BISDN protocol reference model.

task is to provide coordination between all the different planes. No layered structure is used within this plane.

The layer management functions are in a layered structure. This structure performs the management functions relating to the resources and parameters residing in its protocol entities, such as signaling. For each layer, the layer management handles the specific O&M information flow, described in chapter 11.

10.2.2 User plane

The function of the user plane is to transfer the user information from point A to point B in the network. All associated mechanisms, such as flow control, congestion control, or recovery from errors, are included. A layered approach is used with the user plane to identify the different functional components involved in providing services to the user.

10.2.3 Control or signaling plane

A layered structure is also used for the control or signaling plane. This plane is responsible for call control and connection control functions related to setting up and tearing down a connection. These are all the signaling functions necessary to set up, supervise, and release a call or connection. The functions of the physical and ATM layers are the same for the control and user planes. The difference in function occurs at the ATM adaptation layer, or AAL, and in the higher layers. The control plane is discussed in detail in chapter 11.

10.2.4 ATM cell types

Before we go into detail addressing each layer, we need to understand the cell terminology, used in both the physical and the ATM layer. According to ITU-T, "A cell is a block of fixed length. It is identified by a label at the ATM layer of the BISDN PRM."

Many different types of cells are in BISDN protocol layers:

- idle cell
- valid cell
- invalid cell
- assigned cell
- unassigned cell

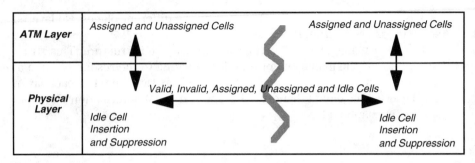

Figure 10.4 Cell types.

Figure 10.4 shows the relationship between the different cells with respect to BISDN physical and ATM protocol layers.

10.2.4.1 Idle cell (physical layer). These cells are used by the physical layer to adapt the ATM cell rate to the transmission rate using a predefined header. The cells are added at the physical layer at the originating end and removed at the physical layer at the terminating end. These cells are not seen by the ATM layer. Thus, they are switched.

10.2.4.2 Valid cell (physical layer). A cell with no header errors and that is not modified by the header error control (HEC) verification process is called a *valid cell*. This cell is present only at the physical layer and has no significance to the ATM layer.

10.2.4.3 Invalid cell (physical layer). An invalid cell is one whose header has errors, has not been modified by the cell HEC verification process, and is discarded by the physical layer. These cells are relevant to the physical layer only.

10.2.4.4 Assigned cell. These cells carry valid information for a service in the upper layers with a valid header. They originate from the ATM layer with appropriate header values for routing. At the destination side, these cells are passed on to the ATM layer on completion of error checking by HEC. These cells are the last cells to be discarded in case of congestion or any other problem in the network.

10.2.4.5 Unassigned cell. A cell that contains no valid information or preassigned header is called an *unassigned cell*. These cells are again originated from the ATM layer or above. They carry predefined header values and payload. Usually, these cells are used for OAM, signaling, etc., where certain values are reassigned to be identified by the switch for special purposes. We address some of these functions again in their respective layers. In the later sections of this chapter, we address each of the layers in reasonable detail.

10.3 Broadband Functional Architecture

The functional reference architecture identifies the network function performed by each layer in the broadband protocol.

A good functional architecture should be designed in a way that provides mobility of function between different physical elements. Figure 10.5 shows the functions of BISDN layers in the BISDN protocol. It gives an overview of

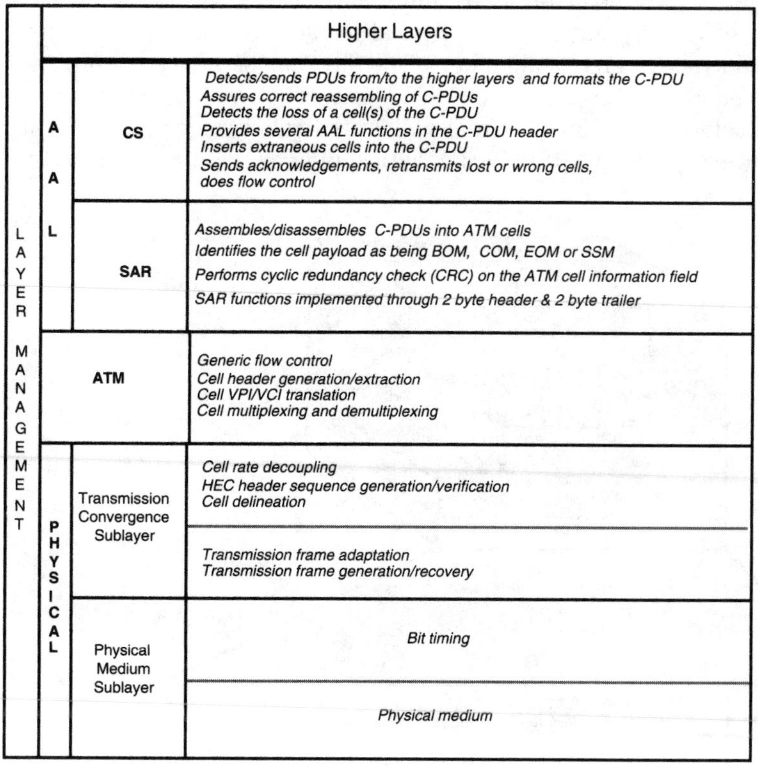

			Higher Layers		
L A Y E R M A N A G E M E N T	A A L	A A	CS	*Detects/sends PDUs from/to the higher layers and formats the C-PDU* *Assures correct reassembling of C-PDUs* *Detects the loss of a cell(s) of the C-PDU* *Provides several AAL functions in the C-PDU header* *Inserts extraneous cells into the C-PDU* *Sends acknowledgements, retransmits lost or wrong cells,* *does flow control*
		L	SAR	*Assembles/disassembles C-PDUs into ATM cells* *Identifies the cell payload as being BOM, COM, EOM or SSM* *Performs cyclic redundancy check (CRC) on the ATM cell information field* *SAR functions implemented through 2 byte header & 2 byte trailer*
	ATM				*Generic flow control* *Cell header generation/extraction* *Cell VPI/VCI translation* *Cell multiplexing and demultiplexing*
	P H Y S I C A L	Transmission Convergence Sublayer			*Cell rate decoupling* *HEC header sequence generation/verification* *Cell delineation*
					Transmission frame adaptation *Transmission frame generation/recovery*
		Physical Medium Sublayer			*Bit timing*
					Physical medium

AAL	ATM Adaptation Layer
ATM	Asynchronous Transfer Mode
CS	Convergence Sublayer
HEC	Header Error Control
SAR	Segmentation and Reassembly
VPI	Virtual Path Identifier
VCI	Virtual Channel Identifier

Figure 10.5 Function of BISDN layers.

the functions of the different layers of the protocol. In this section, we address only the high-level functions of each layer. The details of each layer are given later.

10.3.1 Physical-layer functions

Figure 10.6 shows the physical-layer functions in the BISDN protocol. The physical-layer function again is divided into two sublayers:

- physical medium (PM)
- transmission convergence (TC)

10.3.1.1 Physical medium sublayer. The physical medium sublayer is the lowest layer of the BISDN protocol, and it includes functions that are only physical-medium-dependent. The physical medium provides bit transmission capability, including bit alignment. The physical medium itself provides line coding and, if necessary, electrical-to-optical conversion. The physical medium itself can be of any material such as optical fiber, coax, or even air. ITU-T, however, defined BISDN using optical fiber as the physical medium.

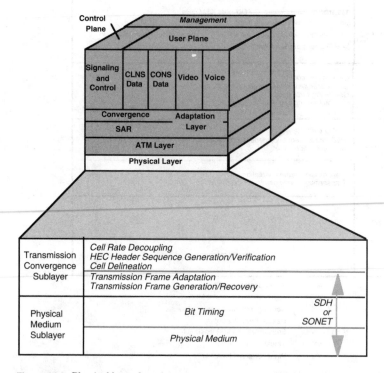

Figure 10.6 Physical layer functions.

In addition to the above functions, the physical medium provides bit timing where generation and reception of waveform is performed.

10.3.1.2 Transmission convergence sublayer. The TC sublayer is the second layer of the physical layer. Figure 10.6 shows the TC sublayer functions. The TC performs the following five functions:

- cell rate decoupling
- HEC header sequence generation/verification
- cell delineation
- transmission frame adaptation
- transmission frame generation/recovery

Each is described in the following subsections.

10.3.1.2.1 Cell rate decoupling. Idle cells are inserted during transmission, and idle cells are removed during reception. This mechanism is called *cell rate decoupling*. The purpose of this mechanism is for the ATM cell rate to adapt to the payload capacity of the transmission system. Other cells are passed untouched. As mentioned earlier, idle cells have no significance at the ATM layer.

10.3.1.2.2 HEC header sequence generation/verification. The HEC sequence is inserted in the 1-byte HEC field in the header of the ATM cell during transmission. At the receiving side, the HEC value is recalculated and compared with the received value. If there are single-bit errors, error correction is performed; otherwise, the cell is discarded. The details of HEC are covered in Section 10.4.

10.3.1.2.3 Cell delineation. Cell delineation enables the receiver to recover cell boundaries. This mechanism is described in ITU-T Recommendation I.432. To protect the cell delineation mechanism from malicious attack, the information field of a cell is scrambled before transmission, and descrambling is performed on the receiving side.

The next two functions are transmission frame-specific. All of the above functions are common to all possible transmission frames.

10.3.1.2.4 Transmission frame adaptation. This adaption is responsible for all actions necessary to adapt the cell flow according to the used payload structure of the transmission system in the sending direction. In the opposite direction, the reverse is done. The frame can be a cell equivalent, such as SONET envelope or DSS PLCP frame. The details of a SONET envelope were described in chapter 9.

10.3.1.2.5 Transmission frame generation/recovery. The lowest of all functions is the generation and recovery of the transmission frame. Its basic function is to generate the required frames so that ATM cells can be mapped. The frame size depends on the transmission speed on the transmitting side. On the receiving side, transmission frame recovery is performed by identifying the frame so that ATM cells can be identified and recovered from the payload envelope.

10.3.2 ATM layer functions

The next layer of the BISDN protocol is the ATM layer. Figure 10.7 shows the ATM layer and its functions in the BISDN protocol.

This layer has characteristics independent of the physical medium. Simply stated, the function of this layer is switching. The functions provided by this layer can be categorized as follows:

- generic flow control (GFC)
- cell header generation/extraction
- cell virtual path identifier (VPI)/virtual channel identifier (VCI) translation
- cell multiplexing and demultiplexing

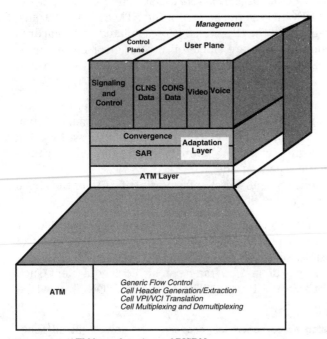

Figure 10.7 ATM layer functions of BISDN.

Each is described in the following subsections.

10.3.2.1 Generic flow control. As mentioned earlier, the GFC function is defined only at BISDN user network interface (UNI) to provide access flow control. It supports control of ATM traffic flow from a customer network or customer premises. GFC can be used to alleviate short-term overload conditions. The specific information is carried in assigned or unassigned cells.

10.3.2.2 Cell header generation/extraction. Cell header generation/extraction is done at the terminating points of the information or payload in the ATM layer. In the transmit direction, after receiving the ATM payload from the adaptation layer (48 bytes), the cell header is added, except for the HEC. The VPI and VCI values that are part of the header are obtained from the service access point (SAP) identifier. In the receive direction, the cell header is extracted, and the payload is forwarded to the AAL layer. Only the cell information field is passed to the higher layer (AAL layer). Here, the VPI and VCI values are translated into an SAP identifier.

10.3.2.3 Cell VPI/VCI translation. Cell VPI/VCI translation is the basis of ATM switching. It is performed at the ATM switching nodes or cross-connect nodes where the VPI and VCI values are translated. In a virtual path (VP) switch, the incoming VPI values within a cell are translated into a new outgoing VPI value. Here, the VCI values within the VPIs are preserved (not changed). In a VC switch, the values of the VPI as well as the VCI are translated. The VP switch is called an *ATM* cross-connect, and the VC switch is called an *ATM switch*. It is not necessary that the VPI and VCI values be translated just because the traffic passes through a switch. One could preserve the values of VPI and VCI end-to-end if necessary.

10.3.2.4 Cell multiplexing and demultiplexing. In the transmitting direction, cells from the individual VPs and VCs are multiplexed into one resulting cell stream. This function is called *cell multiplexing* and is accomplished by changing the VCI or VA value accordingly. The composite stream is normally noncontinuous cell flow. At the receiving side, the cells are demultiplexed into individual cells, and they flow into appropriate VP or VC to the destination.

10.3.3 ATM adaptation layer function

Figure 10.8 shows the AAL layer in the BISDN protocol. The basic function of AAL is the enhanced adaptation of the services provided by the ATM layer until the requirement of the higher layer's services are met. In this layer, the higher layer protocol data units (PDUs) are mapped onto the information field of the ATM cell, which is 48 bytes long. The ATM layer adds a header with appropriate VPI and VCI values to this cell.

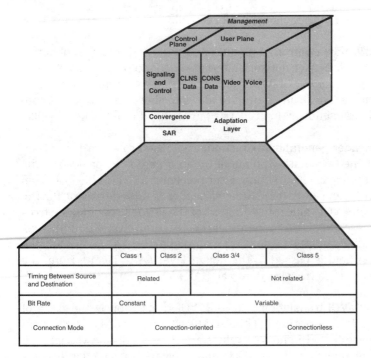

Figure 10.8   AAL layer in BISDN.

To perform its functions, AAL is divided into two sublayers, the convergence sublayer (CS) and the segmentation and reassembly sublayer (SAR).

10.3.3.1 Convergence sublayer. The functions of this layer are service-dependent and provide the AAL service at the AAL-SAP (service access point). Figure 10.9 shows the CS-PDU structure of the convergence sublayer. Details of the CS sublayer are discussed later in this chapter.

Following are the five classes of AAL services:

- AAL 1 or Type 1 constant bit rate (CBR) or circuit emulation
- AAL 2 or Type 2 variable bit rate video and audio
- AAL 3 or Type 3 connection-oriented data transfer
- AAL 4 or Type 4 connectionless data transfer
- AAL 5 or Type 5 high-speed data transfer[1]

---

[1] Not part of the original AAL Type.

To obtain the different classes and services, each of these are classified based on three basic parameters:

- time relation between source and destination
- bit rate
- connection mode

10.3.3.1.1 Time relation between source and destination. Some services have a time relation between source and destination, and in others, no such time relation exists. For instance, voice traffic requires a time relation between source and destination. If a timing problem occurs, the information will not arrive in the same manner as transmitted. For example, say the information transmitted is "Good Morning." If there is no timing sync, the information can arrive as "Morning Good." For data traffic, the information could arrive out of sync with no harm done because no real-time delivery is required. The traffic is reassembled using the sequence number in each 3-PDU and can be delivered in the correct order to the end user. But the integrity of the PDU must be maintained.

10.3.3.1.2 Bit rate. Some services, such as conventional voice traffic, have a continuous bit rate (CBR) and others, such as data traffic, have a variable bit rate (VBR). Of course, if the voice is packetized, it becomes a VBR type of traffic. Video traffic can be either CBR or VBR. In an ATM environment, it is preferred that video be VBR. With statistical multiplexing, real bandwidth savings can be achieved via sharing.

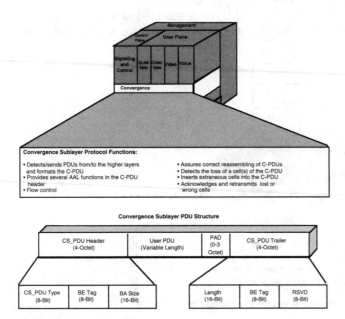

Figure 10.9 CS-PDU structure.

10.3.3.1.3 Connection mode. There are two types of connection modes: connection-oriented and connectionless. In services that are connection-oriented, a connection must be established before any information is transferred. Two examples are frame relay and circuit emulation traffic. The connectionless modes require no prior connection to transfer information. In this service, each packet has the source and destination address so that it can be individually routed in the network. An example of such a service is switched multi-megabit data service (SMDS). Thus, a service typically inherits each one of these characteristics and transfers them across the network.

10.3.3.2 Segmentation and reassembly sublayer. The SAR at the transmitting side segments the incoming higher-layer PDUs into a suitable size so they can fit into the information field of the ATM cell (48 bytes). At the receiving side, the information from different cells is reassembled and passed on to the higher layer as PDUs in the original form. Simply stated, the segmentation process is like cutting a long string into 48-byte pieces for easy handling; reassembly is piecing the string back together. For example, the PDU size for frame relay is different from that of SMDS. Figure 10.10 shows the generic SAR-PDU and the functions of SAR.

10.4 Physical Layer of BISDN Protocol

Figure 10.11 shows the details of the physical layer in the BISDN protocol. This figure is more detailed than the previous one.

As mentioned earlier, the physical layer is subdivided into two layers— the physical medium sublayer and the transmission convergence sublayer.

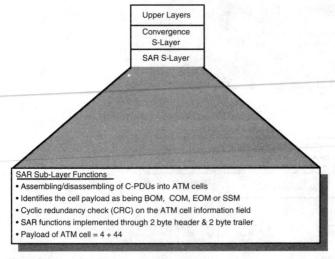

Figure 10.10 SAR-PDU structure and functions.

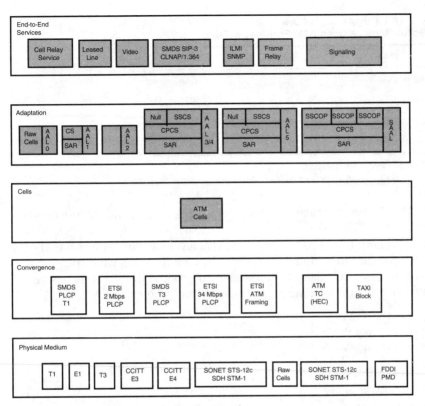

Figure 10.11 Physical layer of broadband protocol.

ITU-T has defined the physical medium as fiber-optics; to accommodate existing physical media for smooth transitions, however, coaxial cables are included. Currently, ITU-T has defined the physical medium interface based on the SONET/SDA for fiber and plesiochronous digital hierarchy (PDH) for coaxial. Using these physical media, ITU-T has defined the interface between customer (user) and the network, known as UNI. Recently, the ATM forum provided the details of the UNI specification which, in some cases, is in addition to ITU-T specifications. Table 10.1 shows the UNI specifications addressed by ITU-T and ATM forums.

The ATM forum, a special interest group formed by manufacturers from computer and telecom industries, has been defined as an additional interface to accommodate and expedite the deployment of ATM in existing interfaces. Some of the potential physical medium interfaces shown in Figure 10.11 are T1 (1.544 Mbps), E1 (2 Mbps), T3 (45 Mbps), E3 (34 Mbps), E4 (140 Mbps), SONET STS-3c / SDH STM-1 (155 Mbps), mapping of a raw ATM cell on any medium beginning (155 Mbps), SONET STS 12c/SDH STM-4 (622 Mbps), and FDDI PMD (100 Mbps).

**TABLE 10.1 Comparison of UNI
Specifications Between ATM Forum and ITU-T**

	ITU-T 1990	ATM forum 1992
45-Mbps DS3	X*	√
STS-1	X*	X**
155-Mbps or STS 3c	√	√
622-Mbps or STS 12c	√	X**

* Not an international standard.
** This rate was not chosen by ATM forum.

Each of these physical medium types has a corresponding transmission convergence layer protocol to act as a translator for different physical media and speeds to the ATM layer. The ATM layer can only process ATM cells of 48 bytes of payload and 5 bytes of header. The following transmission convergences (TC) have been defined to process the different specifications from the physical medium. The different protocols are:

- SMDS PLCP T1
- ETSI PLCP E1
- SMDS PLCP T3
- ETSI PLCP E4
- ETSI ATM framing
- ATM TC (HEC)
- TAXI block

Figure 10.12 shows the mapping of the PM protocol to the TC sublayer protocol. Of the different media protocols, ITU-T has defined three basic physical-layer interfaces for BISDN so far:

- ATM cell-based interface
- SDH/SONET-based interface
- digital interface

10.4.1 ATM cell-based interface

In this interface, a continuous stream of ATM cells are transported without any regular framing. In this cell stream, special cells conveying OAM information concerning the physical layer itself are transmitted. These cells are identified by special header values. These special OAM cells are not delivered to the ATM layer; they have significance only at the physical layer. Table 10.2 shows the values that identify these cells.

These are preassigned values of the cell header (excluding HEC octet). They perform functions such as monitoring, detecting, and reporting transmission errors by calculating an error code over the block of cells between two subsequent physical layer OAM ATM cells. This error code is transmitted in the information field of the physical-layer OAM cells. Figure 10.13 shows the ATM cell stream with OAM cells. However, no external frame exists to carry the ATM cells.

10.4.2 Physical layer for SONET/SDH interface

Here, the ATM cells are carried in a SONET/SDH STS-3c frame as shown in Figure 10.14. They are mapped into an STS-3 payload with a path overhead (POH) pointer indicating the beginning of the ATM cells. ATM

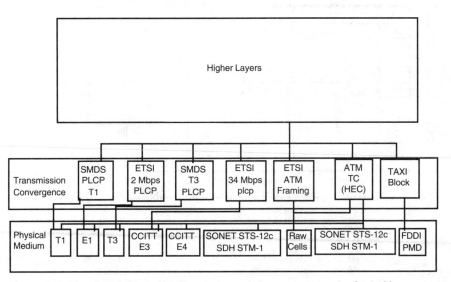

Figure 10.12 Mapping of physical medium to transmission convergence in physical layer.

TABLE 10.2 Preassigned ATM Cell Values

	Octet 1	Octet 2	Octet 3	Octet 4
Reuse by physical layer	PPPP0000	0	0	0000PPP1
Unassigned cells	AAAA0000	0	0	0000AAA0
Physical layer OAM	0	0	0	00001001
Idle cells	0	0	0	00000001

A: Bit available for use by ATM layer
P: Bit available for use by physical layer

(ITU-T/I.361, Table 1. Reprinted with permission.)

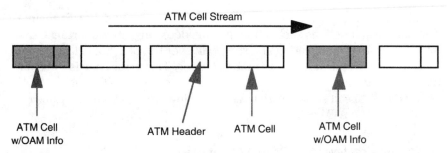

Figure 10.13 ATM cell stream.

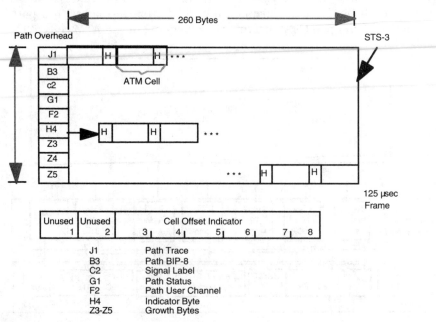

Figure 10.14 Cell mapping on a SONET STS-3c payload.

cell bytes are aligned within a payload but can cross the STS-3 frame. The H4 pointer of the POH is the pointer to point to the beginning of a cell in the payload. At the receiving end, the pointer used to identify the beginning of the cell boundary can be optionally used to help the cell delineation based on the HEC mechanism. The OAM implementation is in accordance with SONET/SDH implementation. The OAM allows frame alignment, error monitoring, error reporting, etc.

10.4.3 Digital interface

This interface was proposed to accommodate the existing asynchronous digital transmission signal as part of the broadband interface to enable easy evolution to BISDN. Existing digital standards such as DS1, DS3, E1, and E3 are used in this interface. The cells are directly mapped onto the digital signal using a function called the physical layer convergence procedure (PLCP). Mapping is defined for asynchronous rates with frame on HEC (not PLCP) format. Figure 10.15 depicts the DS3 PLCP frame format.

We mentioned earlier that five functions are performed by the transmission convergence sublayer. Here, we look into two important functions— header error control and cell delineation and scrambling.

10.4.4 HEC

As the name indicates, the basic function of HEC is to reduce the number of errors that occur in the header. The HEC algorithm is implemented in such a way that this objective is achieved in two stages. Stage 1 detects errors. If the detected error is a single-bit error, the algorithm performs error correction. If multiple bit error is detected, the algorithm discards the cell. Figure 10.16 shows the basic operation of the HEC algorithm.

In normal (detection) mode, if the receiving side detects a single bit error, it enters the correction mode. The error is corrected, and the state at the receiver switches back to detection mode. In the detection mode state,

<1>	<1>	<1>	<1>	<53 Bytes>	
A1	A2	P11	Z6	L2_PDU	
A1	A2	P10	Z5	L2_PDU	
A1	A2	P9	Z4	L2_PDU	
A1	A2	P8	Z3	L2_PDU	
A1	A2	P7	Z2	L2_PDU	
A1	A2	P6	Z1	L2_PDU	
A1	A2	P5	F1	L2_PDU	
A1	A2	P4	B1	L2_PDU	
A1	A2	P3	G1	L2_PDU	
A1	A2	P2	M2	L2_PDU	
A1	A2	P1	M1	L2_PDU	
A1	A2	P0	C1	L2_PDU	Trailer: 13-14 Nibbles

A1, A2	Framing Bytes	125 μs
P11-P0	Path Overhead Integrity	
Z1-Z6	Growth Bytes	
F1	PLCP Path User Channel	
B1	BIP-8	
G1	PLCP Path Status	
M1,M2	SIP Level Counter	
C1	Cycle/Stuff Counter	
PLCP	Physical Layer Convergence Procedure	

Figure 10.15 DS3 PLCP.

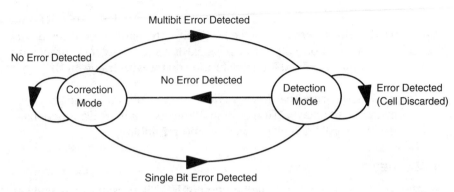

Figure 10.16 HEC algorithm operation. *(ITU-T/I.432, Fig. 3. Reprinted with permission.)*

all cells detected with multiple bit errors are discarded. Headers without errors are examined and forwarded to a higher layer. On the transmission side, the HEC value is calculated using the polynomial generated by header bits (fields other than the HEC field), which are multiplied by 8 and divided by $x^8 + x^2 + x + 1$. The remainder of this calculation is transmitted as the 8-bit HEC field and verified at the receiving end to check the validity of the ATM cells. Figure 10.17 depicts the HEC flow chart.

10.4.5 Cell delineation and scrambling

Cell delineation is the process that allows cell boundaries to be identified. Identification is based on the correlation between the header bits to be protected and the relevant control bits. Figure 10.18 shows a state diagram of HEC-based cell delineation.

The cell delineation algorithm works as shown in the figure. When in hunt state, a bit-by-bit check of the assumed header field is performed. If a correct header is found, the algorithm goes to presync state and the HEC correlation check is performed cell by cell. If δ consecutive correct HECs are found, the algorithm goes to sync state. If not, the system goes to hunt state. The system leaves sync state only if α consecutive incorrect HECs are identified. The default values recommended by ITU-T for α and δ are 7 and 6, respectively.

With α of 7, at 155 Mbps, the ATM HEC will be correct for more than a year with bit error probability of about 10^4. With δ of 6, the same system with the same bit error probability will need about 10 cells, or 28 µs, to reenter sync after the loss of cell synchronization.

This cell delineation method could fail if the HEC correlation were imitated in the information field of ATM cells. In such cases, the information field contents are scrambled, using a self-synchronizing scrambler with the polynomial $X^{43} + 1$. The scrambler is on only in the presync and sync states and is disabled in the hunt state.

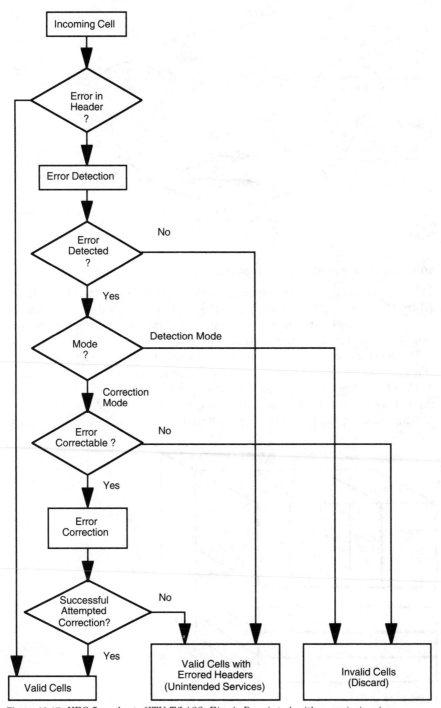

Figure 10.17 HEC flow chart. *(ITU-T/I.432, Fig. 4. Reprinted with permission.)*

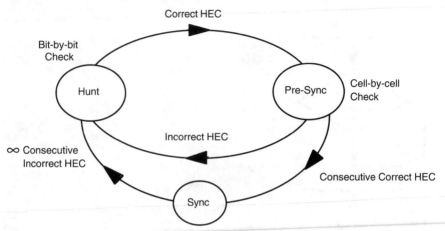

Figure 10.18 Cell delineation state diagram. *(ITU-T/I.432, Fig. 5. Reprinted with permission.)*

10.5 ATM Layer of the BISDN Protocol

The second layer of BISDN protocol is the ATM layer, as shown in Figure 10.19. Five basic functions are performed in this layer. Table 10.3 summarizes the different functions and their corresponding parameters, which were explained in an earlier section.

The ATM layer provides certain services along with its functions. It provides for the transparent transfer of fixed-size ATM layer service data units (ATM-SDUs) between communicating upper-layer entities (e.g., AAL entities). This transfer occurs on a preestablished virtual ATM connection according to a traffic contract. A negotiated traffic contract consists of a

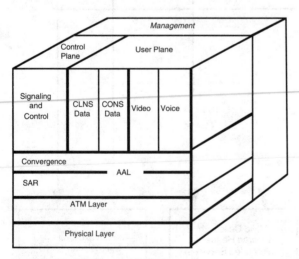

Figure 10.19 ATM layer of BISDN.

TABLE 10.3 Functions Supported at UNI (U-plane)

Functions	Parameters
Multiplexing among different ATM connections	VPI/VCI
Cell rate decoupling (unassigned cells)	Preassigned header field values
Cell discrimination based on predefined header field values	Preassigned header field values
Payload type discrimination	PT field
Loss priority indication and selective cell discarding	CLP field, network congestion state

quality of service (QOS) class, a vector of traffic parameters, a conformance definition, etc. Each ATM end point is expected to generate traffic that conforms to these parameters. We mentioned that transfer occurs on a preestablished virtual ATM connection; two levels of virtual connections exists through which transfer occurs.

Two levels of virtual connections can be supported at the ATM UNI:

- A point-to-point or point-to-multipoint virtual channel connection (VCC), which consists of a single connection established between two ATM VCC end points.

- A point-to-point or point-to-multipoint virtual path connection (VPC), which consists of a bundle of VCCs carried transparently between two ATM VPC end points.

Table 10.4 shows predefined header field values that are defined at UNI by the ATM forum in their UNI specification working document. These values are used to discriminate the ATM cells based on the ATM header values so that processing can be done accordingly.

TABLE 10.4 Predefined Header Field Values

Use	Value[1,2,3,4]			
	Octet 1	Octet 2	Octet 3	Octet 4
Unassigned cell indication	00000000	00000000	00000000	0000xxx0
Meta-signaling (default)[5,7]	00000000	00000000	00000000	00010a0c
Meta-signaling[6,7]	0000yyyy	yyyy0000	00000000	00010a0c
General broadcast signaling (default)[5]	00000000	00000000	00000000	00100aac
General broadcast signaling[6]	0000yyyy	yyyy0000	00000000	00100aac
Point-to-point signaling (default)[5]	00000000	00000000	00000000	01010aac

TABLE 10.4 Continued

Use	Value[1,2,3,4]			
	Octet 1	Octet 2	Octet 3	Octet 4
Point-to-point signaling[6]	0000yyyy	yyyy0000	00000000	01010aac
Invalid pattern	xxxx0000	00000000	00000000	0000xxx1
Segment OAM F4 flow cell[7]	0000aaaa	aaaa0000	00000000	00110a0a
End-to-end OAM F4 flow cell[7]	0000aaaa	aaaa0000	00000000	01000a0a

1: "a" indicates that the bit is available for use by the appropriate ATM layer function.
2: "x" indicates "don't care" bits.
3: "y" indicates any VPI value other than 00000000.
4: Indicates that the originating signaling entity sets the CLP bit to 0. The network may change the value of the CLP bit.
5: Reserved for user signaling with the local exchange.
6: Reserved for signaling with other signaling entities (e.g., other users or remote networks).
7: The transmitting ATM entity sets bit 2 of octet 4 to zero. The receiving ATM entity ignores bit 2 of octet 4.
SOURCE: ATM forum, UNI specification ver. 3.0

Within the ATM header exists the payload type indicator (PTI) field, which has certain predefined values. Table 10.5 shows the predefined cells based on PTI.

The ATM layer routes the cells based on the values of different fields in the header and PTI values. The basic function of the ATM layer from the protocol perspective, however, is in the transmitting direction. The ATM layer utilizes the information received from the higher layers and the management plane to generate the header with certain values and then appends the header to the user information field, which is received from the AAL layer. These cells subsequently are sent down to the physical layer for

TABLE 10.5 Predefined Cells Based on PT Identifier

PTI coding (MSB first)	Interpretation
000	User data cell, congestion not experienced, SDU-type = 0
001	User data cell, congestion not experienced, SDU-type = 1
010	User data cell, congestion experienced, SDU-type = 0
011	User data cell, congestion experienced, SDU-type = 1
100	Segment OAM F5 flow related cell
101	End-to-end OAM F5 flow related cell
110	Reserved for future traffic control and resource management
111	Reserved for future functions

transmission. In the receiving direction, the cells received from the physical layer are disassembled to extract the header information and process it, while the payload is sent to the AAL layer. To take the cell from the higher layer at one end and deliver it to the higher layer at the other end, the ATM layer performs an *ATM connection*. An ATM connection is a transparent connection provided by the ATM layer to the higher layer. It is connected end-to-end through a concatenation of connection elements.

10.5.1 ATM layer connections

Before we explain the ATM connection, one needs to understand the basics of VPI and VCI. Figure 10.20 shows the relationship between a physical path, VPI, and VCI. A physical path can be coax-based DS3 (44.76 Mbps) or fiber-based SONET OC3 (155 Mbps) and above. Within this physical path a certain number of virtual paths (VPs) could exist. The number of virtual paths depends on the number of bits allocated as the VP value in the cell header. Using the VPI value in the ATM cell header, each virtual path is identified. Usually, 12 bits are allocated for the VPI value at the user interface. Having explained the relationship between the physical path and the virtual path, let's look into the relationship between the virtual path and the virtual channel. Within a virtual path there are many virtual channels. The number of virtual channels per virtual path depends on the number of bits allocated to the VCI value in the ATM cell header. Usually, 16 bits are allocated for the VCI value. In an ATM network, the ATM cells are routed using the VP, and VCI values in the header of each cell.

In the real ATM world, two types of ATM connections exist: VCC and VPC. A VC connection is a logical connection between two end points for the transfer of ATM cells, whereas VPC is a logical combination of VCCs. Each VCC is assigned a VCI value, and each VPC is assigned a VPI value. Within a VPC, VC links different from one another exist, and each is differentiated through the use of VCI. On the other hand, VCs belonging to different VPs can possess the same VCI as one used in another VPC. A VC can

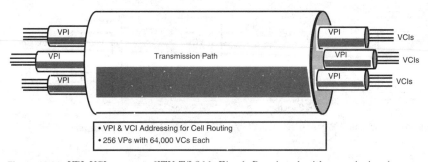

Figure 10.20 VPI, VCI concept. *(ITU-T/I.311, Fig. 1. Reprinted with permission.)*

be completely identified solely on the basis of its corresponding VCI and VPI values. Thus, in an ATM environment, switching of cells is performed based on the VPI and VCI values. The mechanism of switching in an ATM node is explained in Part 4.

If, in an ATM network, the VCI value is not modified but the VPI value is, that function is called a *VP switch* or *ATM cross-connect* or *ATM add-drop*, and its equipment is called *cross-connect equipment* or a *concentrator*. If the function of the equipment is to alter both VPI and VCI values (incoming VPI/VCI values are mapped to outgoing VPI/VCI values), that equipment is called *ATM switch* or *VC switch*. Figure 10.21 shows the concept of VP and VC connections using VP and VC switches. The VP switch corresponds to the add-drop multiplexer and the VP/VC switch corresponds to a normal switching function.

VC or VP connections are defined based on the switching elements used. A VCC refers to a concatenation of VC lines for achieving connection between ATM service access points. In the VC link, the VCI is assigned, gets translated, or is removed. The VCC provided by the ATM switching element can be a permanent or semipermanent connection. The integrity of the cell sequence is ensured within the same VCC. A VCC user is provided with a set of parameters, such as cell delay and cell loss rate, by the network. At the time of VCC setup, user traffic parameters are prescribed through negotiation between the user and the network, and the network monitors these parameters for the duration of the connection. At the user-network interface, four different methods can be used to establish the VCC:

1. *The signaling procedure.* In this connection, setup or release is achieved through a reservation. This method applies to permanent or semipermanent connections.

2. *The metasignaling procedure.* A signaling VC is established or removed through the use of metasignaling VC.

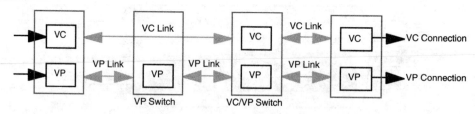

Figure 10.21 VC and VP connections. *(ITU-T/I.150, Fig. 2. Reprinted with permission.)*

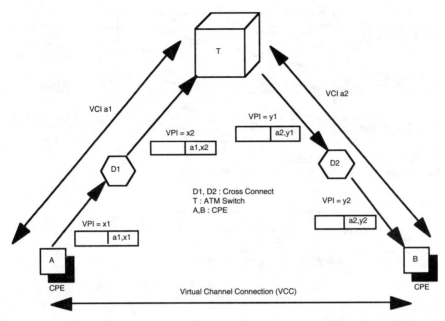

Figure 10.22 Example of VCC.

3. *The user-network signaling procedure.* A signaling VCC is used to establish or release a VCC for end-to-end communication.

4. *The user-to-user signaling procedure.* A signaling VCC is used to establish or release a VCC internal to a VPC preestablished between two UNIs.

Once a VCC connection is established using any of the above methods, ATM cells flow through the network. Each cell header is processed at the network element, such as the cross-connect and ATM switch, which translates the VPI and VCI values to reach their destination. Figure 10.22 shows a typical VCC connection and the location where the VPI and VCI values are translated. *A* and *B* are the CPE, *D1* and *D2* are the ATM cross-connects, where only the VPI values are translated, while the VCI values are preserved. *T* is the ATM switch that performs the translation of VPI and VCI values. Figure 10.22 clearly depicts the concept behind the VCC.

Figure 10.23 shows a blowup of *D1* or *D2* which is a VP switch, and Figure 10.24 shows a blowup of the *T* which is a VC switching function. So far, we have addressed the VC connection and the network elements that participate in the completion of a VC connection. Another type of connection exists called a *VP connection*, or virtual path connection, or simply VPC.

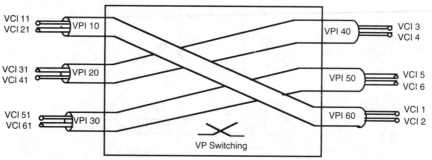

Figure 10.23 VP switching. *(ITU-T/I.311, Fig. 4(a). Reprinted with permission.)*

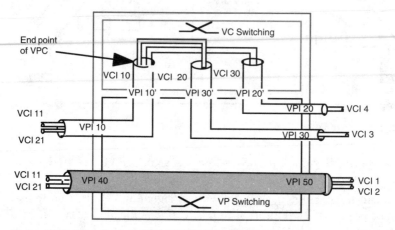

Figure 10.24 VC switching. *(ITU-T/I.311, Fig. 4(b). Reprinted with permission.)*

This VPC refers to the concatenation of VP links for connecting the points at which a VPI is assigned, translated, or removed. A VPC was shown in Figure 10.22. A VPC can be provided through switching equipment and can be permanent or semipermanent. Cell sequence is ensured for each VCC within the same VPC. A VPC between the VPC end points can be established or released in two possible ways:

1. *Without a signaling procedure.* the setup and release of a connection is achieved using a reservation.

2. *By user or network control.* VPIs are assigned in advance by the user or the network service provider.

These are the basic ways through which the ATM layer provides transparent transfer of cells between higher layers at either ends.

10.6   ATM Adaptation Layer of BISDN Protocol

Figure 10.25 shows the ATM adaptation layer of the BISDN protocol. As depicted in the figure, four different classes of AAL exist, each with certain characteristics. Each class has different AAL functions, namely timing, bit rate type, and connection mode. Each of these classes map onto existing services so that they can be adapted to ATM-based broadband networks. In section 10.2 we discussed the functions of AAL.

Certain changes in the AAL classes have occurred from the ones initially proposed by ITU-T. For example, a new AAL class has been included to address the need of high-speed data transfer. In this class, the overhead bytes for a packet are reduced compared to class 3/4. AAL 3 and 4 have also been merged together as a single class. A proposal exists to eliminate class 2, which addresses packetized voice and video. That issue has been postponed, however, until other important issues are resolved. Today's equipment manufacturers are

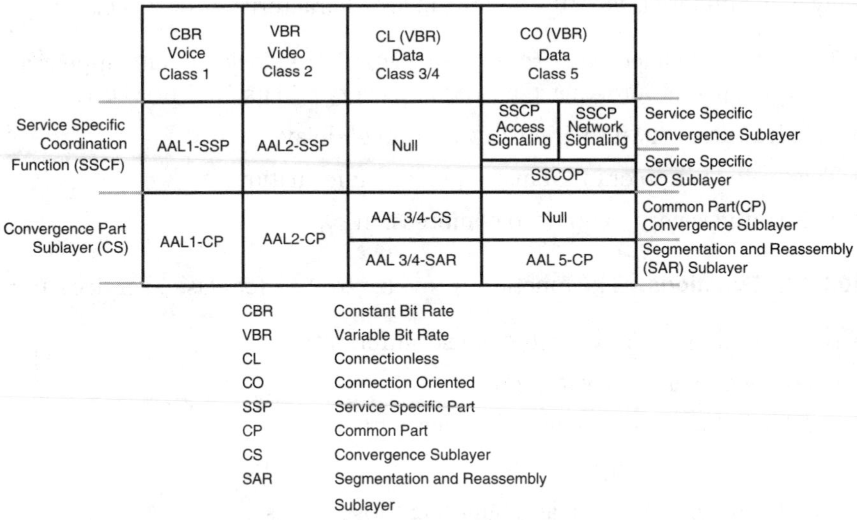

	CBR Voice Class 1	VBR Video Class 2	CL (VBR) Data Class 3/4	CO (VBR) Data Class 5		
Service Specific Coordination Function (SSCF)	AAL1-SSP	AAL2-SSP	Null	SSCP Access Signaling	SSCP Network Signaling	Service Specific Convergence Sublayer
				SSCOP		Service Specific CO Sublayer
Convergence Part Sublayer (CS)	AAL1-CP	AAL2-CP	AAL 3/4-CS	Null		Common Part(CP) Convergence Sublayer
			AAL 3/4-SAR	AAL 5-CP		Segmentation and Reassembly (SAR) Sublayer

CBR	Constant Bit Rate
VBR	Variable Bit Rate
CL	Connectionless
CO	Connection Oriented
SSP	Service Specific Part
CP	Common Part
CS	Convergence Sublayer
SAR	Segmentation and Reassembly Sublayer

Figure 10.25   AAL layer classes.

thus developing equipment based on AAL 1, 3/4, and 5 classes. In addition to these, a proposal exists for AAL 2 for mapping of MPEG 2 video signals into ATM. In this section, we address each of the AAL classes with regard to:

- characteristics
- functions
- services provided
- protocol data unit structure
- mapping of AAL1-PDU to ATM and physical layer

10.6.1   AAL 1 or class 1

Services that have constant bit rate (CBR) use class 1, because it receives and delivers SDUs with constant bit rate from and to the above layers. Along with data, class 1 transfers certain characteristics of CBR, such as the timing information between the source and destination. An indication of lost or errored information is sent to the higher layers if or when failures cannot be recovered. This section addresses the AAL 1 characteristics, services, SDU functions, and how the mapping is accomplished to the ATM and physical layer of the BISDN protocol.

10.6.1.1   Characteristics.   As mentioned above, CBR type of services typically come under class 1. The following are characteristics of this class:

- This traffic is called *isochronous traffic*, where blocks of data appear at known constant intervals, e.g., 193 bits every 125 μs for T1 or DS1.
- This traffic is very intolerant to variation of delay.
- This traffic is intolerant of missequenced information.
- This traffic is very tolerant to compression.

10.6.1.2   Functions.   The functions provided by AAL for class 1 services are:

- segmentation and reassembly of user information
- handling of cell delay variation
- handling of lost and misinserted cells
- source clock recovery
- monitoring for bit errors and handling those errors
- structure pointer generation and detection

10.6.1.3   Services provided.   This class of AAL provides constant-bit-rate service that can be voice, video, or data. It takes incoming structured (bytes) or unstructured (bits) information and maps it into 48-byte payloads to be shipped to the lower layer.

10.6.1.4 Protocol data unit structure. Figure 10.26 shows the format of the SAR-PDU for AAL class 1. Three fields exist, namely the SN (sequence number) field, SNP (sequence number protection) field, and SAR-PDU (segmentation and reassembly-protocol data unit). SN detects the loss or misinsertion of cells. SNP provides error detection, which is done by CRC process, and correction capabilities. The SAR-PDU carries the actual payload of the information to be transported.

10.6.1.5 Mapping of PDU to the ATM and physical layer. Figure 10.27 shows a high-level view of how unstructured CBR DS1 traffic is mapped onto the ATM layer and then to the physical layer. In this example, the physical layer uses the SONET format for carrying the ATM cells.

10.6.2 AAL 2 or class 2

Class 2 addresses the same type of traffic—voice and video—but with different characteristics. Class 2 addresses variable-bit-rate traffic, which is usually data. This type of traffic is not widely used, and the standards for

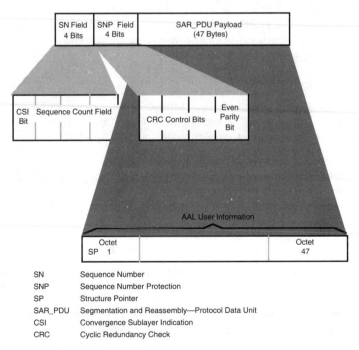

SN	Sequence Number
SNP	Sequence Number Protection
SP	Structure Pointer
SAR_PDU	Segmentation and Reassembly—Protocol Data Unit
CSI	Convergence Sublayer Indication
CRC	Cyclic Redundancy Check

Figure 10.26 AAL 1 SAR-PDU structure.

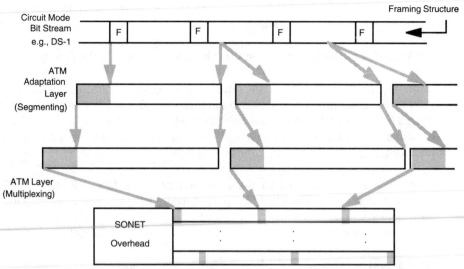

Figure 10.27 Mapping of AAL 1 PDU to physical layer.

this layer have not been well-defined because converting voice traffic to variable-bit-rate traffic has numerous problems. The obvious ones are sequencing the data and preserving the timing. These parameters are very critical for real-time voice traffic, but they are not critical to data. Converting voice to VBR needs additional processing by the upper layer. For example, the upper layer should sense silence during a voice conversation and suppress those data units and transmit only the ones with information. At the same time, the timing integrity between source and destination must be maintained. Thus, the standards body has decided to postpone working on this class of service because it will take a while to be used. There are other basic issues that also need to be defined.

10.6.2.1 Characteristics. Characteristics of AAL class 2 service are the following:

- The traffic has burst characteristics from time to time.

- A time stamp for each packet is required so that it can be reassembled.

- This traffic is very intolerant of missequenced information.

10.6.2.2 Functions. The following are the functions performed by AAL class 2 to enhance the services provided to the ATM layer:

- segmentation and reassembly of user information
- handling of cell delay variation
- handling of lost and misinserted cells
- recovery of source clock at the receiver
- monitoring for bit error and handling of these errors
- monitoring of the user information field for bit errors and possible corrective actions

10.6.2.3 PDU structure. Figure 10.28 shows the AAL 2 protocol structure. The structure consists of three fields, the header, payload, and trailer. The information is carried in the payload. The SN in the PDU header detects lost or misinserted cells. The IT indicates the type of information being carried, such as BOM (beginning of message), COM (continuation of message), or EOM (end of message). In the PDU trailer, LI is used to indicate the number of CS-PDU bytes carried in the SAR-PDU payload field. The CRC protects the PDU against bit errors.

10.6.3 AAL 3/4 or class 3/4

In this section we address AAL 3 and 4 as one service. The initial definition of BISDN was AAL 3 as connection-oriented and AAL 4 as connectionless. The new class is defined as a connectionless-based service, and the services provided by AAL 3 are adopted by the new class called AAL 5.

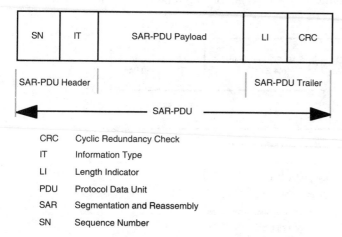

CRC	Cyclic Redundancy Check
IT	Information Type
LI	Length Indicator
PDU	Protocol Data Unit
SAR	Segmentation and Reassembly
SN	Sequence Number

Figure 10.28 AAL 2 PDU structure.

10.6.3.1 Characteristics. The characteristics of class 3/4 are different from that of the other classes. Most of them are an extrapolation of the characteristics of existing VBR traffic. They are:

- burst of information with variable frame length

- delay is not that critical, as in case of class 1 or class 2 service such as voice or video

- because delay is not critical, packets can be resequenced based on the sequence number

10.6.3.2 Services provided. Two models of services are defined for AAL 3/4 class: message mode and streaming mode. Message mode is used for framed data transfer, and streaming mode is suitable for the transfer of low-speed data with low delay requirements.

10.6.3.3 PDU structure. Figure 10.29 shows the AAL 3/4 PDU structure. This PDU has three parts: SAR-PDU header, payload, and trailer. The header consists of fields ST (segment type), SN (sequence number), and MID (multiplexing identifier). The segment type consists of 2 bits and indicates whether the packet is beginning, or continuing, the end of the message with a 2-bit code. The next four bits are for the sequence number. The SN is incremented by one relative to the previous PDU belonging to the same source connection. The next 10 bits (MID field) are used to assist in the interleaving of ATM-SDUs from different CS-PDUs and reassembly of these CS-PDUs at the other end.

The second major field is the payload field, which consists of 44 bytes used for CS-PDU data. If this field is not fully filled, the remaining unused

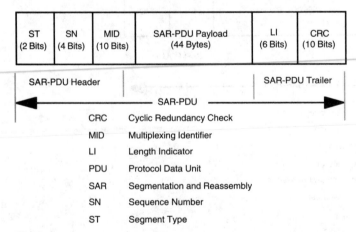

Figure 10.29 AAL 3/4 PDU structure.

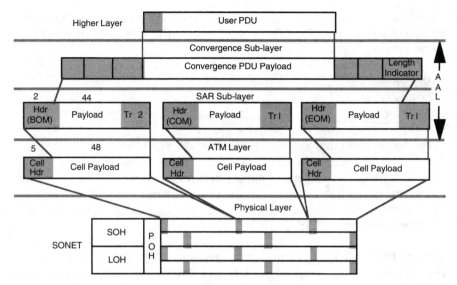

Figure 10.30 Mapping of user data in AAL 3/4.

bits are coded as zeros. The third major segment is the trailer, and it consists of two fields: LI (length indicator) and CRC (cyclic redundancy check). The LI consists of 6 bits and contains the number of bytes from CS-PDU that are included in the PDU payload field. Its maximum value is 44. The CRC is a 10-bit field filled with the results obtained from a CRC calculation performed over the SAR-PDU header.

10.6.3.4 Mapping of PDU to the ATM and physical layers. Figure 10.30 shows the mapping of user data into the ATM and the physical layer of the BISDN protocol. The figure shows how a user's data is segmented into 44 bytes of information in the AAL layer and mapped into the ATM layer of 48 bytes by the addition of 4 bytes, where two bytes are for the header and two bytes for the trailer by the AAL layer.

10.6.4 AAL 5 or class 5

Class 5 service was not part of the BISDN standards initially proposed. It was proposed by the computer and data processing vendors to the standards bodies in the United States in August 1991 for two major reasons:

- AAL 5 has a low overhead compared to AAL 3/4.
- A TCP/IP acknowledgment fits into a single cell with AAL 5, versus two cells in AAL 3/4.

This class is optimized for local usage. The AAL 5 class uses 48 bytes of information in the cell payload, achieved by removing the MID field and reassembling it based on VCI values only. The message type in this class is indicated based on the PT (payload type) field in the ATM cell header. They are as follows:

0X1: EOM
0X0: BOM or COM

Here, X is don't care, i.e., it can be 0 or 1.

The BOM is the first cell of a particular VPI/VCI value with a PT value 0X0, followed by cells with a PT value of 0X1. By using this technique, there is a tremendous performance improvement to be obtained. The receiving end simply has to queue the cells until it encounters the EOM message bit, which is the indication of the last cell. On receiving the last cell, the CRC and LI get checked and are passed on to the higher layers.

10.6.4.1 PDU structure. Figure 10.31 shows the PDU structure of AAL 5. The structure is similar to AAL 3/4 except that no MID field exists in the SAR-PDU structure. In the AAL 5 convergence layer, all cells except the last cell of the packet are completely filled with 48 bytes of data. The CS-SDU trailer has 4 fields: a 1-byte user-to-user indication (UU) field, a 1-byte common part indicator (CPI), a 2-byte field, and a 4-byte CRC field. The UU and CPI fields are currently unused and set to 0.

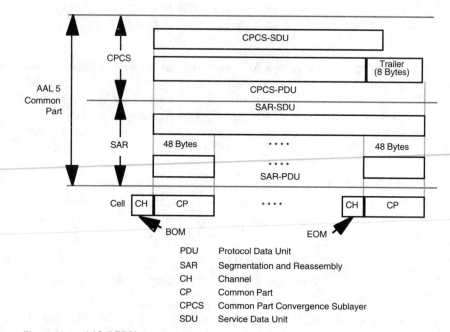

PDU	Protocol Data Unit	
SAR	Segmentation and Reassembly	
CH	Channel	
CP	Common Part	
CPCS	Common Part Convergence Sublayer	
SDU	Service Data Unit	

Figure 10.31 AAL 5 PDU structure.

10.7 Summary

In this chapter, we addressed the lower layers of the BISDN protocol, namely the physical layer, ATM layer, and the ATM adaptation layer, along with their respective functions. Each layer of the BISDN has certain basic functions. The physical layer deals with transport of bits, the ATM layer deals with routing and switching of the ATM cells to the destination, and the AAL layer deals with the adaptation of different protocols, such as frame relay, SDMS, etc. The adaptation layer inherits the characteristics and function of the protocol, such as frame relay, SMDS, etc., so as to make it as transparent as possible to the user who has subscribed to that service. The AAL layer maps the information into the ATM payload so that the AAL layer at the destination can recover the information with the same characteristics with which it was transmitted.

11

BISDN Higher Layers

11.1 Overview

In the previous chapter, we addressed the lower layers of the BISDN protocol reference model: the physical, ATM, and ATM adaptation layers. These lower layers provide the basic functions, such as transmission (transportation of cells), switching (routing of cells), and adaptation of the other protocol-based services, including frame relay, SMDS, etc.

In this chapter, we address the higher layers of the BISDN protocol. The higher layers are the management plane, user plane, and control or signaling plane. The functions of these layers are independent of the lower layers. These layers perform functions to provide the BISDN services and features to the end user. The BISDN protocol uses the existing higher layers, such as the SS7 network for signaling and IN for providing intelligence to the services, and, of course, adapting user protocols to the ATM-based BISDN protocol. This chapter discusses the functions and services provided to the user by these layers.

11.2 Management Plane

This plane performs the management-related functions for the BISDN protocol. This plane is divided into two sublayers: plane management and layer management. Figure 11.1 shows the two management sublayers:

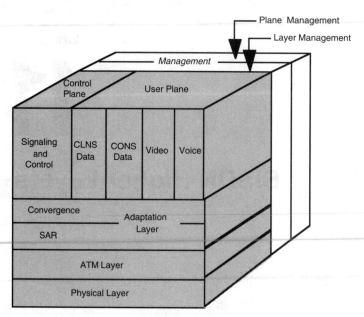

Figure 11.1 BISDN protocol—management plane.

11.2.1 Plane management

This layer performs the management functions related to the whole system. Its main task is to coordinate functions between all planes, including getting the status of information on each plane and informing the other planes of that status.

11.2.2 Layer management

The functions of this sublayer are categorized into layers. Layer management performs the management functions relating to performance, operations, administration, resource management, and parameters for each of the user plane layers. For each of the user plane layers, such as the physical layer, ATM layer, etc., layer management handles specific performance and operations, administration and management (OAM) functions. The initial view on OAM functions is provided in ITU-T Recommendation I.610. In this recommendation, ITU-T defined the basic OAM principle, which is based on controlled maintenance. The OAM provides status, testing, and performance monitoring information for different protocols to prevent errors.

To obtain an optimal functionality with OAM, the following functions have been defined by ITU-T:

- performance monitoring
- defect and failure detection

- system protection
- performance information
- fault localization

Each function is described in the following subsections.

11.2.2.1 Performance monitoring. In normal operation, performance monitoring continuously checks or periodically controls functions to guarantee the provision of maintenance information. The performance information obtained by the relevant performance monitoring mechanisms is transported via cells to the applicable OAM entities, which use this information for long-term system evaluation, short-term service quality control, or to initiate preventive actions if necessary.

11.2.2.2 Defect and failure detection. By continuous or periodic checking of the functions, failures can be detected or made known. If a failure is detected, the necessary actions can be initiated to localize the failure, such as disconnecting the failed equipment from other equipment in the network.

11.2.2.3 System protection. If a failure is detected, the failed entity is excluded from operation, thereby minimizing the effect of the failure on the whole system. This process protects the rest of the system from failure.

11.2.2.4 Performance information. If an entity fails, other management entities are informed in a timely manner on the status of the entity. Status information is also exchanged with other entities. This status information about the failed entity is used by the system in the protection phase so that the system can exclude failing entities. The information is also used by neighboring entities to ensure that a failure message is spread over the entire network. These entities can then update their routing table so as to avoid the failed entity.

11.2.2.5 Fault localization. Internal or external test devices determine the exact location of the failed entity. When the faults are exactly identified, the system enters into the protection phase, where the failing entities are excluded.

To identify the fault exactly, ITU-T has defined five physical hierarchical OAM levels. With each of the levels, an associated information flow exists in the layer management sublayer. These levels are illustrated in Figure 11.2. In this figure, two levels are defined in the ATM layer: virtual channel level, identified by $F5$, and virtual path level, identified by $F4$. The other three levels are in the physical layer. The transmission path level is identified by $F3$, the digital section level by $F2$, and the regenerator section level by $F1$.

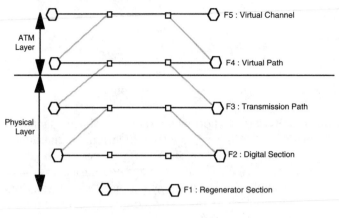

Figure 11.2　OAM hierarchical levels. *(ITU/I.311, Fig. 3. Reprinted with permission.)*

These levels are not present in all parts of the network. In case of a fault condition, the relevant OAM functions are performed on a higher level. These levels are described in the following subsections.

11.2.2.5.1　Virtual channel (F5). Both end points perform virtual channel identifier (VCI) termination functions for a broadband connection. Such a connection comprises several virtual paths. The OAM functions are performed on a VCI level and might provide input to any of the five OAM categories mentioned above. For example, it is possible to conduct performance monitoring on a VCI level using the PTI bits on the ATM cell header. Performance monitoring occurs at the ATM layer in the BISDN protocol.

11.2.2.5.2　Virtual path (F4). Both end points perform virtual path identifier (VPI) termination functions for a broadband connection. Such a connection comprises several physical transmission paths. Again, one of the five OAM categories might be involved in the virtual path maintenance. This virtual path maintenance is performed at the ATM layer in BISDN protocol.

11.2.2.5.3　Transmission path (F3). Both end points perform the assembly/disassembly of the ATM payload and the OAM-related functions of a transmission system. Because cells must be recognized at a transmission path to extract OAM cells, cell delineation and HEC functions are required at the termination point of each transmission path. A transmission path consists of several digital sections. Assembly/disassembly is performed at the physical layer of the BISDN protocol.

11.2.2.5.4 Digital section (F2). Both end points are section termination points. Every digital section comprises a maintenance entity, capable of transporting OAM information between adjacent digital sections. This function is performed at the physical layer of the BISDN protocol.

11.2.2.5.5 Regenerator section (F1). This is the smallest recognizable physical entity for OAM and is located between repeaters. The mechanism provides OAM functions, and the information flows associated with it depend on the respective OAM layer.

11.2.3 Physical-layer mechanisms

At the physical layer, the OAM information flow of F1, F2, and F3 depends on the type of transmission system. In plesiochronous transmission systems (G.702, G.703), the bit error rate per section is monitored via cyclic redundancy check (CRC) by counting the number of code violations. In SONET/ SDH, special bytes in the section overhead (SOH) and path overhead (POH) transport error measurement codes such as BIP-8 (bit interleaved parity). In a cell-based transmission system, OAM is performed by special OAM cells called PLOAM (physical-layer OAM) cells. These cells are valid only on the physical layer and are not passed to the ATM layer. The OAM information is transferred in different ways based on the physical transmission systems defined.

11.2.4 ATM layer mechanisms

At the ATM layer, the OAM information flows of F4 and F5 have dedicated cells used to perform virtual channel (VC) and virtual path (VP) maintenance. These cells can also be used to transport OAM information. Here, the payload type indicator (PTI) bits can be used to identify the OAM cells.

11.2.5 OAM of the physical layer

Let's look into some examples of OAM flow in the physical layer. To achieve end-to-end OAM information flow, different sections of the physical layer must be maintained separately. A possible physical configuration of the network is shown in Figure 11.3.

We see that the F1 flow is terminated by LTs (line termination) and a regenerator, whereas F2 is terminated solely by LTs. The F3 flow requires the recognition of ATM cell streams.

In each of the OAM information flows, different errors can be recognized and assigned to one of the three levels described. For example, in SONET/

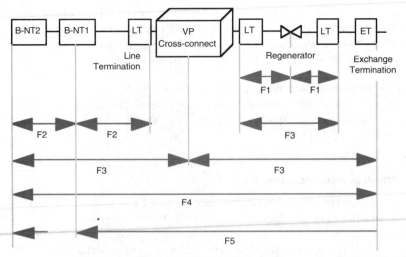

Note: * Termination of F5 at the B-NT1 is for further study

Figure 11.3 Example of a physical configuration and OAM flows at the physical and ATM layers.

SDH-based transmission systems, the following is the OAM information flow identified for each of the different levels:

F1, F2:

- *Loss of frame.* In SONET/SDH, frame synchronization is lost.

- *Degraded error performance.* The quality of the received bit stream is not at an acceptable level, i.e., too many bits are in error, which could be caused by a bad transmission system, an out-of-sync clock, etc.

F3:

- *Loss of cell delineation.* The cell delineation algorithm is no longer in the SYNC state (see Figure 10.19), i.e., the algorithm cannot identify the beginning of the cells.

- *Uncorrectable header.* The header has more errors than can be corrected (i.e., multiple bit errors). This error is detected using the header error control (HEC) mechanism.

- *Degraded header error performance.* There are too many header bit errors. This information can be detected by the HEC mechanism.

- *Loss of H4 pointer.* The H4 pointer of the SONET/SDH is not identified, resulting in an unrecognizable SONET payload.

- *Degraded error performance.* The grade of service is no longer acceptable. This performance level is measured by inserting special OAM cells or by calculating a BIP-8 over the preceding cells, or by using a special pattern in the information fields of the unassigned cells.

- *Failure of insertion and suppression of idle cells.* If too many idle cells are arriving, no useful information can be transported.

If a cell-based transmission system is used, the errors detected could be similar to ones in a SONET/SDH-based transmission system. The following errors are detected in a cell-based transmission system:

F1, F2:

- *Loss of PLOAM cell recognition.* This error occurs when the receiver does not recognize PLOAM cells. Thus, no performance monitoring can be done or provided.

F2:

- *Degraded error performance.* This error is the same as for a SONET/SDH-based transmission system.

F3:

- *Loss of cell delineation*
- *Uncorrectable header*
- *Degraded header error performance*
- *Failure of insertion and suppression of idle cells*

For all the F3 errors, the description is the same as for the SONET/SDH-based transmission system.

11.2.5.1 OAM of the ATM layer. An example of the physical termination points of the ATM layer OAM flows was shown in Figure 11.3. In the figure, an end-to-end virtual path and virtual channel is maintained with the F4 and F5 information flows.

Two possible failures are identified by ITU-T in the ATM layer:

F4:

- *Path not available.* In this case, the virtual path cannot be established and requires a system protection action from setting up a virtual path. This situation can occur if the required bandwidth is not available (end-to-end) or if the number of virtual path connections is exceeded.

F4, F5:

- *Degraded performance.* The ATM cells arriving at the VCI/VPI processing nodes (switching nodes) do not meet performance requirements. This degraded performance can be caused by cell loss, cell insertion, too high a bit error rate in the information field, etc.

ITU-T Recommendation I.610 is only the first document on the maintenance principles of an ATM network. Further work on this subject is continuing.

11.3 User Plane

The higher layers of the user plane contain all service-specific protocols necessary for the completion of end-to-end communications. The higher layer protocols should be independent of the protocols used at the underlying layers. The nonshaded portion of Figure 11.4 shows the different services on top of the BISDN protocol layers. These services have a protocol independent of the protocol used by the BISDN layers. In fact, the higher-layer service-specific protocols can provide services regardless of the lower-layer protocol.

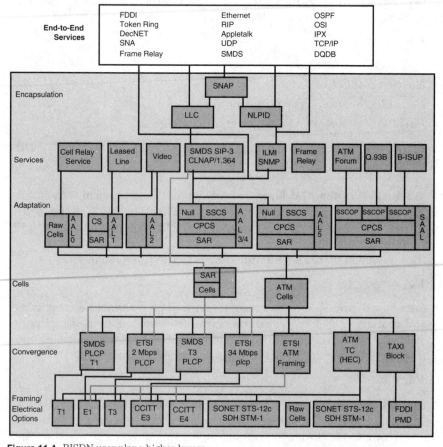

Figure 11.4 BISDN user plane higher layers.

These services typically form end-to-end communications with the BISDN protocol in the network portion. Some of these protocols are addressed in Part 2 of this book (FDDI, DQDB, frame relay, ATM, SMDS). These protocols have functions duplicated by the BISDN protocol layers. Thus, in many cases, some of the protocols' functions are unnecessary. In the long run, if protocols similar to the one used in BISDN, such as ATM, are used for end-to-end communications, the duplicated functions can be eliminated. This increased efficiency will in turn enhance the performance of the network.

11.4   Control or Signaling Plane

The control plane of the higher layers provides signaling message transport and connection/call control capabilities. Signaling specifications (ATM UNI signaling protocol for SVC) are completely defined by ITU-T and are expected to be approved by ITU-T in their SG11 meeting in September 1994. Current commercial ATM implementation is based on permanent virtual circuit (PVC), which is an operator-assisted connection setup. For the customer to receive more flexible services, such as bandwidth on demand, SVCs are required, which calls for the implementation of the signaling protocols. There are two BISDN signaling specifications—Q.2931 (formerly known as Q.93B) and B-ISUP (Q.2761). The standards bodies plan to implement the signaling requirements in two phases or releases. Phase 1 is short-term and Phase 2 is to address the long-term requirements. The ATM forum in its June 1993 working document, ATM User-Network Interface Specification Version 2.2, has defined the Phase 1 requirement for signaling.

In this section, we address only Phase 1 of the signaling requirements definition and the features provided as part of Phase 1. As far as Phase 2 is concerned, nothing has yet been proposed. We do speculate on the type of signaling that might be defined for the long-term, which is in addition to the Phase 1 signaling requirements.

Figure 11.5 shows the protocols required to support signaling in the BISDN protocol. Before we detail the signaling phases, we must understand the need for signaling to provide SVC service, which is simply an automatic call setup, and, at a high level, similar to the regular voice-based phone conversation.

We can now define the procedures for dynamically establishing, maintaining, and clearing ATM connections at the user-network interface. The procedures are defined in terms of messages, and the information elements are used to characterize the ATM connection and ensure interoperability. This implementation agreement is based on a subset of the broadband signaling protocol standards (formerly known as Q.93B; now Q.2931). Additions to these Q.2931 specifications have been made wherever necessary to support capabilities identified by the ATM forum for

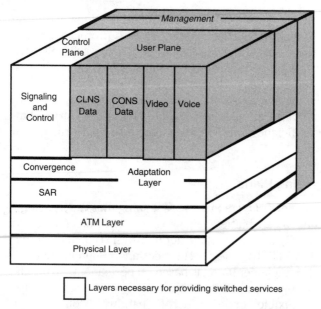

Figure 11.5 BISDN layers required for providing switched services.

early deployment and interoperability of ATM equipment. The primary areas where the standard has been supplemented are to support point-to-multipoint connections, additional traffic descriptors, and private network addressing issues.

The procedures proposed by the ATM forum in its signaling implementation agreement apply to the interface between the terminal or end-point equipment and a public network, referred to as public UNI, and terminal or end-point equipment connected to a private network, referred to as private UNI. Note that the term "*Phase 1* or *Release 1*" refers to the protocol as described in this section.

11.4.1 Reference configuration

Before we go further we need to understand the reference configuration. The protocol specified as part of the ATM forum is valid for the private and public UNI as defined in Figure 11.6. For this UNI, the protocol must be symmetrical, i.e., it must also apply to the interface in the configuration on both ends of the ATM network or even end-to-end. The purpose of a reference configuration for the UNI signaling specification is to list all the elements of an ATM network and the links between them to which this signaling specification applies.

Network elements in this context are:

- end-point equipment (CPE)
- private ATM network
- public ATM network

For the purposes of this section, let's assume a network, public or private, consists of one or more ATM switching platforms under the same administration. The possible reference configuration is illustrated in Figure 11.6. The references to public UNI and private UNI refer to Phase 1 of signaling.

11.4.2 Phase 1 signaling capabilities

In this section, we mention the signaling capabilities defined in Phase 1 at a high level. The basic capabilities supported by the Phase 1 signaling release are the following:

- switched connections
- point-to-point and point-to-multipoint switched connections
- connections with symmetric or asymmetric bandwidth requirements
- single-connection (point-to-point or point-to-multipoint) calls
- basic signaling functions via protocol messages, information elements, and procedures
- class A (AAL 1), class C (AAL 3/4), and class X (AAL 5) ATM transport services
- request and indication of signaling parameters
- VPI/VCI assignment
- a single, permanently defined out-of-band channel for all signaling messages
- error recovery
- public UNI and private UNI addressing formats for unique identification of ATM end points
- a client registration mechanism for exchange of addressing information across a UNI
- multicast service addresses
- end-to-end compatibility parameter identification

Each are described in the following subsections.

11.4.2.1 Switched connections. The purpose of this specification is to support switched connections. The switched connections are established in real-time using certain signaling procedures. These on-demand connections can remain active for a certain amount of time but are not automatically reestablished after a network failure.

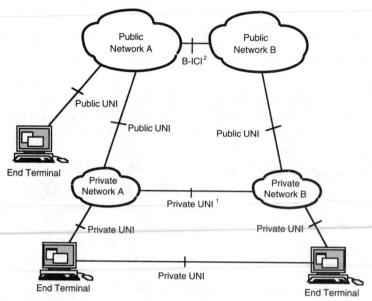

¹Private ATM networks can be connected using the private UNI signaling. Features specific to private network interworking, however, are not a requirement for Phase 1 of the protocol. In releases after Phase 1, such internetworking features can be implemented in a private network-to-network interface (NNI) specification.

²The connection between public networks is outside the domain of the UNI specification. It is addressed by the B-ICI (Broadband Intercarrier Interface) specification of the ATM Forum.

Figure 11.6 Reference configurations.

In contrast, permanent connections are those set up and torn down via provisioning by the service provider. These connections remain established for long periods of time and are automatically reestablished in the event of network failure. The reestablishment is usually done by the service provider. Phase 1 of the signaling deals only with switched connections.

11.4.2.2 Point-to-point and point-to-multipoint switched connections. A point-to-point connection is a collection of associated ATM VC or VP links that connect two end points (see Figure 10.25). The Phase 1 signaling supports point-to-point connections.

11.4.2.3 Point-to-multipoint connection. The point-to-multipoint connection is defined as a connection that is a collection of related ATM VC or VP links with associated end-point nodes. For ease of understanding, we assume the following properties:

- One ATM link, called the *parent link*, serves as the parent in a simple tree topology. This parent node originates the call. When the parent node sends information, all the remaining nodes requesting a connection, called *child nodes*, receive copies of the information.

- In Phase 1, only zero-return bandwidth (i.e., from the child to the parent) is supported.

- In this connection type, the child nodes cannot communicate directly with each other. The only way two child nodes can communicate is by going through the parent's connections.

- A distributed implementation can be used to connect leaves to the tree. For example, each child node can act as a parent node and make connection to other nodes, which then become the child nodes.

A typical point-to-multipoint connection setup is achieved by first establishing a point-to-point connection between the parent node and one child node. After this setup is complete, additional child nodes can be added to the connection by "add party" requests by the parent node. In addition, the Phase 1 signaling specified by the ATM Forum supports the ability of the parent node to have multiple add party requests pending at one time (that is, the parent node does not have to wait for a response from one add party request before issuing the next). The add party response identifies the child that was added (or that failed to get added) so that responses can be paired with requests. Note that the parent node could choose to add child nodes serially (that is, the parent could wait for each add party to be completed before issuing the next), even though the network allows child nodes to be added in parallel.

A child node can be added or dropped from a point-to-multipoint connection at any time after establishing the connection. A new child node can be added to an existing connection via the parent node issuing an add party request. A child node can be dropped from a connection as a result of a request sent by either the parent node or by the child node to be dropped (but not by another child).

This point-to-multipoint connection is different from a multipoint-to-multipoint connection, where every node can communicate with every other node in the call or group. Multipoint-to-multipoint connections are not supported in Phase 1 signaling. Multipoint-to-multipoint connections can be achieved using point-to-multipoint connections in the following ways:

- Each node in a group that wishes to communicate can establish a point-to-multipoint connection to all the other nodes in the group. For example, a group of N nodes, requires N point-to-multipoint connections.

- Each node in the group that wishes to communicate can establish a point-to-point connection to a *multicast server.* The multicast server is the parent node in a point-to-multipoint connection to each node in the group. Any information sent by a node in the group to the multicast server is transmitted back from the multicast server through the point-to-multipoint connection to each of the nodes in the group. For a group of N nodes, this requires N point-to-point connections and one point-to- multipoint connection.

11.4.2.4 Connections with symmetric or asymmetric bandwidth. Point-to-point, bidirectional connections usually have the same bandwidth in both directions, but the specifications allow bandwidth to be specified independent of the other direction. Forward and backward directions can have different bandwidths. The forward direction is from the calling party to the called party, while the backward direction is from the called party to the calling party.

For point-to-multipoint connections, the Phase 1 signaling specification supports only nonzero (identical) bandwidth in the forward direction from the parent node to each child node, and zero bandwidth in the reverse direction from each child node to the parent node.

11.4.2.5 Single connection per call. Here, one and only one connection per call is supported. The single connection can be either a point-to-point or point-to-multipoint connection.

11.4.2.6 Protocol that supports basic signaling functions. The signaling protocol that supports the basic signaling functions at the UNI interface are:

- call setup
- call request
- call answer
- call clearing
- reason for clearing
- out-of-band signaling

11.4.2.6.1 Call setup. This protocol supports the establishment of a connection or call between different parties. The call establishment includes call request and call answer.

11.4.2.6.2 Call request. The function of the call request protocol is to allow an originating party to request the establishment of a call to a certain destination. In this request, the originating party might provide information related to the call. The request might contain information such as destination number, bandwidth required, etc.

11.4.2.6.3 Call answer. The function of the call answer protocol is to allow the destination party to respond to an incoming call request. In other words, it is an acknowledgment. The destination party can include information related to the call. (Rejecting the call request is considered part of the call-clearing function.)

11.4.2.6.4 Call clearing. The function of the call-clearing protocol is to allow any party involved in a call to initiate its removal from an already-established call. If the call is between two parties only, the whole call is removed. This function also allows a destination party to reject its inclusion in a connection or call.

11.4.2.6.5 Reason for clearing. The function of this protocol is to allow the clearing party to indicate the cause for initiating its removal from a connection or call.

11.4.2.6.6 Out-of-band signaling. The function of this protocol is to specify that call-control information uses a channel different from the channels used for exchanging data information between the end parties for carrying control information (i.e., a specific VPI/VCI value is used for the call control signaling channel—SS7).

11.4.2.7 Class A, class C, and class X ATM transport services. These three classes relate to AAL 1, AAL 3/4, and AAL 5 basic services, respectively. Each is described in the subsections following.

11.4.2.7.1 Class A (AAL 1) ATM transport service. Class A service is a connection-oriented, constant-bit-rate ATM transport service. Class A service has certain characteristics, such as end-to-end timing requirements, that might require stringent cell loss, cell delay, and cell delay variation performance. The user chooses the desired bandwidth and the appropriate quality-of-service (QOS) during the setup procedure to establish a class A connection (e.g., voice traffic).

11.4.2.7.2 Class C (AAL 3/4) ATM transport service. Class C service is a connection-oriented, variable-bit-rate ATM transport service. Class C service has no end-to-end timing requirements. The user chooses the desired bandwidth and QOS with appropriate information during the setup procedure to establish a class C connection (e.g., frame-relay type connection).

11.4.2.7.3 Class X (AAL 5) ATM transport service. Class X service is a connection-oriented ATM transport service where the AAL, traffic type (VBR or CBR), and timing requirements are user-defined (i.e., transparent to the network). Class X service is also known as AAL 5 service. The user chooses only the desired bandwidth and QOS with appropriate information during the setup procedure to establish a class X connection (e.g., SMDS type traffic).

Thus, Phase 1 signaling specifications support class A, class C, and class X service. Class D service is not directly supported by signaling. It can be supported via a class X or class C connection to a connectionless server, however.

11.4.2.8 Signaling parameters—request and indication. The Phase 1 signaling specification does not provide support for the negotiation of signaling parameters (e.g., QOS, cell-transfer rate, end-to-end compatibility parameter values). Instead, the sender chooses a value for each parameter to be sent during the connection setup request, and the receiver indicates whether or not the chosen values can be accommodated.

11.4.2.9 VPI/VCI support. The Phase 1 signaling specification supports the virtual path connection identifier (VPCI) to identify the virtual path across the UNI, with the restriction that a one-to-one mapping exists between VPCI and VPI, and hence values beyond 8 bits are restricted.

The following list describes the Phase 1 signaling capabilities with respect to VPIs, VPCIs, and VCIs.

1. It provides for the identification of virtual paths (using VPCIs) and virtual connections within virtual paths (using VCIs).

2. It does not (in Phase 1) include negotiations of VPCIs or VCIs but does not preclude negotiation in future releases.

3. It does not (in Phase 1) include provisions to negotiate or modify allowed ranges for VPCIs or VCIs within virtual paths but does not preclude this in future releases. (Negotiation or provisioning of VPCI/VCI ranges is outside the scope of the signaling protocol for Phase 1.)

11.4.2.10 Support of a single signaling virtual channel. For single point-to-point signaling virtual channels, VCI of 5 and VPCI of 0 are used for all signaling in Phase 1. The association between signaling entities should be permanently established. Metasignaling is not supported in Phase 1. Broadcast signaling using a virtual channel is also not supported.

11.4.2.11 Error recovery. The error recovery capabilities of Phase 1 signaling are the following:

- *Detailed error-handling procedures*: One such signaling entity informs its peer when it has encountered a nonfatal error (i.e., insufficiently severe to force call clearing). Examples of nonfatal errors are message format errors, message content errors, and procedural errors (messages or message contents received in a state in which they are not expected).

- *Procedures for recovery from signaling AAL reset and failure* (and, by extension, from physical layer outages and glitches).

- *Mechanisms for signaling entities to exchange status information for calls and interfaces,* and to recover gracefully if a disagreement occurs; these procedures must operate in error conditions by requesting a signaling entity (i.e., status enquiry).

- *Capability to force calls, VCCs, and interfaces to be disconnected,* either by manual intervention or due to severe error(s).

- *Cause and diagnostic information for fault resolution* provided with call clearing, nonfatal errors, and recovery from errors affecting the whole interface.

- *Mechanisms to recover from loss of individual messages* (e.g., timers and associated procedures).

11.4.2.12 Public UNI and private UNI ATM addressing. Phase 1 signaling supports a number of ATM address formats to be used across the public and private UNIs to unambiguously identify the end points in an ATM connection.

11.4.2.13 Client registration mechanism. Phase 1 signaling supports a mechanism for the exchange of identifier and address information between an end system and a switch across a UNI. The basic capability allows a network administrator to manually configure ATM network address information into a switch port, without having to configure that information into any terminal later attached to that port. Instead, the terminal uses the client registration mechanism to exchange its identifier information for the ATM address information configured in the switch port. The client registration mechanism allows this exchange to take place whenever, for example, the terminal is initialized, reinitialized, or reset.

At the conclusion of the client registration exchange, the terminal has automatically acquired the ATM network address as configured by the network provider, without any requirement for the same address to have been manually provisioned into the terminal. The terminal can then use and transfer its network address as needed by higher-level protocols and applications. This function is similar to the client-server mechanism used in LAN equivalents such as fiber distributed data interface (FDDI), where each station connected to the network can be identified with a unique address.

11.4.2.14 Multicast service addresses. The addressing format contains a field called an end system identifier (ESI) to satisfy the following requirements related to multicasting:

- The addressing schemes allow for multicast service addresses to be distinguishable. For example, when an IEEE 48-bit media access control (MAC) address is used as an ESI value, a multicast address is distinguished by a 1 in the multicast bit of the address.

- An ATM end point can have multiple multicast service addresses (e.g., when multicast is supported on top of the network by a multicast server, the server can have a separate multicast service address for each multicast address it supports, plus its own nonmulticast address).

- The significance of a multicast service address can be restricted to an administrative domain, or it can be global.

- Multicast service addresses can be carried within the called party identifier in a point-to-point connection to a multicast service, and within the calling party identifier in a point-to-multipoint connection originated by a multicast service.

11.4.2.15 End-to-end compatibility parameter identification. On a per-connection basis, the following end-to-end compatibility parameters can be specified:

- AAL type (e.g., Type 1, 3/4, or 5)
- method of protocol multiplexing (e.g., LLC vs. VC)
- for VC-based multiplexing, the protocol that is encapsulated (e.g., any of the list of known routed protocols or bridged protocols)
- protocols above the network layer

11.4.2.16 Example of ATM signaling for point-to-point connection. Having addressed different capabilities of signaling in Phase 1, let's address some of the features of Phase 1 signaling and explain how some of these capabilities operate. Of all the capabilities, the most important are point-to-point and point-to-multipoint connections. We also address these two connections. Figure 11.7 shows a basic ATM network signaling for setting up point-to-point connection.

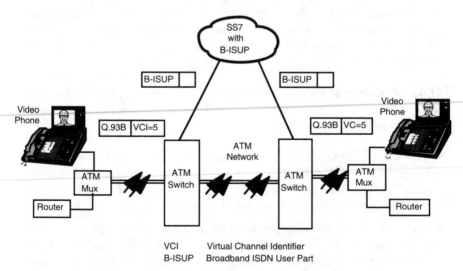

Figure 11.7 ATM signaling in point-to-point mode.

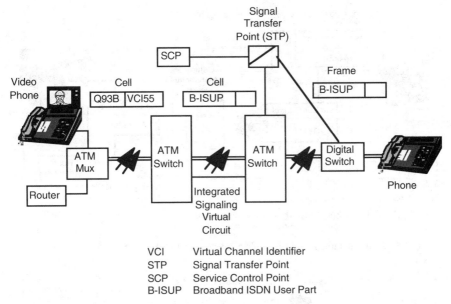

Figure 11.8 ATM signaling in SS7 network architecture.

ATM signaling is based on the existing SS7 protocol used in today's narrowband ISDN. Because the signaling is based on an existing signaling protocol, no new messages are required or created. In this protocol, however, the B-ISUP (Q.2761) (similar to ISUP for ISDN) messages are carried as cells in the signaling VC between the ATM switches. The same message is carried as frames between the STPs and non-ATM switches, as shown in Figure 11.8.

The point-to-point signaling is carried on VC = 5 and VP = 0. The protocol used for access and network signaling is based on Q.2931.

11.4.2.17 Example of ATM signaling in point-to-multipoint connections.
The signaling for point-to-multipoint connections is based on setting up multiple point-to-point connections. In this connection, each virtual path has a separate signaling channel (VC = 1). The signaling is used to establish, monitor/maintain, and tear down signaling VCs. This signaling channel is also used to carry the service profile information between the end points. Figures 11.9, 11.10, and 11.11 show the different stages in setting up a point-to-multipoint connection: connection request, acknowledgment, and connection setup.

In Figure 11.9 (connection request), the ATM CPE equipment requests the signaling VC using signaling channel VC = 1 and VP = 0. The information such as the bandwidth requirement and service profile identifier (SPID) is carried within the cell to the ATM switch via the ATM mux, where the actual call control or connection setup occurs.

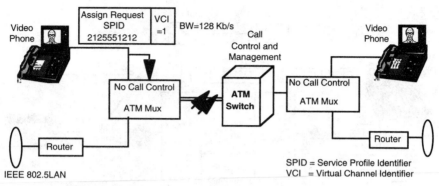

Figure 11.9 ATM signaling in point-to-multipoint (connection request).

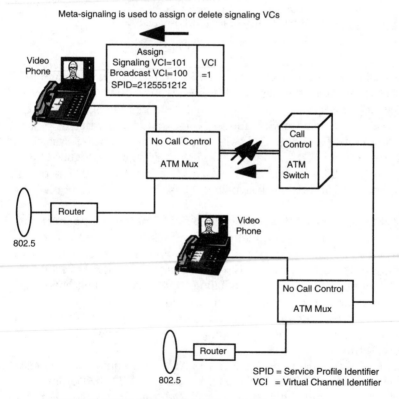

Figure 11.10 ATM signaling in point-to-multipoint (acknowledgment).

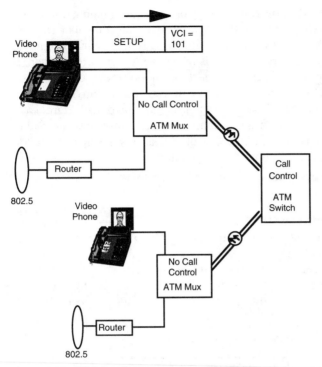

Figure 11.11 ATM signaling in point-to-multipoint (connection setup).

Once the cell with the information is received by the ATM switch, the switch performs call control and forwards the cell to the appropriate child node. Upon receiving the cell, the child node sends an acknowledgment with information on the channels to be used. This process is repeated for every child node in the parent's list, as explained earlier. Figure 11.10 shows an answer or acknowledgment with the cell carrying the appropriate VC, higher bandwidth value.

Once the CPE equipment receives the appropriate channel for the transmission of information, the setup acknowledgment is sent on the appropriate channel (VC number), which is followed by the actual information transfer as shown in Figure 11.11.

11.4.3 Phase 2 of ATM signaling

As mentioned earlier, not much is defined by the standards bodies as to what features are to be incorporated in Phase 2. It is speculated that the signaling features in Phase 2 will be based on a multipoint-to-multipoint connection, multimedia type of call setup, compared to Phase 1 signaling, where even point-to-multipoint connections are achieved by setting up point-to-point connections. Figure 11.12 shows an example of a type of signaling that might be done in an ATM environment. In this example, the call control or connection occurs at the ATM mux level, and a connection request carries multiple connection requests on different channels of different bandwidths, such as low bandwidth for voice and high bandwidth for video traffic. Once this information is received by the network, it sets up the appropriate connection for information transfer, thus enabling a multimedia type of call. Here, each of the calls is independent in terms of channel. Synchronization occurs between different channels in a multimedia call. The issues of synchronization for a multimedia call are still under investigation.

11.5 Summary

In this chapter, we addressed the higher layers such as user plane, management plane, and control/signaling plane of the BISDN protocol and their functions. In the management plane, there are two sublayers—plane management and layer management. The management plane's main functions are related to OAM and performance of the ATM network. In the control or signaling plane, we addressed the features or capabilities proposed for signaling implementation in Phase 1.

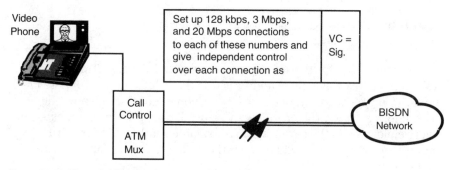

Figure 11.12 Phase 2 ATM signaling.

12

Other Aspects of BISDN

12.1 Overview

In this chapter, we address the BISDN service aspects [I.211], BISDN network aspects [I.311], and BISDN user-to-network interface (UNI) aspects [I.413, I.432]. The BISDN service aspects define the different types of services that ITU-T has proposed for both residential and business needs for various types of bit-rate traffic. In the network aspects, we address the network layering structure, signaling principles, and traffic control mechanisms. In the BISDN UNI, we address the customer network configuration and its different topologies, as recommended by ITU-T.

12.2 Broadband Service Aspects

The broadband service aspects are defined by ITU-T in Recommendation I.211. In this recommendation, ITU-T has defined the different classes of services, taking into consideration features such as:

- services that have capability to increase the flexibility of a connection
- services that have the capacity for flexible bandwidth allocation

In addition, video coding aspects are considered where visual services come into play. In a broadband environment, the capacity available to the user is increased dramatically, enabling a range of services that can be supported. ITU-T has classified the services that could be provided by a broadband network into *interactive services* and *distribution services*, as illustrated in Figure 12.1.

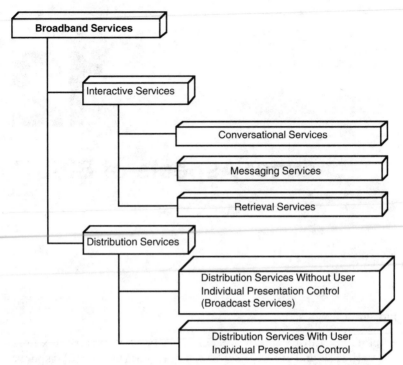

Figure 12.1   BISDN service classification. (*ITU-T I.211/Fig 1. Reprinted with permission*)

12.2.1 Interactive services

Interactive services are those services in which a two-way exchange of actual information (other than control signaling information) occurs between two subscribers or between a subscriber and a service provider. These services include conversational services, messaging services, and retrieval services.

12.2.1.1 Conversational services.   Conversational services provide bidirectional, real-time communication with end-to-end information transfer between two users or between a user and a service provider. These services support the general transfer of information specific to a given user application. That is, the information is generated by and exchanged between users; it is not public information.

This category encompasses a wide range of application types. Table 12.1 divides conversational services into four categories: moving picture (video), sound, data, and document (text). Of these, one of the most important for broadband service is conversational video, such as video telephony service, which is based on video telephones, and desktop video, which uses a computer terminal and video camera.

TABLE 12.1 Conversational Services ITU-T/I.211, Table 1 (reprinted with permission)

Type of information	Examples of broadband services	Applications	Possible attribute values
Moving picture (video) and sound	Broadband[1,2] video-telephony	Communication for the transfer of voice (sound), moving pictures, and video-scanned still images and documents between two locations (person-to-person)[2]: Tele-education Tele-shopping Tele-advertising	Demand/reserved/permanent Point-to-point/multipoint Bidirectional symmetric/bidirectional asymmetric (value for information transfer rate is under study)
	Broadband[2] videoconference	Multipoint communication for the transfer of voice (sound), moving pictures, and video-scanned still images and documents between two or more locations (person-to-group, group-to-group)[2]: Tele-education Business conferencing Tele-advertising	Demand/reserved/permanent Point-to-point/multipoint Bidirectional symmetric/bidirectional asymmetric
	Video-surveillance	Building security Traffic monitoring	Demand/reserved/permanent Point-to-point/multipoint Bidirectional symmetric/bidirectional asymmetric
	Video/audio information transmission service	TV signal transfer Video/audio dialogue Contribution of information	Demand/reserved/permanent Point-to-point/multipoint Bidirectional symmetric/bidirectional asymmetric
Sound	Multiple-sound program signals	Multilingual commentary channels Multiple program transfers	Demand/reserved/permanent Point-to-point/multipoint Bidirectional symmetric/bidirectional asymmetric

Type of information	Examples of broadband services	Applications	Possible attribute values
Data	High-speed unrestricted digital information transmission service	High-speed data transfer LAN interconnection MAN interconnection Computer-computer interconnection Transfer of video information Transfer of video and other information types Still image transfer Multisite interactive CAD/CAM	Demand/reserved/permanent Point-to-point/multipoint Bidirectional symmetric/bidirectional asymmetric Connection-oriented Connectionless
	High-volume file transfer service	Data file transfer	Demand Point-to-point/multipoint Bidirectional symmetric/bidirectional asymmetric
	High-speed teleaction	Real-time control Telemetry Alarms	
Document	High-speed telefax	User-to-user transfer of text, images, drawings, etc.	Demand Point-to-point/multipoint Point-to-point/multipoint Bidirectional symmetric/bidirectional asymmetric
	High-resolution image communications service	Professional images Medical images Remote games and game networks	
	Document communication service	User-to-user transfer of mixed documents[3]	Demand Point-to-point/multipoint Bidirectional symmetric/bidirectional asymmetric

[1] This terminology indicates that a redefinition of existing terms has occurred. The new terms might or might not exist for a transition period.

[2] The realization of the different applications might require the definition of different quality classes.

[3] Mixed document means that a document can contain text, graphic, still and moving picture information, and audio annotation.

Video services simply means that the service includes both voice and picture. Currently, two vendors offer video telephone products. Both operate on a regular phone line (64 kbps). In the future, with better compression algorithms, the quality of video can be improved. Higher bandwidth has already improved the quality drastically.

Another video service in this category is videoconferencing. The simplest form of this service is a point-to-point video conference, which connects conference rooms in two locations. A point-to-point videoconference has additional features such as facsimile and document image transfer and special equipment such as electronic blackboards. Today's videoconference speed ranges from 128 kbps (T1) to 1.554 Mbps (E1). Another type of videoconferencing is the multipoint-to-multipoint service, which allows participants in multiple locations access to a videoconference connection without leaving their workplaces. This application can be accomplished using a videoconference server within the network. Such a system would support a small number (e.g., five) of simultaneous users. In this system, one participant (the one who speaks) would appear on all screens at a time, managed by the videoconference server.

Another type of video service in this category is video surveillance. Video surveillance is not a distribution service because the information delivery is limited to a specific, intended subscriber. This type of service is typically applicable to building security and traffic monitoring. In both cases, users might like to control the camera (change orientation, etc.). Attributes of these services can be on-demand, which can be point-to-point or point-to-multipoint.

The next type of service is audio. In audio (sound), multiple-sound program signals could be transmitted. This service has all the attributes of video services.

As shown in Table 12.1, the next service is data. Examples of data applications that could use this service are the following:

- file transfer in a distributed environment, where file servers are distributed across the network high-speed or large-capacity transmission of measured values or control information

- computer-aided design and manufacturing (CAD/CAM)

- connection of multiple local area networks (LANs), MANs, and WANs distributed across a large region

Finally, there is the conversational transfer of documents. The document could be very high resolution images, with voice annotation, and/or a video component.

12.2.1.2 Messaging services. Table 12.2 shows the messaging services. Messaging services offer end user-to-end user communication, usually between an individual user and a file server for electronic mail, video mail, etc.

TABLE 12.2 Messaging Services ITU-T/I.211, Table 1 (reprinted with permission)

Type of information	Examples of broadband services	Applications	Some possible attribute values
Video and sound	Video mail service	Electronic mailbox service for the transfer of moving pictures and accompanying sound	Demand Point-to-point/multipoint Bidirectional symmetric/ bidirectional asymmetric Unidirectional (for further study)
Document	Document mail service	Electronic mailbox service for mixed documents[3]	Demand Point-to-point/multipoint Bidirectional symmetric/ bidirectional asymmetric Unidirectional (for further study)

[3] *Mixed document* means that a document can contain text, graphic, still and moving picture information, and audio annotation.

Video mail is an enhancement to E-mail. In video mail, all video, text, and voice can be sent simultaneously. Messaging services can also edit, process, and convert the information. In contrast to conversational services, messaging services are not in real-time. Hence, they place lesser demands on the network and do not require that both users be available at the same time. These messaging services are analogous to narrowband services such as X.400 and teletex. They have all the standard attributes of other services.

12.2.1.3 Retrieval services. Retrieval services provide the user with the capability to retrieve information stored in file servers that are, in general, available for public use. This information is sent to the user on-demand; that is, the time at which an information sequence is to start is controlled by the user. This service is called *broadband videotex*. In it, the information is a combination of voice, video, and text.

Narrowband also has a service called videotex. It is a general-purpose database retrieval system that can be used in any network. Table 12.3 shows a list of retrieval services. The videotex information provider usually maintains a variety of databases provided in the system. Examples of videotex services are stock market advisories, yellow pages, etc.

Broadband videotex is basically an enhancement of the existing videotex system. The user will be able to select a variety of sound, images, video, and text. Examples of broadband videotex services are:

- retrieval of encyclopedia information
- results of consumer goods comparisons
- electronic mail-order catalogs and travel brochures with the option of placing a direct order or making a direct booking

Another retrieval service is video retrieval. With this service, a user could order full-length films or videos from a film/video library facility online. Because the provider might need to satisfy many requests, bandwidth considerations dictate that only a small number of different video transmissions can be supported at any time. A realistic service would offer perhaps 500 movies or videos for each two-hour period. Using a 50-Mbps video channel to each subscriber, it would require a manageable 25-Gbps transmission capacity from video suppliers to distribution points. With advances in video compression technology such as MPEG and JPEG coding, however, systems such as ADSL, which run from 1.5 Mbps, are available to provide video over existing copper-based networks.

12.2.1.4 Distribution services. Distribution services are services in which the information transfer is primarily one-way, such as from service provider to broadband subscriber. Information can be broadcast services, for which the

TABLE 12.3 Retrieval Services ITU-T/I.211, Table 1 (reprinted with permission)

Type of information	Examples of broadband services	Applications	Some possible attribute values
Text, data, graphics, sound, still images, moving pictures	Broadband videotex	Videotex, including video Remote education and training Tele-software Tele-shopping Tele-advertising News retrieval	Demand Point-to-point Bidirectional asymmetric
	Video retrieval service	Entertainment Remote education and training	Demand/reserved Point-to-point/multipoint Bidirectional asymmetric
	High-resolution image retrieval service	Entertainment Remote education and training Professional image communications Medical image communications	Demand/reserved Point-to-point/multipoint Bidirectional asymmetric
	Document retrieval service	Mixed documents retrieval from information centers, archives, etc.[3,4]	Demand Point-to-point/multipoint[5] Bidirectional asymmetric
	Data retrieval service	Tele-software	

[3] *Mixed document* means that a document can contain text, graphic, still and moving picture information, and audio annotation.

[4] Special high-layer functions are necessary if post-processing after retrieval is required.

[3] Further study is required to indicate whether the point-to-multipoint connection represents in this case a main application.

user has no control over the presentation of the information or cyclical services, which allow the user some measure of information presentation control.

12.2.1.4.1 Broadcast distribution services. Services in this category are also referred to as services without user presentation control. They provide a continuous flow of information distributed from a central source to all users connected to the network. Each user can access the flow of this information but has no control over the presentation. In particular, the user cannot control the starting time or order of the presentation of the broadcasted information. All users simply tap into the flow of information. A good example of this is the CATV network. Table 12.4 shows the different types of broadcast distribution.

In CATV networks, the signals are broadcast to every subscriber on the network. With broadband communications, these types of services can be integrated with telecommunications services. In addition, higher resolutions and better services can be achieved with the availability of higher bandwidth.

12.2.1.5 Cyclical distribution services. Services in this class are very similar to broadcast services. With these services, the user can individually access the distributed information by controlling the starting point and order in which the information is presented.

Table 12.5 shows cyclical distribution services. The information is actually a one-way broadcast of a video signal. Currently, the information capacity is limited to the available bandwidth.

Examples of information presented by such a system are stock market reports, weather reports, news, leisure information, and recipes. This type of information is currently available through online services such as Prodigy, CompuServe, etc.

Broadband can enhance this service. In the narrowband environment, a very small bandwidth is available. With broadband, a service could use a full digital broadband channel to transmit the information, using text, images, video, and audio. Broadband service can provide low-cost access to timely and frequently requested information, such as an electronic newspaper that uses public networks, or an in-house information system for trade fairs, hotels, or hospitals. A typical system like this can access up to 10,000 pages in one second. Table 12.6 shows characteristics of broadband service in terms of bandwidth utilization.

12.3 Network Aspects

According to ITU-T Recommendation I.327, the information transfer capabilities of a broadband network include:

- broadband capabilities
- 64 kbps-based ISDN capabilities
- user-to-network signaling
- user-to-user signaling

TABLE 12.4   Distribution Services Without User Presentation Control ITU-T/I.211, Table 1 (reprinted with permission)

Type of information	Examples of broadband services	Applications	Some possible attribute values
Data	High-speed unrestricted digital information distribution service	Distribution of unrestricted data	Permanent Broadcast Unidirectional
Text, graphics, still images	Document distribution services	Electronic newspaper Electronic publishing	Demand (selection)/permanent Broadcast/multipoint[5] Bidirectional asymmetric/unidirectional
Video and sound	Video information distribution services	Distribution of video/audio signals	Permanent Broadcast Unidirectional
Video	Existing quality TV distribution service (NTSC, PAL, SECAM)	TV program distribution	Demand (selection)/permanent Broadcast Bidirectional asymmetric/unidirectional
	Extended quality TV distribution service Enhanced definition TV distribution service High quality TV	TV program distribution	Demand (selection)/permanent Broadcast Bidirectional asymmetric/unidirectional
	High-definition TV distribution service	TV program distribution	Demand (selection)/permanent Broadcast Bidirectional asymmetric/unidirectional
	Pay-TV (pay-per-view, pay-per-channel)	TV program distribution	Demand (selection)/permanent Broadcast/multipoint Bidirectional asymmetric/unidirectional

[5] Further study is required to indicate whether the point-to-multipoint connection represents in this case a main application.

TABLE 12.5   Distribution Services with User Presentation Control

ITU-T/I.211, Table 1 (reprinted with permission)

Type of information	Examples of broadband services	Applications	Some possible attribute values
Text, graphics, sound, still images	Full-channel broadcast videography	Remote education and training Tele-advertising News retrieval Tele-software	Permanent Broadcast Unidirectional

TABLE 12.6   Characteristics of Broadband Services

Service	Bit rate (Mbps)	Burstiness*
Data transmission (connection-oriented)	1.5 to 155	1 to 50
Data transmission (connectionless)	1.5 to 155	1
Document transfer/retrieval	1.5 to 45	1 to 20
Videoconference/video-telephony	1.5	1 to 5
Broadband videotex/video retrieval	1.5 to 155	1 to 20
TV distribution	1.5 to 50	1
HDTV distribution	155	1

* Burstiness = peak bit rate/average bit rate.

Figure 12.2 depicts the information transfer capabilities of broadband.

In a broadband network, the information transfer is provided by ATM. As we know, the ATM data unit is a cell of 53 bytes, of which 5 bytes are the header. The header carries the necessary information to identify the cell so it can be routed to its appropriate destination.

ATM is a connection-oriented technique, and a typical connection within an ATM layer consists of more than one link, each of which is assigned an identifier. These identifiers remain unchanged for the duration of the connection. For signaling, the control information is carried on a different connection using a separate identifier (via an out-of-band signaling mechanism).

Although ATM is a connection-oriented technique, it offers a flexible transfer capability common to all services, including connectionless services such as SMDS. The broadband network aspects can be categorized into the following:

- network layering
- signaling principles
- traffic control

Each is described in the subsections following.

12.3.1 Network layering

The layered structure of a broadband network is addressed in ITU-T Recommendation I.311, which is depicted in Figure 12.3. In the figure, BISDN is divided into two categories, the higher layer functions and the ATM transport layer. Because network aspects of BISDN are handled by the ATM transport network layer, we address only ATM transport layer functions in this section.

The ATM transport network layer is split into two parts—the physical layer and the ATM layer. Both are hierarchically structured. The physical layer consists of the transmission path level, digital section level, and regenerator section level.

The transmission path extends between network elements that assemble and disassemble the ATM payload of a transmission system. The digital section extends between network elements that assemble and disassemble continuous bit or byte streams. The regenerator section is a portion of a digital section extending between two adjacent regenerators.

The ATM layer has two hierarchical levels—virtual channel (VC) level and virtual path (VP) level. The VC and VP are used for the switching of ATM cells. Typical applications of VC/VP-based connections are between

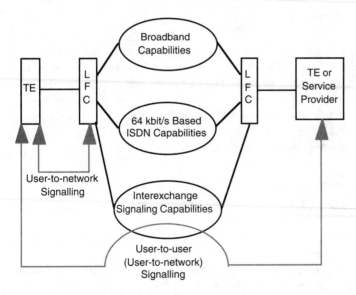

LFC Local Function Capabilities
TE   Terminal Equipment

Figure 12.2 Information transfer and signaling capabilities.

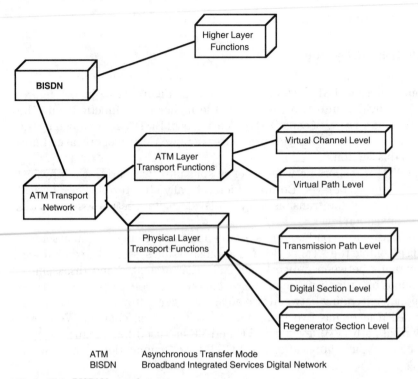

ATM	Asynchronous Transfer Mode
BISDN	Broadband Integrated Services Digital Network

Figure 12.3 BISDN layered structure.

user-to-user, user-to-network, and network-to-network segments of the network. All cells with specific VP/VC connections are transported along the same route through the network. The cell sequence is preserved (first sent-first received) for all virtual connections, which is an advantage of ATM-based switching, where the timing between the cells is maintained from source to destination. The relationship between the different layers of the broadband transport network are shown in Figure 12.4.

12.3.2 Signaling principles

A broadband network follows the principle of out-of-band signaling similar to the one used in N-ISDN using the SS7 signaling network. In BISDN, however, the VC concept provides the means to separate logical signaling channels from user channels.

In chapter 11, we described the signaling features to be implemented in Phase 1 of signaling as part of the control layer function. Here, we give an overview of the different capabilities of broadband signaling. It has been recommended that the existing signaling function according to ITU-T recommendation Q.2931 is to be included as part of broadband signaling. But broadband signaling is based on the ATM transport network, which has been designed to address the increasing desire for advanced forms of communications, such as multimedia services.

In broadband networks, ATM network-specific signaling capabilities are:

- establish, maintain, and release ATM virtual circuit connection (VCC) and virtual path connection (VPC) for information transfer
- negotiate the traffic characteristics using usage parameters for a connection

Other signaling requirements are not ATM-related. Examples are the support of multiple connection calls and multipoint calls.

As mentioned earlier, in a broadband network, signaling messages are conveyed by out-of-band signaling techniques in dedicated switched virtual channels (SVC). Table 12.7 shows different types of SVCs, which are described in the following paragraphs.

Metasignaling channel: One channel exists per interface. This channel is bidirectional and permanent. It is an interface management channel used to establish, check, and release the point-to-point and selective SVCs.

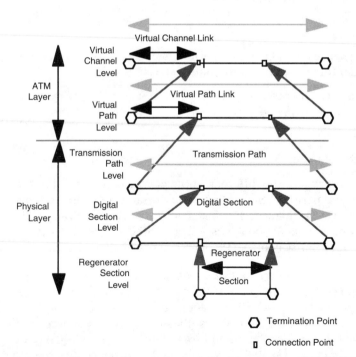

Figure 12.4 Hierarchical layer-to-layer relationship principles. *(ITU-T/I.610, Fig. 3. Reprinted with permission)*

TABLE 12.7 Different Types of SVCs

SVC type	Directionality	No. of SVCs
Metasignaling channel	Bidirectional	One
General broadcast SVC	Unidirectional	One
Selective broadcast SVC	Unidirectional	Multiple
Point-to-point SVC	Bidirectional	One per signaling-end point

General broadcast SVC: This channel is unidirectional, in the network-to-user direction. It sends signaling messages to all signaling end points.

Selective broadcast SVC: This channel is also unidirectional. It sends signaling messages to selected end points. The selective broadcast SVC can also be provided as a network option to address all terminals belonging to the same service profile category.

Point-to-point SVC: This channel is allocated to a signaling end point only while it is active. A signaling end point at the user side can be located in a terminal. In a multifunctional terminal, multiple signaling end points can occur, which is bidirectional. This channel is used to establish, control, and release VCC or VPCs to carry user data.

12.3.3 Traffic control

To provide desired broadband network performance, ITU-T in its I.311 recommendation identified a set of traffic control capabilities:

- connection admission control (CAC)
- usage parameter control (UPC)
- priority control (PC)
- congestion control (CC)

Figure 12.5 shows where the different traffic controls are applied in the reference broadband network. Each segment of the network contains some sort of traffic control mechanism. For instance, Network A has the traffic control mechanisms of connection admission control (CAC), resource management (RM), and priority control (PC) implemented in its network.

12.3.3.1 Connection admission control. CAC is defined as the set of actions taken by the network during call setup phase (or during call renegotiation) to establish whether a VC/VP connection can be accepted.

A connection can only be accepted if sufficient network resources are available to establish the connection end-to-end at its required quality of service. The quality of service already existing in the current network connections must not be influenced by the new connection. Thus, the resources can be bandwidth requested and in terms of average information rate, peak information rate, etc. Two classes of parameters are foreseen to support connection admission control:

- a set of parameters to describe the source traffic characteristics
- another set of parameters to identify the required QOS class

    Source traffic can be characterized by its:

- average bit rate
- peak bit rate
- burstiness
- peak duration

12.3.3.2 Usage parameter control. UPC is defined as the set of actions taken by the network to monitor and control user traffic in terms of traffic volume, cell loss, and cell routing validity. Its main purpose is to protect network resources from malicious and unintentional misbehavior of traffic, which can affect the quality of service of other already-established connections. UPC monitors and detects violations of negotiated parameters. Usage parameter control applies during the information transfer phase only of a connection. Connection monitoring encompasses all connections crossing

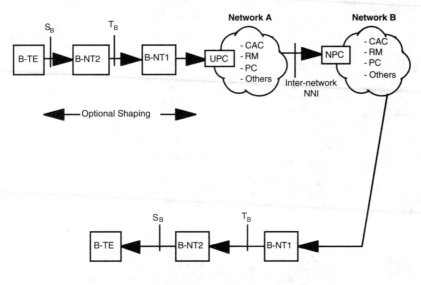

UPC:	Usage Parameter Control	NPC:	Network Parameter Control
CAC:	Connection Admission Control	RM:	Resource Management
PC:	Priority Control	Others:	Spacing, Framing, Shaping, etc.

Figure 12.5 Reference configuration for traffic control.

the user-network interface, including signaling. Usage parameter monitoring includes the following functions:

- checking of the validity of VPI/VCI values
- monitoring the traffic volume entering the network from individually active VP and VC connections to ensure that parameters agreed upon are not violated
- monitoring the total volume of accepted traffic on the access link
- discarding those cells that violate the negotiated traffic parameters. In some cases, the "guilty" connection might be released. Another, less rigorous option is to reroute the violating cells. These cells can be transferred as long as they do not cause serious harm to the network. Thus, the overall throughput of ATM cells might possibly be raised.

Let's illustrate the different UPCs used in a broadband network. Figure 12.6 shows different access network arrangements. The appropriate usage parameter control measures are applied to VCs or VPs at the access point where they are terminated within the network.

In case a, a user is connected directly to a VC switch. Usage parameter

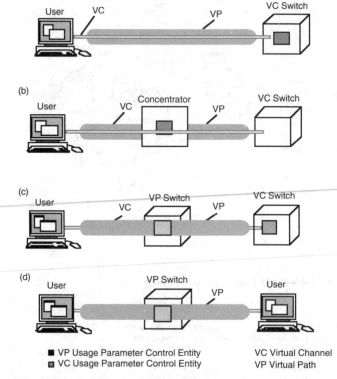

Figure 12.6 Illustration of usage parameter control.

control is performed within the VC switch on each VC before switching it. In case *b*, a user is connected to a VC switch via a concentrator. Usage parameter control is performed within the concentrator on each VC only. In case *c*, a user is connected to a VC switch via a VP switch. Here, usage parameter control is performed within the VP switch on each VP and within the VC switch on each VC. In case *d*, a user is connected to another user via a VP switch. Usage parameter control is performed within the VP switch on each VP connected to the user.

12.3.3.3 Priority control.

Two levels of priority are available in the ATM cell header field called CLP (cell loss priority). These priority classes could be treated separately by connection admission control and usage parameter control. Different buffering mechanisms are used in different switching systems with the two priorities. The mechanisms are common buffer with pushout mechanism, partial buffer sharing, and buffer separation.

In common buffer with pushout, cells of both priorities share a common buffer. If the buffer is full and a higher priority cell arrives, a cell with lower priority (if any is available) is pushed out and lost. To guarantee the cell sequence integrity, a complicated buffer management mechanism is necessary.

In partial buffer sharing, low-priority cells can only access the buffer if the total buffer size is less than a given threshold S_L (S_L < total buffer capacity). High-priority cells can access the whole buffer. By adjusting the threshold S_L, it is possible to adapt the system to various load situations.

In buffer separation, each of the priorities have different buffers. This mechanism is simple to implement, but cell sequence integrity can only be maintained if a single priority is assigned to each connection.

12.3.3.4 Congestion control.

Congestion in broadband networks is defined as a status of network elements (e.g., switches, concentrators, transmission links) in which, due to traffic overload or control resource overload, the network cannot guarantee the negotiated quality of service to the already established connections and the new connection request. Congestion can be caused by unpredictable statistical fluctuations of traffic flows or fault conditions within the network. For example, a user or users can use more resources than they have requested at the time of connection setup negotiations.

12.4 User-Network Interface Aspects

One of the essential parts of the BISDN network is the customer network. It is sometimes referred to as customer premises network (CPN) or subscriber premises network (SPN).

Figure 12.7 shows the reference configuration of BISDN UNI, which was extended from ISDN UNI as described in ITU-T Recommendation I.411. This ISDN configuration was general enough to be considered for BISDN

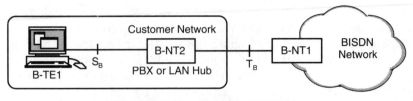

B-NT Network Termination for BISDN
B-TE Terminal Equipment for BISDN

• Functional Groups: B-NT1, B-NT2 and B-TE1
• Reference Points: T_B and S_B

Figure 12.7 Reference configuration of the BISDN UNI.

UNI configuration. The BISDN UNI consists of the following:

- Functional groups

  ~B-NT1: Broadband-Network Terminal 1

  ~B-NT2: Broadband-Network Terminal 2

  ~B-TE1: Broadband Terminal Equipment 1

- Reference points: T_B, S_B.

The B-NT1 performs line transmission termination and related OAM functions. The B-NT2 can be a PBX or LAN interface that performs multiplexing and switching of ATM cells. The reference point T_B is the interface between the CPN and the public network. A number of customer network categories exist, depending on the aspects of the customer network, such as environment, number of users, or topology. The major categories of broadband customer network are residential customer network or business customer network.

Residential network. In this network, a small number of users use broadband services mainly for entertainment purposes. In this case, no internal switching capabilities are necessary within a customer network.

Business network. This category can be divided into the subcategories of small, medium, and large businesses. The subcategory is based on the number of users in each environment. For example, in the case of small business, the number of users varies from 1 to 20; their requirements are similar to residential users, except that the end use is not for entertainment.

Two requirements exist for the above CPN: service requirement and structural requirement.

Service requirements include the consequences of supporting these services. Each of the services has different requirements, such as bit rate, quality, grade of services, and delay variation. The customer network should be designed in such a way to handle a wide variety of services. For instance, to carry switched video services, the customer network should be able to handle high bandwidth, high-quality video signal, which requires certain levels of delay and grade of service (GOS).

Structural requirements are the physical arrangements of the network elements. The structure should be flexible, modular, reliable, and provide maximum performance at low cost. Flexibility deals with the ability to cope with system changes in case of failure or other outages. There are four points: adaptability, expandability, mobility, and interworking. Modularity is the ability to have flexible structure and add additional equipment to the existing network configuration. Reliability is the ability to perform without any mishaps or errors in traffic under extreme environments. Reliability can also sometimes require redundancy (backup) of the customer network components. All these requirements enhance or maintain the performance at a reduced cost.

Figure 12.8 shows the physical configurations typically used in a customer network. The configurations can be the following:

- star configuration
- dual-bus medium (shared-medium) configuration
- combination of star and shared-medium configuration

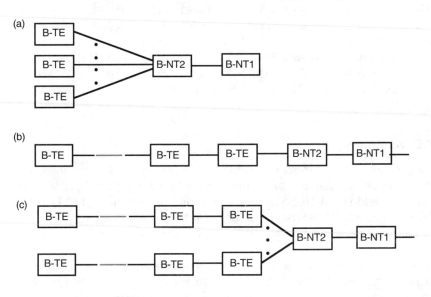

B-NT Network Termination for BISDN
B-TE Terminal Equipment for BISDN
Dashed Box Means the B-NT2 May or May Not Exist

Figure 12.8 Physical configurations of customer premises network.

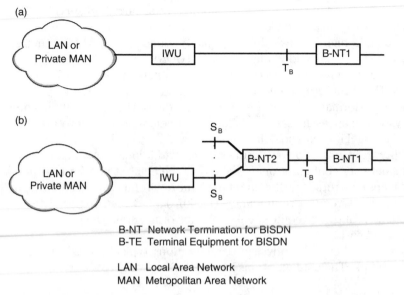

B-NT Network Termination for BISDN
B-TE Terminal Equipment for BISDN

LAN   Local Area Network
MAN Metropolitan Area Network

Figure 12.9 Interworking LAN/MAN with BISDN.

These configurations are typically used for LANs. To connect different LANs located in geographically dispersed areas, one needs interconnection. Figure 12.9 shows a typical example of the interworking of LANs via a MAN and the interworking of multiple MANs with a broadband network. The interworking unit (IWU) was created for this purpose. The IWU can be attached directly to a B-NT1 at the reference point T_B or the LAN/MAN can be connected via an IWU through S_B or B-NT2 (if present) as shown in Figure 12.9.

12.5 Summary

In this chapter, we addressed the different aspects of BISDN, such as service, network, and the user or customer-to-network interface. In service aspects, we addressed the different categories of service that ITU-T has defined in its recommendation: interactive and distribution services. The interactive services were subdivided into conversational services, messaging services, and retrieval services, while the distribution services were subdivided into distribution without user presentation control and distribution with user presentation control. The network aspects were divided into three subcategories: network layering, signaling principles, and traffic control. The final section covered aspects related to the user or customer-to-network interface, which discussed the requirements and topologies of customer networks.

Broadband ATM Switching and Transmission

In this part, we look into the different components that constitute the broadband transmission and switching systems. We know by now that ATM is the switching technology, and SONET/SDH is the fiber-based transmission system for future BISDN. Here, we examine the building blocks of the ATM switching system, its working principles, and the different types of ATM switching fabrics currently in use.

13

ATM-Based Broadband Switching

13.1 Overview

For broadband networks, since ATM is the switching technology, we can now turn our attention to ATM switching concepts, including the basic architecture, requirements, principles, and building blocks. Numerous books and papers have been published that address issues of switch design to increase the speed, capacity, and overall performance of ATM switching systems. Appendix B lists numerous reference materials. This chapter gives an overview of ATM switching systems.

13.2 ATM-Based Switching

Switching systems developed for conventional voice or data networks are not directly applicable for the broadband system. Hence, a compromising switching technology, asynchronous transfer mode (ATM), had to be developed and adopted for broadband-based networks. Two major requirements impact the definition of broadband ATM switching systems:

1. high-speed interfaces (50 Mbps to 2.4 Gbps) to the switch, with switching rates up to 80 Gbps in the backplane

2. statistical capability of the ATM streams passing through the ATM switching systems

To meet these requirements, the ATM switches had to be completely different from conventional switches. A large number of alternatives exist for ATM switches based on different factors addressed in this chapter.

First, however, we need to understand the basic ATM switch. A generic ATM switch can be divided into two categories—*hardware architecture* and *software architecture*—which are subdivided in turn. Each is explained in the architecture sections that follow.

13.2.1 Hardware architecture

The hardware architecture of an ATM switch is divided into two main components: the switch core, which performs the switching of cells, and the switch interface, which performs the external input and output functions. Figure 13.1 depicts the ATM switch and its physical interfaces. Each interface is connected to the switch core through two ATM core interfaces, one for input and other for output.

13.2.1.1 Switch interface. The switch interface adapts between the access devices and the ATM switch core whenever needed. The bit rate and format of the ATM cells are adapted by the switch interface to fit the switch core. In addition, almost all functions that handle ATM cell labeling reside in the switch interface, including virtual path identifier/virtual channel identifier (VPI/VCI) assignment, addition of routing information to cells, and discarding cells if needed. The switch interfaces can be designed for various bit rates. Currently, 1.5 Mbps, 50 Mbps, 155 Mbps, 622 Mbps, and 2.4 kbps interface speeds have been developed for different physical-layer protocols. The virtual path (VP) switching and virtual circuit (VC) switching are simultaneously supported by the switch interfaces, which results in valuable cell-routing flexibility.

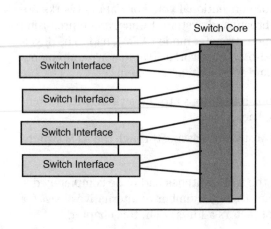

Figure 13.1 Generic ATM switch.

To summarize, a switch interface performs optical-to-electrical signal conversion, cell synchronization, header translation, and insertion and extraction of routing information.

13.2.1.2 Switch core. The switch core is a space switch that supports both point-to-point and point-to-multipoint connections. It is equipped with large buffers at the outputs or inputs to cope with variations in the asynchronous cell flow. The decision to put buffers at the input or output or both is an aspect of ATM switch design and varies depending on vendor methodology.

The switch core comprises of three different functional units:

- concentrators
- multiplexers
- switching matrix

These three can be cost-effectively combined to meet the various capacity requirements for a switching node. Switch capacity can be increased by using additional switch cores. Figure 13.2 shows a switch core containing all three types of units.

13.2.1.2.1 Concentrators and multiplexers. Devices with a lower bit rate at the switch interface are connected to the switching matrix via a concentrator to better utilize the incoming link connected to the switch matrix. Thus, a concentrator aggregates the lower, variable bit rate traffic into a higher bit rate so that the switching matrix can perform the switch at a standard interface speed.

Multiplexers are used when switching matrix interfaces are at a higher link speed than the switch interface. In this process, the cells from a number of interfaces are multiplexed into a single cell stream. The output cell stream is at the cell rate required by the switching matrix.

13.2.1.2.2 Switching matrix. The incoming cell streams are passed through the switching matrix. No switching occurs in the concentrators and multiplexers; the switch core always contains the switch matrix. Many types of switching matrices exist, some of which are discussed in Section 13.6.

13.2.2 Software architecture

The ATM switch is controlled and supervised by software. The software architecture in an ATM switch is usually divided into three functional categories:

- handling of traffic management
- handling of operation and maintenance
- handling of system functions

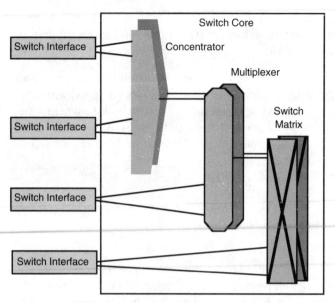

Figure 13.2 Switch core of the ATM switch.

Each of these categories is further divided into seven functional areas:

- connection handling
- configuration management
- fault management
- performance management
- billing management
- security management
- system functions

An essential requirement when designing software for the ATM switch has been the partitioning of the software into hardware-dependent and hardware-independent components. Software blocks have well-defined interfaces to other software blocks and entail as little interdependency as possible. This modular approach is a prerequisite for simplified, systematic upgrading of the switch to introduce any new functionality.

Each of these seven functional areas controls the ATM switch. Each performs certain functions related to a call. For example, performance management provides statistics on the performance of the network, on a VP- or VC-connection basis. In addition, statistics such as delay and throughput are available. Very little work has been done so far in software architecture of the ATM switch.

Usually, the ATM cells in the switch must be transported from an input to one or more outputs. The switching from input to output can be combined with concentration, expansion, multiplexing, and demultiplexing of the ATM traffic. From a functional point of view, an ATM switch is the same as a packet switch. The main difference between the ATM switch and the packet switch is the switching speed, processing of the packets or cells, and the packet size.

13.3 ATM Switching Principle

Having seen the basic architecture of the ATM switch, let's go through the principle behind it, which is shown in Figure 13.3. Here, the incoming ATM cells are physically switched from an input I_n to an output O_q while the cell's header values are translated from an incoming value β to an outgoing value δ. Each incoming and outgoing link has unique header values, but identical headers can be found in different links (e.g., X on link I_q and I_n). The translation ta-

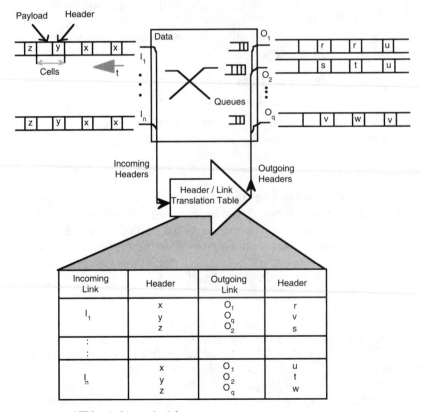

Figure 13.3 ATM switching principle.

bles map the incoming header value to the outgoing header value. For example, the cells with a header value of x on incoming link I_1 are switched to output O_1, with their headers translated (switched) to value r. Similarly, all cells with header value x in link I_n are switched to output O_1 with header value u.

In this switching system, however, it is possible that two cells of inputs I_1 and I_n arrive simultaneously at the ATM switch and are destined for the same output (O_1). In such a case, the cell cannot be put on output at the same time, and the switch must buffer the cells that cannot be served. Buffering is typical of an ATM switch.

13.4   ATM Switching Requirements

As mentioned earlier, a broadband network must be capable of transporting all types of information, ranging from low-bit-rate basic voice to high-bit-rate high-definition video. Each of these services has different requirements in terms of bit rate, characteristics, temporal behavior (constant bit rate or variable bit rate), semantic transparency (cell loss, bit error rate) and time transparency (delay, delays jitter). These different service requirements must be met by the broadband ATM switches.

To address these services, the following are classified as the basic requirements for an ATM switch:

- ability to handle different information rates
- capability to do broadcast and multicast
- ability to have high performance in terms of bit error rate, delay, and throughput

13.4.1   Information rates

Because the information rates of the different services are very diverse, a large number of information rates must be switched in broadband switches. These rates range from kbps to values of 150 Mbps. Currently, ATM switches have a maximum bit rate of about 150 Mbps per port. Thus, the internal backplane switching capacity of an ATM switch can vary from 2 Gbps to 80 Gbps, depending on the number of ports in a switch.

13.4.2   Broadcast and multicast

In typical voice and packet switches, only point-to-point connections are available because information must be switched from a logical input to a logical output. In broadband networks, however, additional requirements arise because of the new services. As mentioned in chapter 12, some services are distributive in nature—requiring the broadcasting of information from one source to all destinations—and others are multicast—providing the information from one source to selective destinations. CATV distribution is an example of broadcast. Both multicast and broadcast facilities require trunk circuits to subscribers.

13.4.3 Performance

Typically, the performance of a switch is characterized by throughput, call blocking probability, bit error rate (BER), and switching delay. In an ATM switching environment, however, two additional parameters are also important, namely cell loss probability or cell insertion probability and jitter on the delay.

In ATM switches, as in conventional switches, the performance characteristics are based on the technology and dimensioning of the system. We addressed the different technology drivers for broadband in chapter 2.

In this section we address the three parameters of connection blocking probability, cell loss or cell insertion probability, and switching delay.

13.4.3.1 Connection blocking probability

Although ATM switching is based on ATM cells, it is inherently connection-oriented. Thus, a connection must be established between an input and an output interface through the switching matrix, which does not mean that ATM switch implementation is internally connection-oriented.

Connection blocking is determined as the probability that not enough resources are available between the input and output of the switch to guarantee the quality of all existing connections as well as the new connection. Thus, if enough resources (i.e., guaranteed bandwidth) are available on the switch for all the connections, no blocking occurs. The number of connections depends on the internal design of the switch because resources in the switch must be allocated internally for every new connection. The blocking probability of the switches is determined by the dimensioning of the switch, such as the number of internal connections and the load (or bandwidth utilized) on those connections.

13.4.3.2 Cell loss/cell insertion probability. In ATM switches, there could be instances of too many cells destined for the same link. The consequence of this situation is that particular link receives more cells than it can handle, forcing cells to be discarded or lost. The probability of losing cells should be kept within specified limits to ensure a high semantic transparency. Typical values for cell loss probability for ATM switches are in the range of 10^{-8} to 10^{-11}. In other words, approximately only one cell per billion should be lost (10^{-9}).

13.4.3.3 Switching delay. Switching delay is the time taken to switch an ATM cell through the switch. Typical values mentioned for the delay in ATM switches range between 100 and 1000 μs, with a jitter of a few 100 μs. This delay is negligible compared to the switching delay of 20 ms that is found in a packet switch.

13.5 ATM Switch Building Blocks

An ATM switching network or fabric comprises basic building blocks called *switching elements*. Switching elements consist of queues and switching matrices. First, we focus on the building blocks, then we look at the different switching matrix types.

ATM switching elements are typically small. Currently, elements vary from two inputs and two outputs at 150 Mbps each to a maximum of 32 inputs and 32 outputs at 150 Mbps. The number of ports varies depending on the interface speed of the switch. It also depends on the technology used and the level of integration achieved by the switch manufacturer. In this section, we discuss the queues of switching elements. Typically, the switching element acts as a statistical multiplexer, which resembles queuing. A cell is queued when two cells arrive at two inputs destined for the same output. Depending on the architecture of the switching element and on internal speed, the cell can be queued at the input, the output, or internally in the switching element (i.e., between two switching matrices). Depending on the queue location, a switching element is called *input-queued switching element*, *output-queued switching element*, or central queued switching element. Each is described in the following sections.

13.5.1 Input-queued switching element

Figure 13.4 shows the switching element with an input-queuing mechanism. In this approach, the contention problem is resolved at the input. Each input port consists of a dedicated buffer to store the incoming cells until the arbitration logic determines that the buffer (queue) can be served. On clearing the queue, the cell goes to the switching transfer medium, which transfers the ATM cells from the input to the output without any internal contention or blocking. The arbitration logic decides which input port is to be served. This logic can be the basic round-robin algorithm, or any complex selection algorithm that uses the information switch as quality of service (QOS) to select the queue to be processed. The switching matrix in the switching element with input queues transfers the selected cells (the number of cells is less than N) from the N input ports to the N selected output ports during one cell time. In this phase, the queue size depends on the switch algorithm. Typically, the queue size ranges from 320 to 640 cells. If

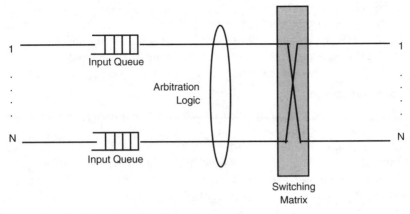

Figure 13.4 Switching element with input queues.

the incoming cells exceed the queue length, the switch drops the lower-priority cells from the queue.

13.5.2 Output queuing

Figure 13.5 shows a switching element with output queues. A collision can occur if several input cells are hunting for the same output. This problem is resolved by using queues at each output of the switching element. Each output consists of a dedicated buffer that stores multiple cells that might arrive during one cell time. To ensure that no cell is lost in the switching matrix, the matrix must transfer the cells at a rate of N times the speed of input ports before the cells arrive at the output queue. The system must then write N

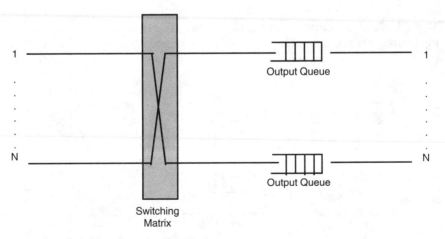

Figure 13.5 Switching element with output queues.

cells in the queue in one cell time. Here, no arbitration logic is required in the switching elements because all the cells can arrive at their respective output queues. The control of the output queues is based on a simple first in first out (FIFO) logic, ensuring that the cells remain in the correct sequence.

13.5.3 Central queuing

Figure 13.6 shows the switching element with central queuing and multiple switching matrices within the switching element. In central queuing, the queuing buffers are not dedicated to a single input or output port of the switching matrix but shared between all inputs and outputs. In this case, each incoming cell is stored directly in the central queue. Every output selects the cell destined for it from the central queue using FIFO.

13.6 ATM Switching Matrix or Network

We mentioned one of the building blocks of the switching network in the previous section. The other is the switching matrix. Switching matrices typically have a large number of input and output ports. They can typically be classified into two major groups, *single-stage switching matrix* and *multistage switching matrix*, as shown in Figure 13.7.

13.6.1 Single-stage switching matrix

A single-stage matrix is characterized by a single stage of switching elements connected to the input and outputs of the matrix. Two types of single-stage networks are described in this chapter: shuffle exchange and extended.

13.6.1.1 Shuffle-exchange switching matrix. The shuffle-exchange switching matrix is based on a perfect shuffle permutation, which is connected to a stage of switching elements. Figure 13.8 shows an example of a shuffle-ex-

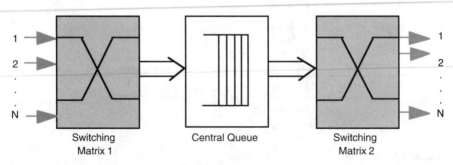

Switching   Central Queue   Switching
Matrix 1   Matrix 2

Figure 13.6 Switching element with central queuing.

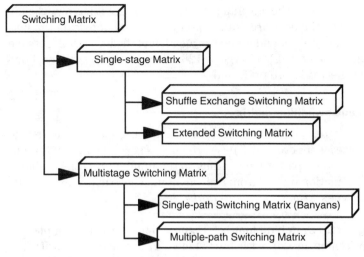

Figure 13.7   Classification of switching matrix.

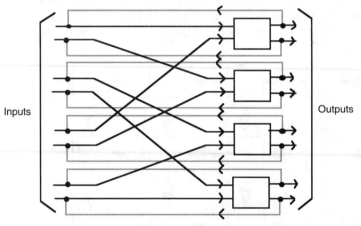

Figure 13.8   Example of a shuffle-exchange switching matrix.

change network. In this matrix, a feedback mechanism is used to reach an arbitrary output from a given input. Thus, a typical cell can pass through the matrix several times before reaching its proper destination. This switching matrix is also called a *recirculating switching matrix*. This matrix requires only a small number of switching elements, but the performance is very poor because of the feedback mechanism.

13.6.1.2 Extended switching matrix.   A switching network with $k \times k$ switching elements is shown in Figure 13.8. In this matrix, any size switching matrix can be implemented. The advantage of this type is that it has a negligible cross-delay because the cells are buffered only once while crossing the network. The cross-delay is dependent on the location of the input. It is possible to form a single-stage network as large as 128×128. For large systems, however, multistage networks are preferred.

13.6.2 Multistage switching matrix

To avoid the drawbacks of the single-stage switching matrix, multistage switching matrices were created. These are built of several stages interconnected in a certain link pattern, according to the number of paths available to reach the destination output from a given input. These matrices are classified into two types: *single-path* and *multiple-path*.

13.6.2.1 Single-path switching matrix. Figure 13.9 shows an example of a single-path switching matrix with four stages. In a single-path switching matrix, only one path exists to reach the destination from a given input. These networks are sometimes called *Banyan* networks. Routing is very simple because only one path exists to reach the proper output. *Delta networks* are a special type of Banyan networks. Here, each output port is identified by a unique destination address that allows for simple routing of

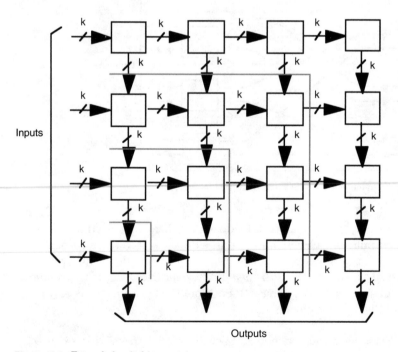

Figure 13.9 Extended switching matrix.

the cells to the destination. Hence, this network is sometimes called the *self-routing* network. The matrix in Figure 13.10 is called a Delta-2.

The Delta-2 switching matrix with four stages has the topology of a baseline switching matrix. The thick lines indicate the path from input 5 to output 13 (binary destination address 1101). The switching element in each stage processes one bit of the 4-bit address. Each bit can have one of two values, 0 or 1. The switching element routes the cells in each stage depending on the value of the bit. In this example, the bit is 1 in the first stage. On arriving at the first stage, the switching element reads the value of the first bit if the bit is 1, as is the case in this example. The cells are routed via the down link. If the value of the bit is 0, as in the case of the second bit, the switching element routes the traffic via the up link. In this way, each cell is routed to its destination point.

13.6.2.2 Multiple-path switching matrix.

In a multiple-path switching matrix, many alternate paths exist for the destination output from a given input. This type of matrix has the advantage of reducing or avoiding internal blocking. The internal path is determined in most cases during the connection setup phase, and all the cells of that connection use the same internal path set at the time of connection. If FIFO is provided at each switching element, cell sequence integrity can be guaranteed, and no resequencing is necessary. There are two types of multiple-path networks: *folded* and *unfolded*. In folded networks, all inputs and outputs are located on the same side of the switching network, and the network's internal links are operated in a bidirectional manner. In unfolded networks, the inputs are located on one side and the outputs on the opposite side of the network. The internal links are unidirectional, and all cells must pass through the same number of switching elements.

Figure 13.11 shows an example of a multipath interconnection (MIN) switching matrix. This switching matrix is realized by adding a baseline switching matrix with reversed topology. The baseline switching matrix topology is shown in Fig. 13.10.

13.7 ATM Cell Processing in a Switch

The main functions of an ATM switching node are VPI/VCI translation and transport of cells from the input to the appropriate output. Earlier, we showed cell processing using a table-controlled principle. There are actually two principles to address these tasks: self-routing or table-controlled.

13.7.1 Self-routing principle

In self-routing, the VPI/VCI translation must be performed at the input of the switching element, as shown in Figure 13.12. After translation, the cell

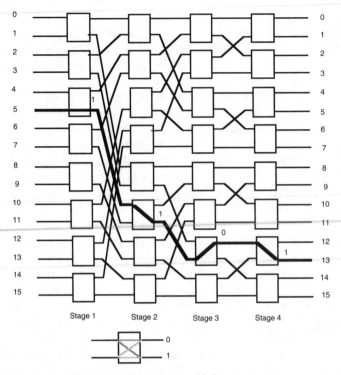

Figure 13.10 Delta-2 switching matrix with four stages.

is extended by an internal identifier that precedes the cell header. A cell header extension requires an increased internal matrix speed. Once the internal identifier is attached to the cell, it is routed via the self-routing principle described in section 13.6.2. Each connection, (incoming to outgoing) has a specific switching matrix internal identifier. This identifier is unique to that switching matrix and cannot be reused at another switching matrix. In a point-to-multipoint connection input, VPI/VCI is assigned a multiple-switch internal identifier, thus enabling cells to be copied and routed to different destinations, depending on the identifier.

13.7.2 Table-controlled principle

In table-controlled switching, the VPI/VCI cell header is translated in each switching element into a new header. Figure 13.13 shows this type of cell translation. During the connection setup phase, the contents of the tables are updated. Each table entry consists of the new VPI/VCI and the number of the appropriate output port or link.

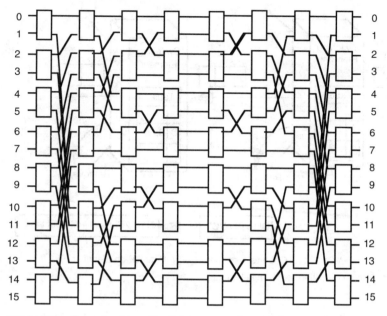

Figure 13.11 Example of a multipath interconnection switching matrix.

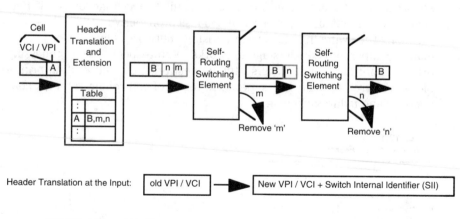

VCI Virtual Channel Identifier
VPI Virtual Path Identifier
SII Switch Internal Identfier

Figure 13.12 Self-routing switching elements.

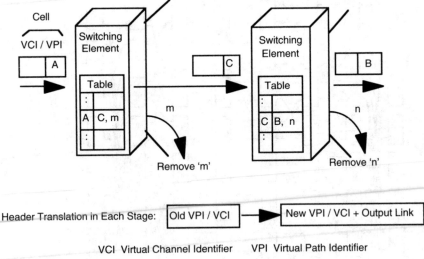

Header Translation in Each Stage: | Old VPI / VCI | → | New VPI / VCI + Output Link |

VCI Virtual Channel Identifier   VPI Virtual Path Identifier

Figure 13.13 Table-controlled switching elements.

13.8 Summary

In this chapter, we covered broadband ATM switching. We addressed the requirements and building blocks, such as different types of queues and switching matrices. Among the different queuing mechanisms, central queuing requires the lowest memory capacity because of buffer sharing. Table 13.1 shows the queue sizes (in cells) for switching elements of different sizes and different queue locations, assuming an average load of 85 percent at each input and a permissible cell loss probability of 10^{-9}.

Among the switching matrices, multistage networks have better performance compared to single-stage networks. With respect to cell processing principles for large multistage switching networks, the self-routing principle is preferred because it is superior in terms of controlling complex connections and failure behavior.

TABLE 13.1
Queue size requirements

Type	Size	
	16×16	32×32
Central queue	113	199
Input queue	320	640
Output queue	896	1824

14

Broadband
Transmission Network

14.1 Overview

In this chapter, we address how information is carried in a broadband transmission network. In the ATM switching environment, it is assumed that all information is placed into the payload of ATM cells. However, not all information is in the form of ATM cells in a broadband transmission network. The broadband transmission network should be designed in such a way that the traffic from the existing digital asynchronous network can also be carried, along with the new ATM traffic. The basic function of a transmission network is to transport information from the originating point to the destination without loss of information or delay. In addition to these responsibilities for a transmission network, others include the following:

- generation of cell (packetizer)
- transmission of cells and existing isochronous traffic
- multiplexing and demultiplexing of cells and other existing isochronous traffic
- cross-connecting of cells and existing isochronous traffic
- switching of cells and existing isochronous traffic

To meet these functions, ITU-T has selected SONET/SDH as the transmission protocol. SONET/SDH, however, is not the only protocol that can transport ATM-cell existing isochronous traffic.

To deploy any successful network, including a broadband one, the most critical component of the whole network is the transmission system, which includes the physical cable. Unless the transmission systems are in place, no switching or any other system in the network can be used. Many of the technological advances in areas such as switching, intelligent networks, etc., were possible because of the advances made in transmission systems, such as fiber-optics, fiber-terminals, etc. Fiber, as technology and equipment, is available and being deployed around the world.

In this chapter we examine the transmission system for the broadband network and address its components and their functions. We also study the broadband reference architecture that brings together the different components. Both local and trunk network architectures are also explained in this chapter.

14.2 Broadband Transmission Functional Components

A transmission system consists of many components, each having a function to perform related to the transport of the information from one point to another. Here, we look into some of the functional components of SONET/SDH and ATM, the transmission and switching technologies for future BISDN.

14.2.1 SONET multiplexer

A SONET multiplexer has p inputs and one output, as shown in Figure 14.1a. An input tributary operates at the STS-3nc/OC3nc rate, which depends on the individual tributary. The SONET multiplexer synchronously multiplexes the input tributaries into an output stream operating at the STS-3m rate. This output rate is equal to the sum of the input tributary rates. SONET multiplexers can also be cascaded back-to-back. It is not necessary that the input be STS-3nc/OC3nc (frame-carrying concatenated ATM cells); in fact, the input can carry existing digital signals such as DS1 or DS3, mapped alongside ATM cells. DS3 is mapped directly onto an STS-1 payload, where DS1 is mapped onto a

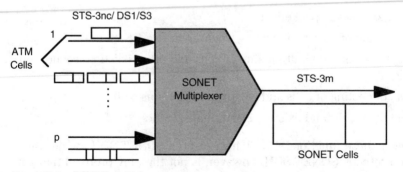

Figure 14.1a SONET multiplexer.

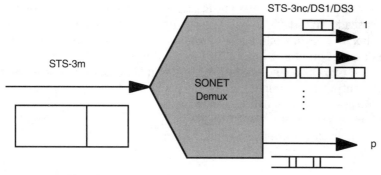

Figure 14.1b SONET demultiplexer.

VT-1.5. Thus, within an STS-1 payload, 28 VT-1.5s can be accommodated in an STS-1 frame. In this multiplexer, the inputs i to p can be STS-1, DS1, or DS3. All these can be present as part of the input. Thus, the output STS-3nc/OC3nc frame consists of the combined payload of DS3, DS1, and ATM cells, enabling all the different traffic to be carried on one transmission network.

14.2.2 SONET demultiplexer

The SONET demultiplexer performs the inverse function of the SONET multiplexer. Figure 14.1b illustrates it. The SONET demultiplexer synchronously demultiplexes the single input stream into p output streams, which correspond to the input stream of the associated SONET multiplexer. Thus, if the input consists of a combination of STS-3nc, DS1 and DS3, the output consists of exactly the same elements. Figure 14.2 shows a cascade of multi-

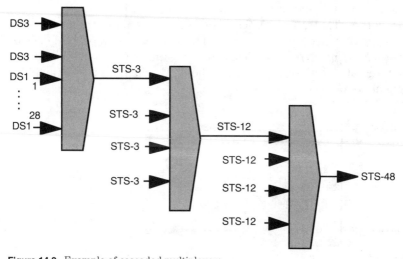

Figure 14.2 Example of cascaded multiplexers.

plexers, where lower-speed inputs are multiplexed back-to-back to higher-speed in a cascade. In today's equipment, the users do not need to buy multiple multiplexers because one piece of equipment can be configured with any combination of input and output interfaces. For example, one can combine four DS1s, two DS3s, and one OC3 as input and one OC48 as output. In this case, the output frame might not be fully packed because of its very high speed, but future expansion can be easily incorporated by adding an additional input interface. In a typical transmission terminal, a SONET multiplexer and demultiplexer are packaged together.

14.2.3 STS-3nc cross-connect

An STS-3nc has p inputs and q outputs, as shown in Figure 14.3, where all of them operate at the same rate. It can connect any input to zero, one, or more outputs, thereby duplicating the input signal at those outputs. The relative timing between STS-3nc streams might or might not be preserved as the streams pass through the cross-connect. The basic function of this cross-connect is to regroup different VT's and STS-3nc to the appropriate port, which is destined for a single location.

14.2.4 ATM concentrator

An ATM concentrator has p inputs and one output. Each input tributary operates at the STS-3nc or DS3 rate. The ATM concentrator receives incoming ATM cells at a variable bit rate at each input. Cells designated as unoccupied are deleted. Occupied cells are buffered at the input until they can be merged with other cells into the output stream. The output stream operates at the STS-3mc rate, illustrated in Figure 14.4. The output rate depends on the ATM traffic characteristics within the input tributaries and on the degree of concentration desired. This concentration of ATM cells is called *statistical multiplexing*. This concentration of cells can be stand-alone or integrated with others, such as the ATM switch, as part of an ATM switching node.

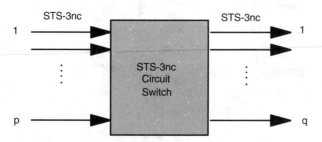

Figure 14.3 STS-3nc SONET cross-connect.

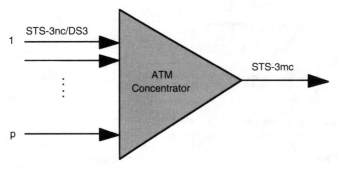

Figure 14.4 ATM concentrator.

14.2.5 ATM expander

The ATM expander performs the inverse function of the ATM concentrator, as shown in Figure 14.5. An incoming STS-3mc stream of cells is put on the appropriate output port with appropriate empty cells to pad the output variable information rate. Usually, an ATM concentrator is cascaded with an ATM expander to perform the bidirectional function for the ATM cells.

14.2.6 ATM switch

An ATM switch typically consists of a concentrator, expander, multiplexer, and demultiplexer to adapt the different rates at the input ports of the ATM switch. As shown in Figure 14.6, p input ports and q output ports are in an ATM switch, where $p > q$. Input and output ports operate at various rates, ranging from DS1 to STS-3mc. The header of the incoming ATM cell is examined to determine the output port to which the cell must be routed. Information in the header is used as an index into a table that contains routing and translation information. The details of the ATM switch were discussed in chapter 13.

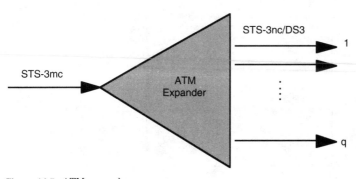

Figure 14.5 ATM expander.

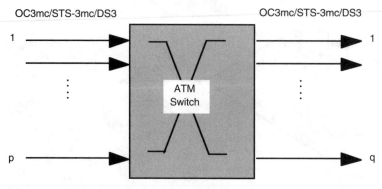

OC3mc/STS-3mc/DS3   OC3mc/STS-3mc/DS3

Figure 14.6 ATM switch.

14.2.7 Service module

A service module is part of the network, but performs its function at a layer higher than the ATM layer (Figure 14.7). For BISDN protocols, these layers are the adaptation layers, management layer, and control layer. Examples of service modules are the following:

- a service module that receives user-network signaling messages, performs call processing functions, and controls the switching matrix table

- a service module that receives multimegabit connectionless data, possibly performs the segmentation and reassembly functions, examines the level-3 header, and routes the datagram to the destination or to another connectionless data service module

- a multicast bridge that receives incoming cells, copies the cells, and routes the copies of the cell to multiple destinations.

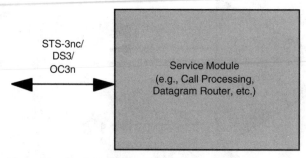

STS-3nc/
DS3/
OC3n

Service Module
(e.g., Call Processing,
Datagram Router, etc.)

Figure 14.7 Service module.

14.2.8 Interworking units

Interworking units provide the adaptation functionality from narrowband to broadband. The existing digital asynchronous transmission system maps to a synchronous transmission system. A typical example is the mapping of DS3 to STS-1 payload or optical carrier interface (OCI) payload or DSI to V1.5 payload in the SONET frame as shown in Figure 14.8.

14.3 Broadband Transmission Functions

One of the basic functions provided by any transmission network is the transfer of information from the source to the destination without error. Because of the high speed and usage of fiber-optics to transfer information in broadband networks, very stringent performance requirements in terms of bit error rate become necessary. In broadband transmission networks, ATM cells carry the information in their payload. It is not necessary, however, that information be carried in the ATM cells. Standard bodies have defined standards such that information from different existing transmission systems, such as digital asynchronous system (DS1, DS3, E1, E3), can carry ATM cells. Of course, ATM cells can also be carried in a SONET/SDH network. The broadband transmission network typically uses some or all of the functional components mentioned in Section 14.2. The transmission functions related to transfer of ATM cells are

- generation of cells
- transmission of cells
- multiplexing and demultiplexing of cells
- cross-connecting of cells
- switching of cells

Functions related to the switching of ATM cells have been studied in chapter 13. The other functions are addressed in this chapter.

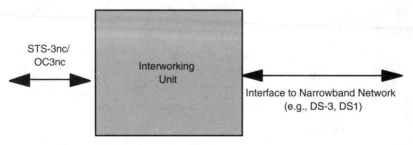

Figure 14.8 Interworking unit.

14.3.1 Generation of cells

Cell generation in a broadband network can occur in two ways:

- In a broadband/ATM customer premises equipment (CPE), which is a broadband terminal that has the capability to generate ATM cells.
- If the CPE is a nonbroadband/ATM terminal, the network must convert the information into ATM cells and vice versa.

In the case of a broadband/ATM terminal, the CPE has an adapter card that segments the information into 53-byte ATM cells with appropriate VPI/VCI values. The adapter card generates speeds ranging from 1.5 to 155 Mbps, enabling the end-to-end services to have the same protocol (including the intermediate nodes). There are numerous advantages to broadband terminals, such as reduced delay due to protocol conversion from one network to another, ease of operations, administration, and maintenance (OAM), etc. Figure 14.9 shows a typical broadband terminal.

In the case of a conventional or non-ATM terminal, the adaptation of information must be performed in the network. The network component that performs this function is called an *ATM packetizer/depacketizer*. The principle here is as simple as cutting a long ribbon into evenly sized pieces. Figure 14.10 depicts this concept.

This function will initially be located in the backbone network due to economics, and will later move to the customer premises to realize a complete ATM-based broadband network.

14.3.2 Transmission of cells

In principle, ATM cells can be transported on any transmission system. The only requirement is guaranteed bit independence so that there are no restrictions on cell information contents. To accommodate the transport of ATM cells in existing systems, standards must be developed that enable the

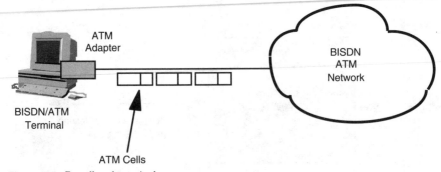

Figure 14.9 Broadband terminal.

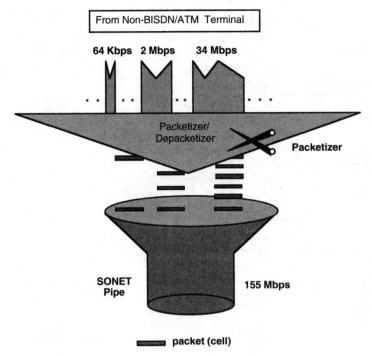

Figure 14.10   ATM packetizer/depacketizer.

user to transmit ATM cells. In fact, the ATM forum's main objective is to develop a user-network interface (UNI) and network-network interface (NNI) for existing systems, in addition to the proposed BISDN standards by ITU-T. To achieve network synchronization on a synchronous transmission network, care must be taken that clocks are synchronized throughout the system. Bellcore, in its technical reference TR-NWT-000253, Issue 2 ("SONET Transport System Common Generic Criteria"), has detailed the requirement of the clock so that synchronization can be maintained in a SONET environment. Figure 14.11 shows how synchronization is maintained in a BISDN network.

In BISDN, network synchronization is obtained from a reliable, stable clock source, which has an accuracy of Stratum 2 or 1[1]. All the elements in the network are directly synchronized to the single clock. Because it is not possible to have a direct link from a single clock to all the network elements, the timing information is usually extracted from the neighboring node (the

---

[1]The Stratum 1 clock is the most accurate clock, whose oscillation remains constant over a long period of time.

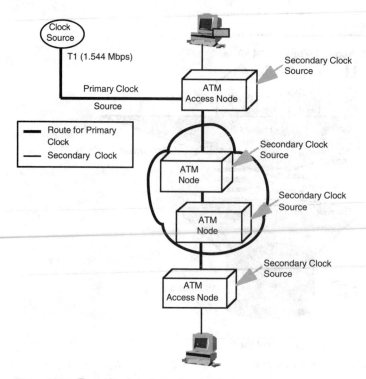

Figure 14.11 Example of clock synchronization.

node that has a direct connection) in the network. This node then transfers that timing to the next node in the line. All nodes in the network are thus synchronized. Because timing is an important element in a synchronous network, care must be taken to have a backup. Thus, in a BISDN network, each node must have some arrangement to fall back on the secondary clock, if the primary clock fails or loses synchronization for a short period. Typically, the secondary source is internal to a node. In some cases, however, it might be another external clock source. Timing is one of the more important features required by synchronous transmission system equipment. In general, no equipment can be sold without incorporating good timing features, such as primary and secondary clocks. In the real world, large pubic networks get their reference timing from a nuclear source, which is supposed to be the most stable and reliable source. All other timing clocks are always in exact accordance with that nuclear source.

14.3.3 Multiplexing and demultiplexing of cells

A typical multiplexer multiplexes several signals that originate from different BISDN customers on to a single access line. Usually, the signal sources in a multiplexer vary from DS1 to OC3, which operate at various line speeds. For example, the input signal can be asynchronous or synchronous, and the output can be SONET, but on the input side, asynchronous traffic is mapped onto synchronous traffic (SONET), which might be OC1/STS-1. These input signals are multiplexed together to form higher SONET rates (OC48/STS-48).

In an ATM multiplexer, all incoming idle cells are sorted out. The ATM traffic can be thus concentrated, which is performed by an *ATM concentrator*. The achievable degree of concentration depends on the traffic characteristics and the requested quality of services. The device that performs the reverse of the ATM concentrator is called the *ATM expander*. This type of multiplexing is statistical.

14.3.4 Cross-connecting of cells

An ATM cross-connect is a VP switch that can flexibly map incoming VPs onto the outgoing VPs and thus enable establishment of VPCs through the ATM network. The cross-connect can also concentrate ATM traffic and perform management functions of the physical and ATM layer of BISDN. Typical use of this cross-connect is as an access vehicle to the ATM switching backbone network. A typical ATM switch can be configured as a regular switch or a cross-connect. In ATM terminology, an ATM switch can be configured as a VC switch or a VP switch.

14.4 Broadband Reference Network Architecture

A reference architecture model for the broadband network is a conceptual division of functions that must be performed to transport user information through the network. It is a basic tool for studying alternate network architectures in both local and trunk segments of the network. In many instances, some of these functions can be merged together or completely ignored. The reference architecture is just a logical segmentation of functions; the functions can be in any part of the network. The reference architecture model that is depicted in Figure 14.12 consists of the following layers:

- customer premises node (CPN)
- distribution/drop plant
- remote multiplexer node (RMN)
- subfeeder plant
- access node (AN)
- feeder plant
- local exchange node (LEN)
- interexchange plant
- transit exchange node (TEN)

A set of functions are performed corresponding to each layer. Depending on individual architecture realizations, some functions in a particular layer might not be implemented, or, in some cases, an entire layer can be eliminated.

CPN provides conversion between the various service interfaces and the integrated user-network interface. It is assumed that single-mode fiber/multimode fiber in the distribution/drop plant carries information to and from a single user. Fibers used in the subfeeder, feeder, and interexchange plant carry information to and from many users.

RMN and AN guarantee that only information destined for a single customer flows downstream over a distribution fiber. Upstream, the RMN and AN prevent information from one customer adversely affecting information of another. The LEN provides interconnections among the customers served by that LEN by performing switching. The TEN provides interconnections and switching among the LENs served by that TEN. Because narrowband networks will continue to exist for many years, internetworking units (IWU) are necessary at the local exchange and at transit exchange nodes to provide the necessary conversions and adaptations for interconnecting the narrowband networks. The IWUs can be separate modules connected to the broadband switching systems or integrated into broadband

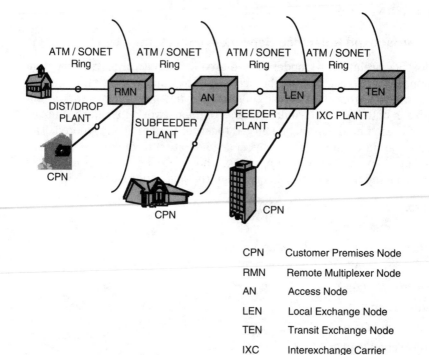

CPN	Customer Premises Node
RMN	Remote Multiplexer Node
AN	Access Node
LEN	Local Exchange Node
TEN	Transit Exchange Node
IXC	Interexchange Carrier

Figure 14.12 Broadband reference network architecture.

TABLE 14.1 Function of Broadband Reference Network Layers

Network Layer	Functions
Customer premises node	ATM and SONET multiplexing, NT2 functions
Distribution/drop plant	Physical layer functions (transmission)
Remote multiplexer node	SONET/ATM multiplexing/demultiplexing and ATM concentrator and expander
Subfeeder plant	Physical layer functions (transmission)
Access node	STS-3nc cross-connect, ATM concentration/expansion, traffic segregation, ATM switching, SONET multiplexing/demultiplexing
Feeder plant	Physical layer functions (transmission)
Local exchange node	STS-3nc cross-connect, ATM concentration/expansion, traffic segregation, ATM switching, SONET multiplexing/demultiplexing, higher-layer functions, internetworking with other networks
Interexchange plant	Physical-layer functions (transmission)
Transit exchange node	STS-3nc cross-connect, ATM concentration/expansion, traffic segregation, ATM switching, SONET multiplexing/demultiplexing, higher-layer functions, internetworking with other networks

switching systems. Table 14.1 shows the different functions performed by each network layer.

14.4.1 Local network architecture

Conceptually, the simplest realization of a BISDN local network is the star topology, where one access line exists per customer. This topology is used in today's telecommunication networks. The local network is part of the broadband reference network architecture. Figure 14.13 shows the local network topologies in the BISDN reference network architecture. Other topologies, such as rings and double stars, are also recommended. For broadband networks, ring topologies are considered because of the need for survivability in the local network.

14.4.2 Trunk network architecture

Figure 14.14 shows examples of a typical trunk network architecture. In the first example, the cross-connect is connected to an ATM switch on one side and a SONET fiber linear terminal (point-to-point terminal) on the trunk side in each location. SONET transmission and ATM switching combine to form the typical BISDN trunk network. The cross-connect or multiplexer might be the interface between the BISDN/ATM switch, which switches ATM cells, and the SONET terminal, which multiplexes the ATM and conventional traffic in that location into a common, fiber-based technology.

In the second example, shown in Figure 14.15, the linear fiber terminal is replaced by a ring terminal. There are many advantages in using the ring terminal. Foremost among them is the survivability of the traffic between two

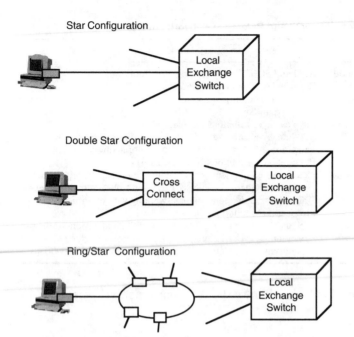

Figure 14.13 Example of local network topologies.

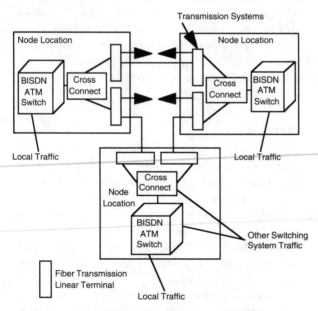

Figure 14.14 Example of trunk network.

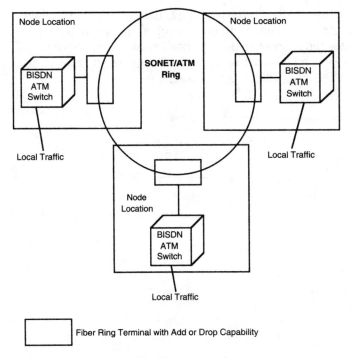

Figure 14.15   Second example of trunk network.

locations; there is always an alternate route. Alternate routes are very valuable in the case of a long-distance backbone network, where a huge amount of traffic is carried on each trunk between two points. In this network, if a link fails, the traffic is automatically routed via another route to its destination. The only traffic lost is the local traffic near the failure. It is expected that ring trunk architecture will be preferred in broadband transmission networks.

14.5 Summary

In this chapter, we addressed the broadband transmission network, its different components, and their functions. The transmission network plays an important part in the success of BISDN deployment. Unless the transmission network is successfully implemented, the overall success of the network is questioned. In fact, any service provider who is in the business of providing communication services spends a lot of time and money to ensure that the transmission network is fail-proof.

For example, service providers are deploying SONET rings instead of SONET linear terminals. In the former, if any link fails, the network automatically reroutes the traffic, thus providing 100 percent survivability, whereas in the latter, human intervention is required to reroute the traffic. In the United States today, most of the backbone networks are fiber-based, and some are already migrating toward SONET ring-based transmission networks.

ATM Environments

We now change our focus from BISDN to ATM exclusively because ATM is the most important subset of BISDN. ATM can be used as a transmission technology as well as the switching technology in BISDN. This part provides an overview of the usage of ATM technology in different environments, namely local area networks, wide area networks and public networks, such as local exchange carriers (LECs) and interexchange carriers (IECs). The application of ATM technology in other environments is also explained in Part 5, such as cable television, where needs and applications differ drastically.

15

ATM in Local Area Networks

15.1 Overview

So far, we have seen different technologies, their architecture, and the concepts behind them, including ATM. Starting with this section, we see how ATM technology can be applied to different environments, such as local area networks (LANs), wide area networks (WANs), public networks, and other networks such as CATV. In this chapter, we address the use of ATM in LANs, first by analyzing the current technology options available for LANs and why ATM has become the talk of LAN vendors as the future technology. We then address the different ways through which one can migrate to ATM in the LAN environment. Last but not the least, we see what applications are driving ATM in the LAN environment.

15.2 LAN Requirements

In today's LAN environment, the requirements are changing so rapidly that no unique solution or technology exists to address all of them. Before we talk about the different technologies, let's see what the requirements are of today's and next generation's LANs. We have classified the requirements of a LAN environment into three major categories:

- unlimited growth path for higher speed
- guaranteed quality of service (QOS)
- enterprise-wide uniform network management

15.2.1 Unlimited growth path for higher speed

Unlimited growth path means that the access speed of the workstations to the servers should be capable of being easily increased without modifying the network or system. Currently, access speeds of even 10 Mbps are more than sufficient. In the future, however, one might need at least 150-Mbps access speed to provide multimedia services. If the user must replace the existing system in order to meet immediate needs, then it has to be in such a way that it can handle future need for multimedia applications also. If a 150-Mbps access speed link is installed now, the user could request higher-speed access when necessary to access the required multimedia information from the server with negligible response time. To achieve this capability, the network and technology should be in place before the need arises.

15.2.2 Guaranteed quality of service

The issue of quality of service (QOS) for each user and each service arises when a network handles different services, such as voice, data, and video. Of course, each service has a different QOS. The new technology should be designed to handle different levels of QOS based on user requirements such as delay, response time, throughput, cost, etc.

15.2.3 Enterprise-wide uniform network management

The concept of enterprise-wide uniform network management becomes simple if everyone uses the same equipment across the entire enterprise network because the management function of all the equipment is the same and thus easy to manage. The same concept of network management becomes very complex, however, if equipment is supplied by multiple vendors with different architectures using different management systems. The big problem then becomes system integration, which is true even in today's network environment. The new technology should therefore have good network management features to adapt to this heterogeneous network management.

15.3 Technology Options for LANs

Having discussed the requirements, let's now look at the different technologies available for the LAN environment. Each of the technologies has its own advantages and disadvantages, depending on its usage, such as the type of application, performance, throughput and latency requirement, geographic bound, and cost.

We analyze four different technological categories applicable for LAN environments:

- switched Ethernet
- fiber distributed data interface (FDDI)
- 100 base-VG (voice grade)
- ATM

Table 15.1 shows a comparison of these four technologies. They are further discussed in the following subsections.

15.3.1 Switched Ethernet

Switched Ethernet is available today and primarily used for data, but it has limited multimedia support. The performance of the switched Ethernet is higher than that of the shared Ethernet, and it has a low latency because of the dedicated bandwidth. This technology is used only in the LAN environment. One of the factors favoring switched Ethernet is the cost. Switched Ethernet is preferred to shared Ethernet because of the requirement of dedicated access. The drawback is, of course, the dedicated access to the workstation or terminal.

15.3.2 FDDI

FDDI is used only for data in very high-performance workgroup communities. Currently, a number of vendors offer FDDI products. The performance

TABLE 15.1 Comparison of LAN Technologies

	Switched Ethernet	Shared FDDI	100 Base-VG	ATM
Product availability	1992–1993	Since 1990	1994	1993–1994
Application drivers	Client-server WG	LAN backbone	Client-server WG	Backbone, Interconnect
End-user performance	Higher than shared Ethernet	Higher than shared Ethernet	Higher than shared Ethernet	Very high
Geographic scalability	Limited to WG	WG, bldg	Limited to WG	LAN=MAN=WAN
Bandwidth scalability	Limited per user, high for aggregation	None	None	Very high per user, high for aggregation
Cost	Lowest	High	Same as Swt. Ethernet	Initially higher than FDDI, expected to drop to Swt. Ethernet
Standards maturity	NA	Stable	Just starting	Evolving

of FDDI is much better than that of Ethernet and token ring-based networks, but its disadvantage is that it is limited to shared media only. Again, this is a LAN-only application with very limited use in WAN. Currently, the cost of FDDI is extremely high but is expected to come down as newer technologies enter the market.

We discussed the FDDI technology in detail in chapter 3. As the name indicates, it is a fiber-based network. Currently, a more cost-effective coax-based network has been developed as an alternative to FDDI called CDDI (coax distributed data interface). Many developments have occurred in FDDI, such as FDDI II, which is capable of handling isochronous traffic (voice). FDDI II was developed to compete against a growing number of new technologies entering the LAN arena and is capable of handling all types of traffic, including ATM.

15.3.3 100 Base-VG

100 Base-VG is a relatively new technology. It was standardized in 1992, but no products are currently available. This technology is primarily used for data application and some multimedia. Its performance is superior to FDDI. Again, this is a LAN-only technology, and its cost is expected to be comparable to switched Ethernet.

15.3.4 ATM

Currently, numerous ATM demonstrations are performed today by a handful of vendors, and lots of others have announced the introduction of ATM product lines. The biggest advantage of this technology is that it is suitable for the transmission of data, voice, video, and image. The method of transportation of these was mentioned in chapter 8. This technology has the highest performance, unlimited bandwidth, and lowest latency. It is available in LAN, MAN, and WAN environments. It is currently expensive but is expected to be comparable to currently available technologies. A detailed review of ATM technology was also discussed in chapter 8.

15.4 LAN Environment

LAN environments can be categorized based on the following:

- topology of network
- media type used in network
- access method to network

Each is discussed in the subsections following.

15.4.1 Topology of the network

Topology addresses how the nodes or terminals are connected with each other. Currently, most LANs are either in a bus or ring topology, as shown in Figures 15.1 and 15.2, respectively. The reason these two topologies are used predominantly in a LAN environment is that both operate on the principle of sharing the transmission medium. In the early years of LAN, the ob-

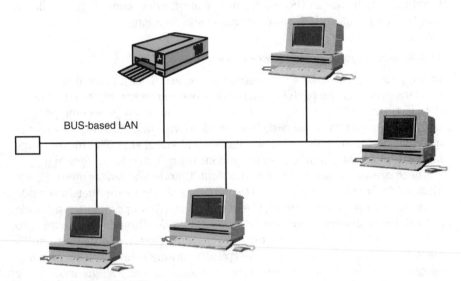

BUS-based LAN

Figure 15.1 Bus-based LAN topology.

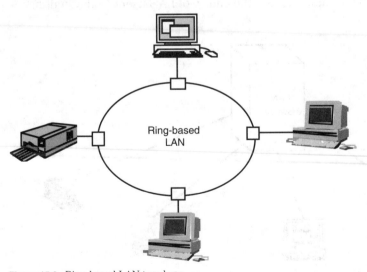

Ring-based LAN

Figure 15.2 Ring-based LAN topology.

jective was to connect multiple users located within a building. Today, a growing need exists for dedicated bandwidth to access new applications (such as multimedia) where voice, video, and data are integrated. In this environment, the bandwidths to each user can no longer be shared, but must be dedicated. New technology is therefore needed to provide a dedicated bandwidth to each user, enabling the user to access information at a high speed. The topology that offers such dedicated bandwidth is something already in use in the telecommunication environment. It is called a star topology. Figure 15.3 shows a simple star topology.

15.4.2 Media type used in network

In the past, copper cables connected computers. Today, coaxial cables are used instead because of the reduction in cost and the ability to carry more data in terms of bits per second. Fiber-optics has recently become popular because of its ability to carry high-speed data with negligible loss or error in information for long distances and its small physical size. The future usage of fiber-optic cables in a LAN environment is predicted to be very high.

There are two types of fiber-optics—multimode and single-mode fibers. Multimode fibers are made of plastic, whereas single-mode fibers are made of glass. In a LAN environment, the plastic fibers that operate in multimode are relatively cheaper than single-mode fibers. Multimode fibers are also designed to carry information error-free over a reasonable distance sufficient for the LAN environment. A typical multimode fiber can carry information error-free for a distance of three miles, whereas a single-mode fiber can carry information error-free for 25 miles; a LAN usually covers less than three miles. Thus, multimode is the choice of LAN users. Today, multimode

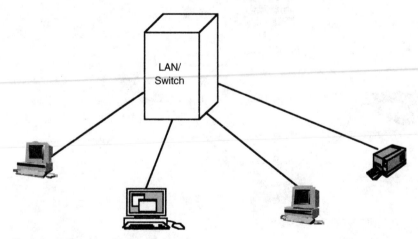

Figure 15.3 Star-based LAN topology.

fibers are comparable in cost to coaxial cable, making them even more attractive in LANs.

15.4.3 Access method to network

In LAN environments, two types of access methods are used, carrier sense multiple-access/collision detection (CSMA/CD) and token passing.

CSMA/CD is an enhanced version of the Aloha protocol developed by the University of Hawaii. CSMA/CD is a technique where all the stations listen on the transmission line. A station that wishes to transmit does so only if it detects the channel is idle. This procedure is called carrier sensing (CS), and the access strategy used is termed a CSMA scheme. Collisions can still occur because stations are physically displaced from one another, and two or more stations might sense that the channel is idle and start transmitting, causing a collision. Once stations detect collision (CD), they transmit a special jam signal to notify all other stations about the collision on the link. In other words, think of CSMA/CD as two polite people who start to talk at the same time. Each politely backs off and waits a random amount of time before starting to speak again.

Another access method is token passing. Here, each device on a LAN receives and passes the token. The device with the token has the right to use the single channel on the LAN. The key to remember is that a token-passing or token-ring LAN has only one channel, which is a high-speed channel that can move a lot of data. It can, however, move only one conversation at a time. In other words, it is like a musical ball game, where a ball is passed from one person to another until the music stops. Then, whoever has the ball must pick an instruction card and act accordingly. In token ring, the ball is the token, and whoever gets the token can transmit data on the ring.

The disadvantage of these access methods is that only one terminal can access the network to communicate at one time. Hence, these methods are not suitable when more than one user tries to use the network. Thus, dedicated access comes into play, where each terminal has a dedicated line connected to a LAN hub, as shown in Figure 15.3. To evolve from the existing bus or ring-based LAN environment into something completely different, such as a switched hub-based network (e.g., telephone public network), however, is a daunting task. One must go through careful planning before such a migration. Numerous problems exist with the existing architecture and topology. Because users at the time of deployment did not have much choice, they had to use what was available. Today, there are about 15 million LAN ports installed, and this amount of usage was not expected by the original designer of the technology. Today, the difficulty arises when one must rearrange the network so that when people move from one place to another, they do not see any change in their networking environment (same user name but different domain).

15.5 Architecture Alternatives

ATMs in LAN environments are basically implemented in existing products, such as routers and hubs. Figure 15.4 shows a typical ATM-based router. The functionality of such a product is basically multiprotocol routing plus ATM, especially in traffic management. The interfaces to the router include all the LAN interfaces, such as HSSI, DSI, DXI, etc.

Different architecture alternatives are available with the presence of ATM in LAN:

- router-centric view
- ATM-centric view
- balanced view
- LAN emulation

Each is discussed in the subsections below.

15.5.1 Router-centric view

In the router-centric view architecture, as shown in Figure 15.5, the backbone network is router-based, and the ATM is an access vehicle. In this architecture, the backbone routing between domains is performed by the router, and the local switching functions are performed by the ATM switch in the network.

15.5.2 ATM-centric view

In an ATM-centric view, as shown in Figure 15.6, the ATM nodes serve as the backbone, and the router becomes the access vehicle to the network.

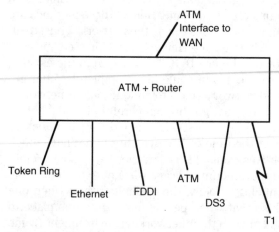

Figure 15.4 ATM as backbone.

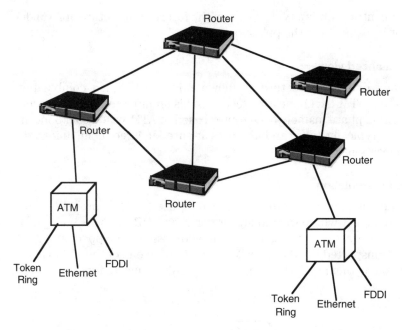

Figure 15.5 Router-centric view.

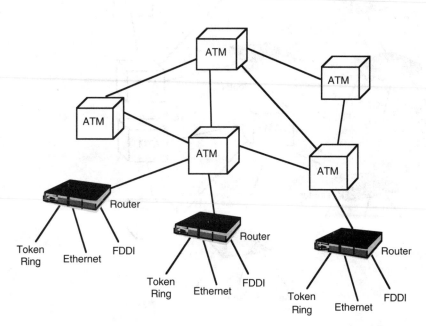

Figure 15.6 ATM-centric view.

The ATM-centric view is a typical view of the future architecture, if ATM deployment is expected in the public network.

15.5.3 Balanced view

The balanced view architecture, as shown in Figure 15.7, is a combination of router and ATM-backbone architecture. This architecture usually occurs in a transition phase, namely from router-based to ATM-based backbone architecture. Typically, these architectures are used in larger companies with multiple locations.

15.5.4 LAN emulation

The LAN emulation, as shown in Figure 15.8, uses ATM to provide virtual LAN interconnect where the intelligent peripheral (IP) address encapsulation occurs. Here, the router-generated IP addresses are mapped onto ATM cells and transported across the ATM network using one of the ATM virtual channel connecting networks, thus enabling LAN connectivity.

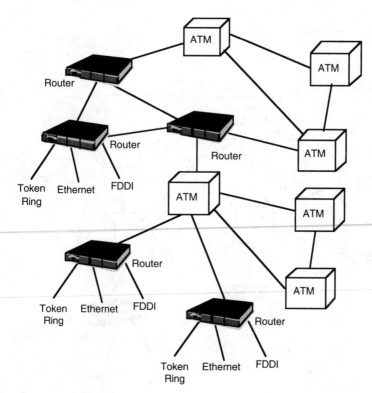

Figure 15.7 Balanced view.

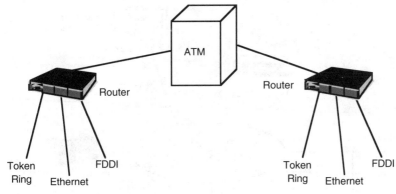

Figure 15.8 LAN emulation.

15.6 Transition to ATM-Based LAN

Today, the whole industry is excited about ATM, especially the LAN industry. Why is there this craze? The answer lies within. During the past five years, the basic technology for bridges and routers has changed drastically. Simple bridges gave way to learning bridges and then to routing bridges. Single-protocol routers have been replaced by multiprotocol routers that offer both bridging and routing. The latest trend includes the migration of these to ATM-based systems.

One of the most difficult decisions for a company or individual to make is whether or not to migrate from an existing technology to a new one, such as ATM. Many questions arise when addressing such a migration. If we assume the new technology is ATM, then the following questions should be answered:

- How will the new ATM feature work in LANs?
- What is the best way to evolve to ATM from today's Ethernet, token ring, and FDDI?
- How will ATM interoperate with existing LANs and WANs?
- What happens to the existing hubs and routers in use?

As far as the first question is concerned, the new feature of switched virtual circuit (SVC) will enhance the existing LAN features. To address the rest of the questions, we propose a three-phase evolution path from the current environment. These phases are based on customer inputs such as the following:

- Existing equipment should not be replaced before it could be written off.
- The new technology should be a more cost-effective alternative.
- The new technology should have an evolution strategy to handle additional developments.

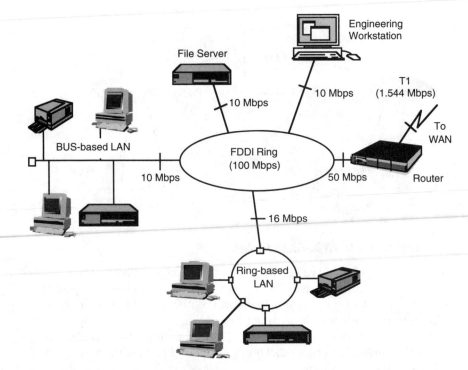

Figure 15.9 Today's LAN environment.

Figure 15.9 shows today's LAN environment. This environment is typical in a corporate network, where FDDI serves as the backbone connection between many departmental LANs based on Ethernet, token ring, etc., with varying speeds. We assume this to be the starting point for our transition. Figure 15.9 also shows the typical interface speed to the different networks, which ranges from 1.544 to 100 Mbps.

Figure 15.10 shows phase 1 of the transition to ATM-based LANs. Here, the FDDI is being replaced by an ATM switch to provide a switched environment compared to FDDI's bus-based shared architecture. The purpose of a switched environment is to provide dedicated connection between LANs on demand. Here, the routers are used to communicate to the outside world. In some instances, FDDI will become one of the access networks to ATM switch, either as a new LAN or as a replacement for one of the existing LANs, such as Ethernet or token ring.

Figure 15.11 shows phase 2 of the transition. Here, an ATM switch is used to connect to the outside world at the interface. The results of ATM deployment in this phase are the following:

- improved performance, scalability, and management of multiple LAN environments
- cell switching used in the backbone network
- ATM backbone becomes a competitive alternative to FDDI in similar networking environment.

This phase is typically applied in a university campus environment, where the LANs are scattered all around the campus. Figure 15.11 also shows the typical interface speed at which different LANs interwork. In this phase, the ATM adapters become increasingly cost-effective, enabling workstations to be connected directly to the ATM switch via a 100-Mbps interface. New applications start to proliferate as ATM migrates to the desktop.

Figure 15.12 shows phase 3 of the ATM deployment in LAN. Here, the token ring or bus LANs disappear, and all the terminals or workstations are directly connected to the ATM switch. The LAN environment is completely cell-based, with switched architecture. In this environment, the interface

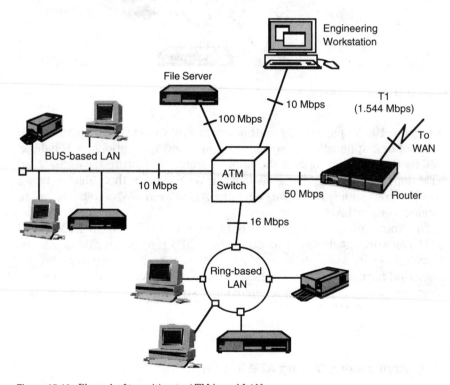

Figure 15.10 Phase 1 of transition to ATM-based LAN.

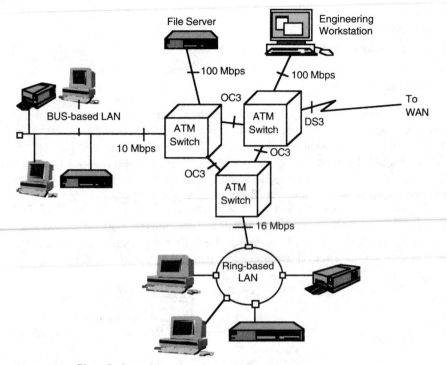

Figure 15.11　Phase 2 of transition.

runs from 100 Mbps and higher, thus opening a window of opportunities in terms of new applications, especially video-based applications, to the desktop. By this time, not only is the LAN environment mature for widespread ATM deployment, but ATM availability in WANs and in the public network also becomes widespread, thus enabling ATM-based WANs to be set up to connect remote LAN sites or servers.

The above phases are not the only ways to evolve to an ATM-based LAN. Networks at different locations have different needs and demands. Thus, users can skip a phase or adopt a different architecture than the one discussed here. The transition here is more of an example than a real deployment scenario. Usually, a deployment strategy is based on many factors, such as cost, technology maturity, applicability, and current LAN environment.

15.7　Applications Driving ATM in LAN

So far, we have seen the requirements for LAN users and the technology to meet those requirements. We have also seen an example of evolving from an existing LAN to an ATM-based LAN and different architecture alternatives to provide a smooth evolution. All these are fine, but what applications will users need to evolve to an ATM-based LAN?

The first application is multimedia. Studies have shown, however, that the main application for an ATM LAN is not multimedia but rather the reinventing of internetworking. Although internetworking is not theoretically an application, it consists of LAN applications such as file transfer, E-mail, remote database access, and other applications that do not belong to this category. Today, internetworking is out of control as we build router-based internets that are too big, too complex, and too diverse to be well-managed. As mentioned in the evolution of existing LAN to ATM-based LAN, the best place to start is with routers, then with a step-by-step repositioning of these routers as traffic fades at the periphery of an ATM network. The reason for starting with routers is that they are the biggest market segment and the most profitable from an equipment vendor's perspective. Other areas do not seem to be a good starting point for ATM-based LANs.

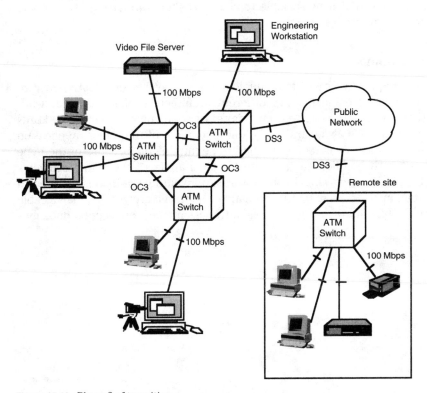

Figure 15.12 Phase 3 of transition.

Compared to router-based internetworking, ATM networking offers many advantages over today's internets:

- Local ATM nodes perform switching better than today's LAN hubs because LAN hubs broadcast information to all LAN ports. In a star-wired configuration, broadcasting is unnecessary and a waste of resources. In ATM, the traffic is routed only to the required nodes, thus saving the resources for other uses.

- ATM switches offer the promise of multiple "virtual LANs" in a single switch, making moves, additions, and changes a simple matter of adjusting workgroup membership routers at a network management console. These virtual LANs can be extended across several switches to provide access to remote servers with almost the same throughput and delay as if they were local.

In short, the need for reinventing internetworking will be the crucial application for ATM LAN. But one must make the change only after careful analysis and planning. To make this change, it is essential that ATM be a standard and that it be scalable to large sizes, operate at high speed, and meet diverse requirements.

15.8 Summary

In this chapter, we saw the technology options available for LAN environments today. We looked into different architecture alternatives and addressed how one can evolve from today's environment to an ATM-based LAN environment. We also clarified the misconception of the "killer" application for ATM, which all along has seemed to be multimedia because of the hype created by the public. But, in reality, the killer application is simply reinventing LAN internetworking. In other words, reinventing the LAN is to redesign the LAN in such a way that it can handle the traffic growth in a manageable manner obtained by using the latest advances in many areas of technology.

16

ATM in Wide Area Networks

16.1 Overview

Before we address the role of ATM in wide area networks (WANs), we need to define WAN and its aliases. WAN is also known as a corporate network, enterprise network, or private network. It is typically an extension of LAN outside the building or campus, over the public or private links to other LANs in remote buildings or cities. A basic difference between WAN and LAN is that WAN uses common carrier lines or private lines and LAN does not. The connection from a LAN to a WAN is typically made through a device called a channel service unit/data service unit (CSU/DSU), which is connected to a router on the LAN side. The CSU/DSU connects the digital line (T1) to a data communication device, such as a router. The basic function of a CSU/DSU is to convert LAN protocol to a public network digital protocol, such as T1 or T3.

As we address the role of ATM in WAN, we look into some of the technology requirements demanded by WAN. Typically, more than one technology can meet the requirements of WANs, such as a router, frame relay, SMDS, and ATM, described in Part 2 of this book. Because ATM has been selected as the future broadband switching technology, we address the transition from today's TDM- (router) based environment to an ATM-based environment. Here, the transition is only an example of how today's WAN can be evolved. In reality, the evolution path depends on various factors, such as optimum cost, (which can vary from business to business), technology availability, etc. It is speculated that early ATM services will appear in the WAN environment for private network applications. In effect, corporations will use ATM to connect scattered LANs together into a large corporate network. Initial applications expected in ATM-based WAN are data, and later, video and voice.

16.2 ATM-Based Hubs

Before we address ATM in a hub environment, we must first understand what is meant by a hub. A *hub* is typically an interface point for different LANs within a building or a campus that enables the LANs to be connected. The hubs typically consist of routers that handle the function of routing between the LANs within a building and to remote LANs via the WAN. Figure 16.1 shows a typical router-based hub backbone environment, where the router enables interaction between the LAN switch in a campus or to the outside world via a WAN. The outside world interaction is done by the CSU/DSU, which performs protocol conversion (i.e., from LAN protocol to TDM-based public network protocol). The equipment is currently designed for data traffic only, however. To handle voice and video, additional equipment, such as a PBX and video codec, are required to be connected to a TDM device. Additional WAN lines are also required because effective sharing is not possible. Thus, with TDM technology, the inherent disadvantage is the waste of bandwidth or the inef-

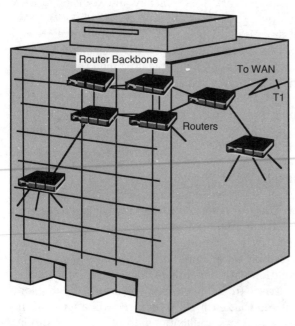

Figure 16.1 Router-based hubs.

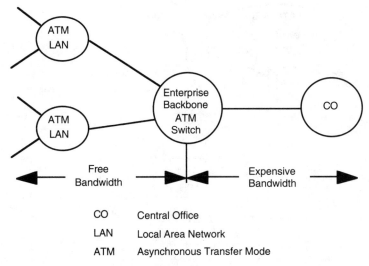

CO   Central Office
LAN   Local Area Network
ATM   Asynchronous Transfer Mode

Figure 16.2 LAN, WAN bandwidth cost.

fective use of bandwidth. In a WAN environment, effective utilization of available bandwidth must be achieved compared to a LAN environment because the bandwidth is free in LANs, whereas the bandwidth is very expensive in WANs. Every effort is thus made to effectively use available bandwidth.

Figure 16.2 illustrates this concept. ATM in the WAN environment is the only technology that fulfills the requirement of effective bandwidth utilization. By replacing the routers in Figure 16.3 by an ATM switch, one can completely migrate to an ATM-based hub environment, thus enabling the network to migrate to a switched environment.

In addition to efficient bandwidth utilization, ATM enables users to integrate voice, data, and video on the same network, thus creating more efficient shared use of the available bandwidth. With ATM in the WAN network, other benefits also arise, such as easy evolution to higher speed bandwidth on demand, conformance with global standards, etc. Figure 16.4 shows an example of the savings in bandwidth when migrating from today's TDM-based WAN environment to an ATM-based WAN environment. The percentage of utilization shown in the figure for the TDM environment is based on today's router-based private WAN network. The reason for such resource wasting is that the network is designed based on busy-hour (BH) utilization, which is usually only 30 percent of a day's traffic. With this type of calculation, a corporation usually ends up having much more bandwidth than typically used.

Thus, ATM in a WAN environment effectively uses resources, which is why experts predict that WAN is the first place where ATM can be tested

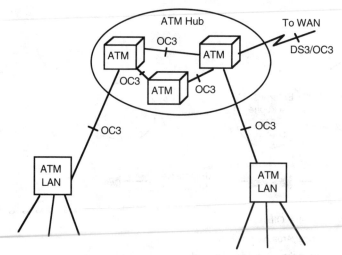

Figure 16.3 ATM-based hubs.

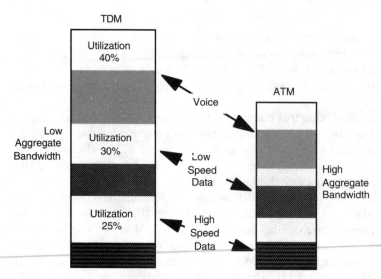

Figure 16.4 Comparison of bandwidth efficiency between TDM and ATM.

and deployed. Numerous vendors are currently manufacturing ATM products applicable for the WAN environment.

16.3   WAN Technology Requirements

WAN is a network that typically covers a vast geographical area. It thus creates certain technical requirements, some of which are the following:

- standards compliance
- switching/routing capability
- performance
  ~throughput
  ~relay through node and network
  ~nodal reliability/availability
- protocol handling support
- relative cost of technology
- network management support

Each is discussed in the following subsections .

16.3.1 Standards compliance

Standards compliance allows the WAN to carry multiprotocol traffic easily and to possibly construct a network from multivendor equipment. It also allows vendor equipment with different internal designs to interact as if they were of the same protocol and from the same vendor. Compliance helps service providers and large corporations get competitively priced equipment from different vendors. The reason service providers and large corporations prefer compliance is that they do not need to depend on one vendor who then controls their network by controlling the cost of equipment. Thus, providers and corporations actively participate in the standards-making process both at national and international levels. Standards compliance is one of the reasons ATM has become a reality around the world.

16.3.2 Switching/routing capability

Universally, all traffic must be routed or switched to arrive at its destination. Without switching, connectivity would be impossible. Switching and routing thus allows a very robust and fault-tolerant network. This capability enables the network to effectively utilize its resources with many users. Switching, along with transmission, is a very important component of the network. In a WAN environment, however, the transmission systems are usually leased from public service providers, depending on the routes required for connectivity. Switching, on the other hand, can be publicly or privately owned, each with its own pros and cons. If the public carrier controls the switching, then the managing of the network is also done by the public carrier because the details of usage, such as billing, are available in the switch itself. In this case, the traffic characteristics are known to the public carrier. If the corporation controls the switching, the traffic information remains within the corporation. The problem here is that the corporations need additional personnel to run and manage the network, which becomes difficult if the corporation's primary area of business is not related to telecommunications or data communications. Additional personnel is also difficult for small companies with many remote offices.

16.3.3 Performance

Performance is another key driver in any network architecture. It is the only indication that a network is healthy or that it is running out of capacity. The network must inform the network operator when it is time to upgrade the network capacity to handle additional users or traffic. This type of information is provided using certain parameters that keep the network operator informed about network status. The most commonly used parameters are

- throughput
- delay through the node and link
- nodal reliability/availability

Each is described in the following subsections.

16.3.3.1 Throughput. Throughput is the key criterion that can affect the performance of any network. The network's aggregate throughput is dependent on the individual nodal throughput; hence, it is important to compare and contrast it. Throughput is one of the performance parameters that end users experience in their network and often complain about. Throughput is the total amount of traffic that can be carried in a unit of time; it is usually measured per second during the busy hour. The higher the throughput, the higher the performance. In a typical network, the throughput is roughly equal to 70 percent of the node or link capacity. In fact, nodes and links are designed to do exactly that as a precautionary measure to prevent the node or link from going down in case of overload. The throughput determines the amount of traffic that can pass through a node or a link in any given unit of time. In a broadband network, the switch node throughput runs about 20 to 100 Gbps.

16.3.3.2 Nodal and link delay. Delay through the network is probably the single most important parameter for end users. Once a WAN is put in place and users try to access a service at a remote LAN, they don't understand why it takes so long to access the information. What they forget or don't care about is that the server is at a remote location and the interconnection is at a low speed.

One alternative that can lower user complaints is to use a larger bandwidth to interconnect nodes, but this approach is obviously very expensive. A second alternative is that the technology support a larger bandwidth. The third alternative is to select a nodal technology that matches the application environment with minimal delays through the nodal equipment.

Nodal equipment delays are a major contributing factor to nodal delays in the WAN. Nodal delay occurs because the user puts in more traffic than the node can handle, causing the node to queue the traffic. If the queue limit is exceeded, traffic is lost (cells are discarded). The delay seen by the user is really the queuing delay at each node. The other delay is the link delay.

One delay no one can prevent is the transmission propagation delay caused by natural phenomena. It is the time taken for an electrical signal to propagate from point A to point B—usually about 10 μs to 20 μs. In a broadband network, even a microsecond is a large delay. With fiber-optics and information traveling at light speed, this delay is reduced, but in a WAN network design, this delay is critical. When delay adds up, it becomes significant.

16.3.3.3 Node and link reliability/availability.

The nodes and links are the fundamental building blocks of any network. Their unavailability negatively impacts productivity. When designing a WAN, one must ensure that the individual transmission and nodal components provide a fairly high degree of availability (usually 99.9 percent). In a WAN environment, redundancies have usually been built into the switch by the vendors because of the market needs. The node, otherwise known as a switch, must be up and running all the time. Equipment vendors, especially for broadband equipment, provide one-to-one, hard-swap backup, including the power source to the node. In addition to availability, reliability is critical. Here, every interface to the switch has a backup interface to meet the strict standards set forth for switching nodes. Link availability and reliability is critical, because the links are usually leased from the public carriers. The carrier usually guarantees the link availability as part of the lease agreement. In fact, today's public network provider guarantees the availability of transmission at all times.

16.3.4 Protocol handling support

Most WAN applications require protocol support of varying degrees. Eventually, the network and the transport-level protocols must be supported if the application is to successfully run on the network. Because a proliferation of protocols exist in any network, one must be careful to accommodate a wide variety. Typically, a careful balance between the native protocol support and the use of a gateway for other protocols is required. In a LAN environment, there are many nonstandard protocols. In a public environment, however, most of the protocols are as per the specifications defined by the standards body at the national or international level. This connection of LAN via public leased lines requires additional equipment to accomplish protocol conversion. The problem with protocol conversion is the additional overhead involved, which in turn reduces the effective throughput or utilization of the expensive WAN bandwidth.

16.3.5 Relative cost of the technology

When implementing any new technology, the "mother of all evil" is the relative cost of the technology. Cost is always a key consideration for any network design. Cost has many components—typically a fixed and variable cost. The biggest cost factor is usually variable cost, which is where most of the users look for savings from a new technology. In today's wide area network, the transmission links are leased. The lease agreement is usually on a yearly basis. Thus, the user agrees to pay for the leased line regardless of the use of those facilities. With new technology, the links can be used on demand with varying bandwidth, and the user pays for the usage only, enabling the users to use the expensive transmission link only when needed. Thus, new technologies such as ATM look cost-effective in spite of the initial fixed cost of the technology.

16.3.6 Network management support

As mentioned earlier, a WAN is a network covering a vast geographical area, which is where the network management feature offered by the equipment vendors plays a key role. Network management is the ease with which the network can be operated on a daily basis by a smaller staff, resulting in overall operational savings. With the advent of network management standards, such as simple network management protocol (SNMP), service providers can manage different network equipment from different vendors using common network management systems. These network management systems enable the network operator to manage the network from a remote location. The network management software also provides a large amount of statistical information about the network, including performance, usage, traffic patterns, and growth in traffic. In some cases, most of the network problems, such as isolating the faulty device, can be accomplished from the network control center.

16.4 Transition to ATM-based WAN

In this section, we address what the best step-by-step migration path is from today's router/TDM-based WAN to an ATM-based WAN. The reasons for migrating to ATM in a WAN environment are the following:

- need for more aggregate bandwidth
- need for on-demand bandwidth
- need for efficient use of available resources, especially the expensive ones
- single network management system

We assume the target WAN is an ATM-based solution from today's router/TDM-based solution. It is not necessary that an ATM-based solution be the target WAN solution, however; it depends on the current and future user requirements. In fact, the end users can migrate from any existing technology to any target technology as long as the technology meets their objectives. For example, a user who has currently deployed frame relay technology can also evolve to an ATM-based network if a need exists for higher bandwidth applications. In this case, the evolution path might or might not be the same as the one proposed here. The transition to ATM is thus an example of a typical WAN environment transition to an ATM-based WAN environment.

The evolution is based on the assumption that frame relay and SMDS are still in the early stages in terms of widespread deployment as the backbone in corporate networks. Also, that frame relay and SMDS cannot handle isochronous (voice) traffic.

We propose three phases of transition from a router/TDM environment to an ATM-based WAN environment. Figure 16.5 shows a typical router/TDM-based WAN environment. In this example, assume that there are three locations in two different cities connected via routers at each location. The different locations are connected via leased T1 private lines whose bandwidths are 1.544 Mbps. In one of the links, there is more than one T1 leased line.

The following is the topology of the links:

- Between City A, location 2 and City B campus are two T1 leased lines
- Between City A, location 2 and City A, location 1 is one T1 leased line
- Between City A, location 1 and City B campus is one T1 leased line

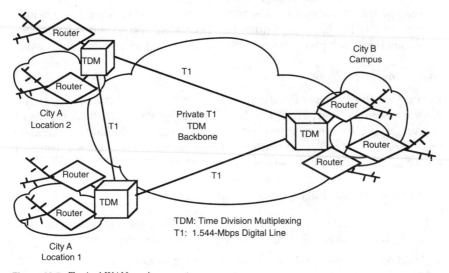

Figure 16.5 Typical WAN environment.

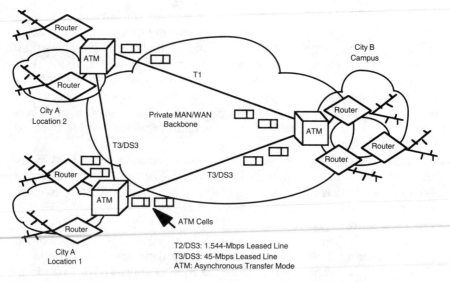

Figure 16.6 Phase 1 of transition.

The link capacities are decided based on the usage between the two points during a busy period. Thus, during a non-busy period, the capacities are not utilized or are underutilized. This network is only for data traffic; a separate network exists for voice. If the need for data traffic is less than T1, some voice can be multiplexed at the TDM box. With ISDN technology, some sharing occurs between data and voice traffic, but not to the extent where one gets real cost savings due to integration.

Some of the limitations of this network are limited backbone bandwidth (limited to T3), low bandwidth utilization (average), and no network management facilities. Each phase of the evolution should address these limitations.

16.4.1 Phase 1

Figure 16.6 shows phase 1 of the proposed evolution. The connection to the WAN is via an ATM switch situated at different locations; dedicated lines of various capacities are required between the locations. Initially, these lines are expected to be T1. All traffic within a premise terminates through a router at the ATM switch, which switches the traffic between the routers for both intra- and interpremises traffic.

The following are the benefits of migrating to phase 1:

■ *Premises traffic consolidation.* All the traffic (data) is consolidated at the ATM hub, which does intra- and interpremises switching.

■ *Backbone upgrade (depending on traffic demand) from T1 to T3.* Once the ATM hub is developed, the backbone network can be easily upgraded from T1 to T3 when the demand arises because the ATM hub already has the capability.

- *WAN traffic consolidation (voice, data, and video) via ATM switch.* With the help of the ATM hub, integrating different traffic becomes easy.

16.4.2   Phase 2

Figure 16.7 shows phase 2 of the WAN network evolution. In this phase, the backbone network speed is upgraded from T3 to OC3. Not only is the capacity of the link increased, but it is also moved from the electrical to the optical domain. At the same time, public ATM service becomes available from service providers. ATM also impacts other traffic, such as voice, where the PBX will be connected to ATM switches, which is where the initial cost savings begin to appear from traffic integration.

As mentioned earlier, backbone link sizes are designed based on the busy period usage. In this phase, capacities not used during the non-busy period are used by others because with ATM, the user sets up a connection when the traffic needs to be transmitted. The customer is billed only for the usage period. This benefit is in addition to the benefits obtained by consoli-

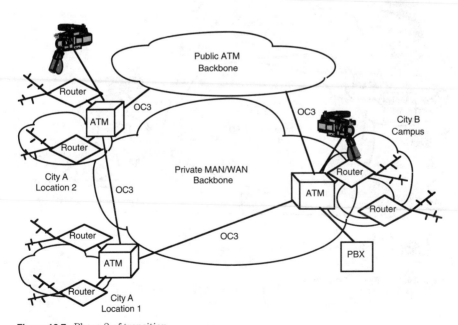

Figure 16.7 Phase 2 of transition.

dating the voice, data, and video traffic. The following are the overall benefits of migrating to this phase:

- total ATM solution in the WAN environment
- appearance of benefits from the consolidation of data, voice, and video
- easy link-speed upgrade and migration to a more reliable optical domain

16.4.3 Phase 3

Figure 16.8 shows phase 3 of the WAN evolution to a complete broad-
band/ATM-based solution. In this phase, the public network has evolved
considerably to provide true BISDN services. The BISDN network becomes available, i.e., SONET, ATM, IN, and all the AAL layers (for adaptation of existing services) are deployed in the network. The backbone
network is fully optical and has 100 percent redundancy. The real benefits of traffic consolidation are achieved. Users in the environment use
the expensive WAN facilities in the most cost-effective manner. New services can be added without restriction from the technology or available
facilities. Video becomes a way to communicate, such as desktop video,
video mail box, etc.

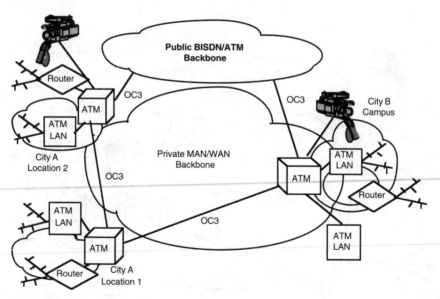

Figure 16.8 Phase 3 of transition.

16.5 Summary

No other network uses ATM more effectively than WAN/enterprise/corporate networks. One of the main driving forces is the need for more bandwidth and better utilization of the same link to boost performance of the network, while protecting the current investment. The driving force behind ATM in WAN is the need to access scarce resources at remote locations that are beyond one's reach. In this chapter, we addressed the transition from existing router-based WAN technology to ATM-based technology, taking into account the issues mentioned above.

17

ATM in Public Networks

17.1 Overview

The most visible area for ATM technology is the public networks. Most of the major networks around the world are public (available to anyone who needs service) and their objective is to provide unbiased communications service to anyone. In the United States, these networks are categorized by regulatory bodies into local exchange carriers (LECs) and interexchange carriers (IECs). LECs provide the local access services, whereas IECs provide long-distance services (interstate/inter-LATA). LATA is an acronym for local access transport area. In most other countries, telecommunication networks are controlled by the Government under the department of Post, Telegraph, and Telephones (PTTs). It is speculated that these countries might follow the footsteps of U.S. public network environments in terms of opening up the networks for competition and private ownership, which leads to the deployment of new technologies to meet market needs.

In this chapter, we address the role of ATM in public local access and backbone environments. In the United States, the public telephone network is segmented into local exchange networks, otherwise known as LECs, owned and operated by the regional Bell operating companies (RBOCS) and independents. These carriers are monopolistic in their serving area and hence regulated by the Government. For example, in the United States, the LECs are regulated by the governmental body called the Federal Communications Commission (FCC). The IEC is simply the long-distance carrier network, which is where many carriers compete with each other for market share.

Each of these carriers have their own traffic share, network, switching and transmission requirements. Because these carriers carry all types of traffic (voice, video, and data), they want to evolve their networks so that one network can handle all types of traffic. To carry all traffic on a single network, the carriers need switching and transmission systems that can handle such demands in terms of different bit rates and characteristics, which is where ATM and SONET come into play. Here ATM provides switching, and SONET provides transmission systems. Thus in this chapter we look into the role that ATM plays in these network environments. We then propose a strategy for the public backbone network's transition from today's network to an ATM-based one.

17.2 Public Network Requirements

For ATM to be deployed in public networks for voice, video, or data, it must meet all requirements put forward by the regulatory bodies, such as the FCC in the United States. In a competitive environment like the United States, the carriers usually not only meet the basic requirements but go farther with their own requirements, which are more stringent than those imposed by the regulatory bodies to differentiate themselves from other service providers in terms of features offered. This act in turn puts the burden on the switch manufacturers to design network equipment (switching and transmission) that can meet these higher requirements.

In addition to the WAN requirements mentioned in chapter 16, (standards compliance, switching/routing capability, performance—throughput, delay through the node, nodal reliability/availability—protocol handling support, relative cost of the technology, and network management support) the public network has the following additional requirements:

- ubiquitous access
- bit error rate (BER) of 10^{-9} or better
- blocking probability of 0.01 or better
- survivability
- redundancy and protection
- downtime as specified by regulatory bodies
- economic efficiency compared to existing technologies
- flexibility for growth, higher speed, switch size, etc.

In this section we describe these requirements. Not all of these requirements can be achieved from day one, but an evolution plan is necessary to meet those requested by the carriers and customers.

17.2.1 Ubiquitous access

The public carrier must provide nationwide coverage for the backbone network. In the United States, a long-distance carrier must have a point of presence (POP) in every local access transport area (LATA). In the case of a local access provider (regulated service provider), every carrier must provide a uniform fixed basic cost of service to anyone who wants telephone service. Additional costs are subsidized through a complex process.

17.2.2 Bit error rate

With today's fiber-optic technology, carriers not only carry huge amounts of data at the speed of light, but the data is error free. In fact, at one time the carrier who deployed the first fiber-optic network advertised that it could guarantee data integrity 99.99 percent of the time. Accurate data is all related to what is called *bit error rate* (BER).

$$\text{BER} = \left(\frac{\text{Number of bits received in error}}{\text{Total number of bits received}} \right) \times 100$$

BER is defined as the percentage of received bits in error compared to the total number of bits received. It is usually expressed as a number to the power of 10. Currently, the BER requirement for a public network is 10^{-9} or better, which means that there should be only one bit error for every one billion bits or less.

17.2.3 Blocking probability

In a telephone environment, if a telephone call cannot be completed, it is said to have been blocked. *Blocking* is a fancy way to say that the caller received a busy signal. A call can be blocked at many different locations in the network. It can be blocked because of the unavailability of the network ports or lines or because of the receiver is using the line. As far as the service provider is concerned, blocking probability refers to the former. A *blocking probability* is a unit that measures the number of calls blocked or measures the network's ability to handle the offered traffic. It is defined as the probability that a call can be blocked in a given period of time, usually during the busy hour. In the public network, blocking probability is 0.01 (1 percent) or less. Thus, less than one percent of the calls made during the busy hour are blocked. For example, if there are 100 calls in a busy hour and the blocking probability is one percent, one call is blocked and 99 calls get through the network. A 1 percent blocking probability is one of the main requirements for a voice-based network. Therefore, regardless of the type of network, this requirement must be met for voice applications. In the case of data, no real-time transfer is required; thus, data needs a completely different set of requirements. Blocking probability is used to measure the grade of service (GOS) provided by the carrier. GOS varies for the different types of networks, data, voice, or video.

17.2.4 Survivability

Survivability simply means that every node and link has a standby backup. If a node fails, the traffic is rerouted via other nodes. If a link fails, all the traffic on that link is rerouted. It is very complex and expensive to design and deploy such a system because the carrier must decide how much (capacity) backup is required.

One can achieve survivability in many ways, such as using rings in the transmission network. In broadband communications using SONET as the transmission standard, SONET rings provide automatic restoration if a cable is cut, thus providing survivability. For switching nodes, redundancy is provided by having a completely redundant interface, switching matrix, and power source for the switches.

17.3 ATM in Local Access Networks

In chapter 12, we mentioned the BISDN reference architecture along with local access networks. We also proposed several local access network alternatives. The voice-based telephone architecture is usually star or double-

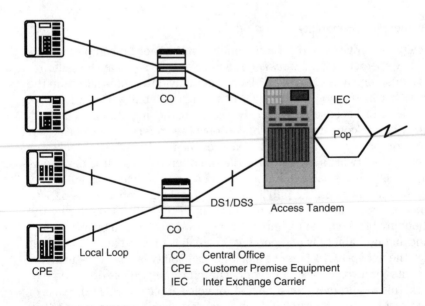

Figure 17.1 Telecommunications network access architecture.

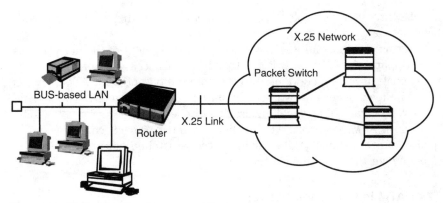

Figure 17.2 Computer network access architecture.

star, and the data-based packet network is also usually star. Figures 17.1 and 17.2 show typical local access architectures for telecommunications networks and computer networks, respectively.

In the figures, the ATM switch is a potential candidate to replace the central office, access tandem switches in the telecommunications network and a packet switch in the public computer network. To deploy the ATM switch and take advantage of services ATM is capable of providing, however, the existing local loop infrastructure must be upgraded from twisted-pair to either coax or fiber. In the trunk network, however, ATM can be deployed immediately because fiber/coax is already there. Standards have also been defined so that coax at DS3 can be used as an ATM interface.

The use of ATM locally varies, depending on the type of customer it serves. In a telecommunications environment, customers are typically categorized as residential or business. The respective networks are referred to as residential and business networks. Each of these addresses a different set of applications. Although ATM can be deployed in any of these logical segments of the public local access network, it is generally deployed in the business network to provide service to those customers.

Deploying ATM as an access switch for residential customers is not currently justifiable because of the voice traffic demand. Future services, especially video-on-demand (VOD), will probably justify deploying ATM in the residential network. To provide such a service, the existing telephone network must first be upgraded. A CATV network can deploy ATM switches and provide VOD services and is addressed in chapter 18, but a public telephone network must upgrade the local loop plant to provide the enhanced services. With new compression technologies, such as asymmetrical digital subscriber loop (ADSL), using MPEG2 coding, VOD can be provided on a limited basis, depending on the loop length, which is the portions between residence and the CO and identified to be about 18 ft. Several studies have been conducted on architecture alternatives for deploying fiber-optics in the residential broadband network.

In terms of opportunities, however, no other segment of the network has more opportunity for ATM switches than the local access network. Currently, many trials are being conducted to evaluate the potential for ATM services in public local access networks. The initial services targeted are for data; voice will be carried as usual on circuit-switched networks. For voice to be carried on ATM networks, ATM technology must prove it can provide the quality of service available today in a circuit-switched network, which will happen over time as the ATM service matures.

17.4 ATM in Backbone Networks

Backbone networks are also called long-haul or long-distance networks, either at the national or international level connecting subnetworks. They consist mostly of long-haul transmission systems and large switching systems carrying large amounts of traffic. The U.S. IECs came into existence in 1984 after the breakup of AT&T. These telecommunications networks provide voice services across interstate, inter-LATA boundaries (LATA is a logical division of regions). In the 1960s, the first data network, called ARPANET, was introduced, and today in the United States it has evolved to what is called NSFNET (National Science Foundation Network) or, more popularly, the Internet. Initially, the Internet was a research network connecting universities across the United States and in other countries. Recently, however, this network has opened up for commercial purposes so that anyone can subscribe or access it.

As global businesses expand, the need for reliable high-speed communication for voice, data, and video increases. An upgrade of the existing network then becomes necessary. ATM meets this need now and in the future. For example, as of 1993, three carriers in the United States had announced their first ATM service, and others are following. Once ATM service proves itself, more carriers will slowly migrate their voice traffic to the ATM backbone.

ATM's initial application is to provide large businesses with wide area LAN connectivity. Here, the IECs can bypass the local carrier and connect directly to the backbone network. Many ATM trials occurred in 1993, and numerous tests are still ongoing to quantify various applications by almost every public carrier around the world.

Figure 17.3 shows a typical backbone network. The network consists of two layers; one is the backbone switching system and other is the concentrator or multiplexer. The latter layer is where the traffic from the access networks is collected and offered to the backbone. Initial deployment of ATM is as an overlay onto the switching layer.

This backbone can be compared to any backbone network in the world usually controlled by the Government. In the United States, the voice net-

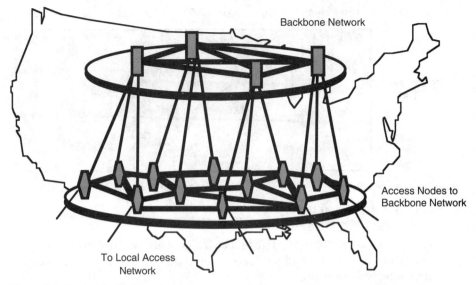

Figure 17.3 Typical backbone network.

work consists of two layers—the access layer, called POP, which is present in every LATA, and the backbone switching layer. The equipment present at each POP is simply a concentrator or multiplexer. The traffic collected from the local access network is offered to the backbone switching layer, which performs the switching.

The public data network also has a similar hierarchy where the POP is replaced by a regional network connecting the universities. The traffic from the universities is then offered to the regional networks, which in turn offer the traffic to the backbone NSFNET, to be switched to other regional networks or to the international network. This network is pure data based on packet-switching technology. It is expected that ATM will be first deployed on a wide scale in this network.

17.5 Public Network Trends

The public network is becoming the mother of all networks—one of the most important networks around the world. With privatization occurring around the world, it has become the battleground to deploy the latest technology in transmission and switching. In this section, we see the trends in the public network in the areas of traffic, technology, BISDN, and the international public network.

As mentioned before, current public networks are dominated primarily by voice traffic. Although data traffic is carried on a separate network, its traffic is negligible when compared to voice. Figure 17.4 shows traffic

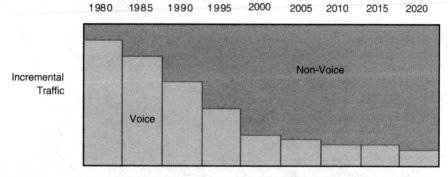

Source: Stephen Flemming, Northern Telecom

Figure 17.4 Public network traffic trends.

trends. One can see from the figure that traffic is shifting from voice to non-voice (video and data) traffic. It is speculated that by the end of the decade, data traffic will dominate.

No other industry has seen more changes in technology than telecommunications. Figure 17.5 shows the trend in technology over time. The initial technology that dominated the public network was analog, followed by digital. The next generation is ATM. Figure 17.5 shows how the new technology slowly replaces the older technology.

Before BISDN can be deployed in the public network, the backbone technologies must be deployed. The two most important technologies for BISDN are SONET for transmission and ATM for switching and access functions, illustrated in Figure 17.6. The first technology to be deployed in the network is SONET. The next step is to provide ATM backbone switching, and the last is to deploy ATM at the access switches.

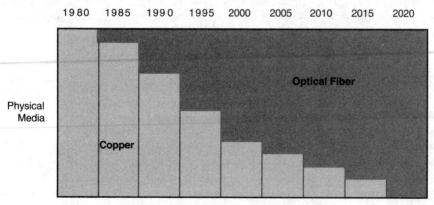

Source: Stephen Flemming, Northern Telecom

Figure 17.5 Public network technology trends.

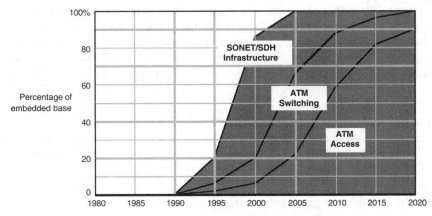

Source: Stephen Flemming, Northern Telecom

Figure 17.6 BISDN capability deployment in embedded base.

Once SONET and ATM are deployed in the public network, the next stage is to deploy BISDN, currently perceived as the ultimate in telecommunications/data communications technology. Figure 17.7 shows a pyramid of technologies to be deployed before BISDN is realized. The bottom of the pyramid is the common transport infrastructure. For broadband communications, SONET/SDH fiber-based transmission technology has been selected and is currently being deployed around the world. The next layer of the pyramid is the broadband switching infrastructure with ATM, which is

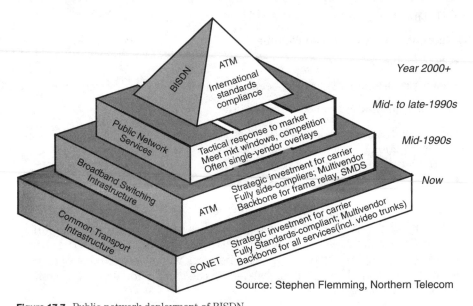

Source: Stephen Flemming, Northern Telecom

Figure 17.7 Public network deployment of BISDN.

in the early stages of trial and deployment. (It is expected to be deployed in mid to late 1990s.) The next level to provide public network services using the available network equipment. The topmost layer, which is the target, is the deployment of the intelligent network (IN). IN is the third component of the BISDN to be deployed with international standard compliance. It is expected to happen by the beginning of the next century.

With a true broadband network deployed across all segments of the public network, the public network view will be completely changed. Figure 17.8 shows one such view. Here, the user plane consists of most of the existing customer premises equipment (CPE) used in residential and business environments. The access part brings the user traffic to the public network through any means available today. Then, depending on the user community of interest (COI), the traffic is divided into different segments. All traffic will be connected via the communication plane, however, where the broadband technologies and existing technology will be present.

17.6   Transition to ATM-Based Public Networks

Since public networks are categorized into local and long-distance, as far as the initial deployment of ATM is concerned, we will focus mainly on backbone networks. Broadband/ATM in the backbone network offers numerous immediate advantages:

- broadband bandwidth

- scalable bandwidth

- improved network response time

- high quality of service

- easy provisioning of services because of unified network management

- efficient utilization of network resources

- increased economics by
  ~aggregation of voice, data, and video
  ~reduction in operating costs (OAM)
- maximum flexibility

Since we mentioned that the initial services provided by an ATM-based network will be data, let's assume the path to true ATM/broadband will evolve from today's packet-switched network, similar to NSFNET in the United States. In other words, the ATM network will initially be deployed in a data network, and voice will be carried on the conventional telecommunications network.

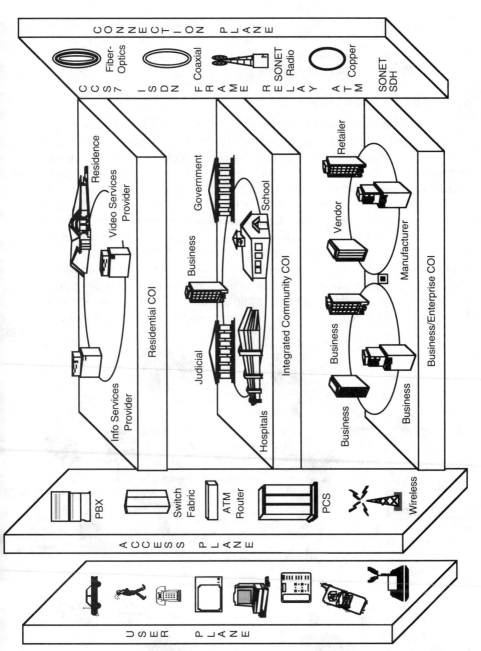

Figure 17.8 A public network view.

The transition to a broadband/ATM network is assumed to take place in four phases from today's X.25-based, packet-switched network to the SONET/ATM-based BISDN. Before describing the four phases, let's look at today's environment of public networks.

17.6.1 Today's environment

Figure 17.9 shows a typical X.25 packet-switched network with an interface dialup of 56Kbps from the LAN. The X.25 network is designed to carry data with 100 percent data integrity. To meet this objective, the X.25 network performs error correction and detection for every packet at every node. Numerous disadvantages exist with this system, such as the very slow transmission speed and unnecessary processing. For example, the X.25 network processing of a packet is performed by as many as three layers of the OSI protocol: the physical, data link, and network layers. The reason for such processing overhead is that at the time the X.25 network was designed, the transmission system was not reliable, and the objective was to deliver the information error-free regardless of the time to deliver. This need has changed; today, the information not only must be delivered error-free, but it also must be delivered fast (depending on the user's request).

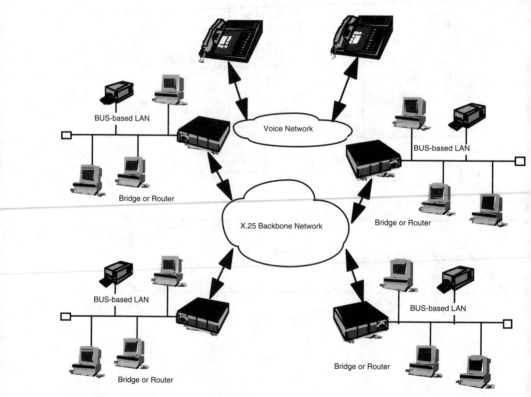

Figure 17.9 Today's packet-switched network.

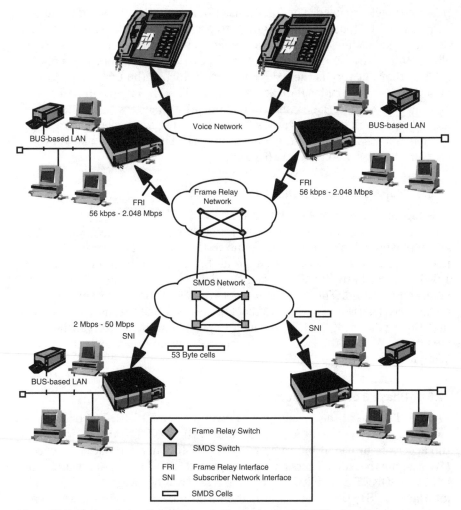

Figure 17.10 Fast packet-switched network (frame relay and SMDS), phase 1.

17.6.2 Phase 1 of the evolution

Figure 17.10 shows phase 1 in the evolution of the packet-switched network. Here, the X.25 switches are replaced with frame relay switches, which are extensions of X.25 switches. This replacement meets the need for data services of less than T1 speed. For high-speed data services, a switched multimegabit data service (SMDS) network is deployed to provide connectionless services for speeds greater than T1.

The frame relay operates in one of the modes of X.25 called *virtual circuit mode*, where setting up the connection is required before data can be transmitted. In the frame relay network, access speeds up to 1.544 Mbps (T1) and 2.048 Mbps (E1) are possible. Here only two layers of the OSI protocol stack are used. The functions of the network layer are done at the CPE itself, which has the intelligence to process the packets. In the initial design of X.25, the CPE was not assumed to be as intelligent as it is today. The details of frame relay protocol were covered in chapters 4 and 5.

The other network is the SMDS network, which is the first public broadband service. It provides data services up to T3, or 45 Mbps. In SMDS service, all network interfaces, such as customer interface, switching interface, and internetwork interfaces are defined. Here, the frame relay network can form an access network to the SMDS network. The details of SMDS protocol were discussed in chapters 6 and 7.

17.6.3 Phase 2 of the evolution

The second phase of the evolution is shown in Figure 17.11. In this phase, the SMDS switches no longer form the backbone. Instead, ATM switches are deployed, which act as a backbone to the switched network. The ATM network provides a similar virtual circuit but with bandwidth ranging from 1.544 Mbps (T1) to 622 Mbps (OC12), enabling a wide range of services to be accommodated. The frame relay networks and the SMDS networks become access networks to the ATM backbone at various speeds. Here, for the first time, voice traffic at T1 or higher speeds can be tested on the ATM backbone. Most of the traffic, however, will still be on an existing circuit-switched network.

17.6.4 Phase 3 of the evolution

The third phase of the evolution is shown in Figure 17.12. The broadband switching is introduced (ATM switching with SONET transmission), and the access is provided by frame relay and SMDS network-based services. (We assume that not all carriers have migrated to SMDS.) Some switches do not have SONET capability but do have ATM switches with standard digital interfaces (DS1, DS3). The information in the backbone, however, is carried in ATM cells, which are switched by the ATM/broadband switch. These cells ride on the SONET transmission system. Most carriers will in fact evolve to this phase in the very near future (depending on the availability of technology) and will remain at this stage for a long time until true BISDN is justified across the network.

17.6.5 Phase 4 of the evolution

In phase 4, the complete or true BISDN network comes into existence, illustrated in Figure 17.13. All the switches in the network adopt the BISDN protocol, and the interfaces are in ATM/SONET. No protocol conversion or adaptation is necessary because the whole network is based on

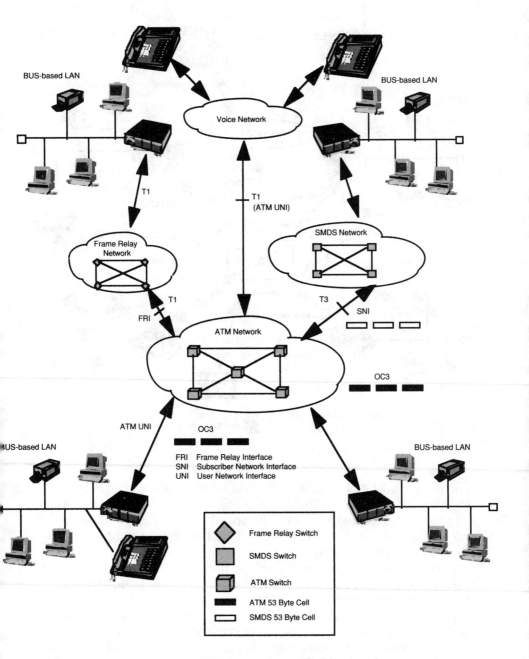

Figure 17.11 Fast packet-switched network (ATM, frame relay, and SMDS), phase 2.

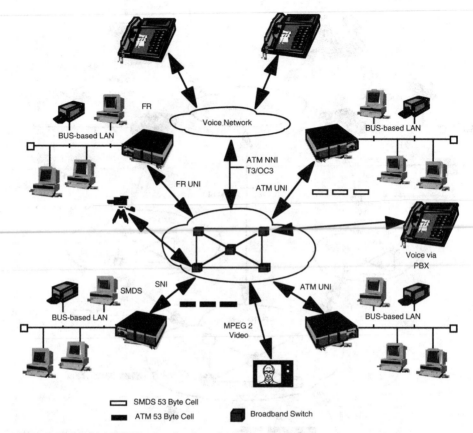

Figure 17.12 Initial BISDN network, phase 3.

one protocol, BISDN ATM/SONET. This network takes advantage of all
the benefits defined for BISDN, and it has the capability of proving that
all present and future services can be provided without any protocol conversion. At this time, IN will be mature and deployed to utilize the underlying network infrastructure.

In fact, many long-distance carriers are already on the verge of migrating
from phase 1 to phase 2. The most realistic architecture is phase 3; hence,
the final stage of most networks will be phase 3. To achieve true integrated
voice, video, and data, we propose phase 4, the true BISDN network, where
voice, data, and video become one with all the intelligent network features.
Although Phase 4 is far from reality in most cases, one should not ignore
these options as a desirable goal.

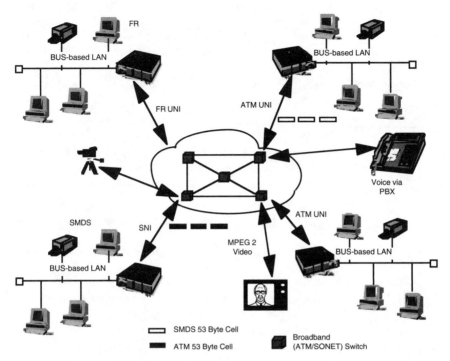

Figure 17.13 BISDN network, phase 4.

17.7 Summary

In this chapter, we addressed the role of ATM in public networks. One of the early deployments of ATM will be in the backbone network, with some local-access exceptions in the case of large businesses. ATM/broadband will be used initially as an overlay network to the existing voice network. ATM's initial traffic is expected to be data. If ATM proves itself for voice traffic, a slow migration of voice will occur. In the transition section, we mentioned the different phases of transition. It is not necessary that all public networks go through this transition; some networks could bypass some of the phases. The transition shown here is only an example of the evolution of the existing network to an ATM/SONET-based BISDN one.

18

ATM in CATV Networks

18.1 Overview

In this chapter, we address the applicability of ATM in networks other than the traditional networks discussed so far, the public telephone network (voice) and computer network (data). The reason for mentioning only these two so far is that they are the most popular networks around the world. In this chapter, we address a network designed for one-way video distribution—the community antenna television (CATV) network. Because CATV is introduced in this book for the first time, we look into it in more detail than other networks.

There are other networks in the United States, such as alternate access providers, who provide local service within a metropolitan area and compete against local exchange carriers. Another area where ATM is being considered is in the wireless data network, where data is in the ATM cell format to be carried on a consistent protocol throughout the network when ATM is deployed across the network. We do not discuss alternate access or wireless data because it is beyond the scope of this book.

CATV provides broadcast video over coaxial cable. In the United States, about 60 percent of homes have cable service, with 98 percent having access to cable service. CATV has become valuable because of its ability to provide video entertainment services to residential subscribers. From the network infrastructure perspective, CATV has become valuable because of its coax in the last mile of the network to every home. Thus, it has the capability to provide broadband video services. That last mile of coax accounts for 50 percent of the total network investment. CATV is not as widespread around the world as the other networks we have mentioned, except in western countries.

18.2   CATV Architecture

Different CATV architectures are used around the world today. Some have evolved from the others; there are three basic types:

- tree-and-branch
- switched-star
- tree-and-bush

Each is described in the following subsections.

18.2.1   Tree-and-branch architecture

This physical architecture is capable of delivering 64 channels of distributed video service, as shown in Figure 18.1. In this network, the signals from the different sources to be delivered to the subscriber are gathered at a central headend. The headend is equivalent to a central office switch in the telephone network. Typical sources of these signals are satellite earth stations, off-air antennas, video tape playback, and super trunking, which provide delivery of signals from studios at remote locations.

At the headend, the various video sources at different radio frequencies are combined into a single broadband signal and transmitted over a single coaxial cable. This coaxial cable undergoes branching through power split-

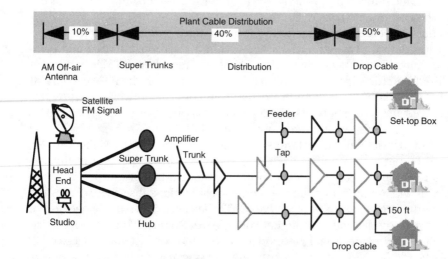

Figure 18.1   Tree-and-branch architecture.

ters until it passes down each street in a community. Broadband amplifiers are required every 1000 to 2000 feet to overcome losses from cable branching and transmission. The biggest concern of the CATV system is its limited capacity, and that its service quality deteriorates due to noise, modulation distortion, and reliability. The current generation of analog coaxial CATV equipment consists of simple unidirectional transmission devices with limited telemetering capability for monitoring.

The electronic components in the network include the trunk and feeder amplifiers, which maintain network signals at adequate power levels to ensure signal quality, and the converter box on the subscriber premises. The passive components include the coaxial cable connectors, signal splitters, and signal taps for the drop. The network requires a power station about every two miles of cable to multiplex electrical power onto the coaxial cable. These stations drive the electronics in the trunk and feeder portions of the network, and the customer power drives the converter box. The network does not provide backup power. Hence, if commercial power fails, the cable system is inoperable, and cable services are lost.

Because every subscriber receives all channels carried by the cable system, controlling access to services such as pay-per-view becomes a key technical issue. One approach employed by roughly half the industry is to install inexpensive coaxial filters that trap signals at the drop, blocking the portion of the bandwidth occupied by the premium channels. When cable service was first offered, whenever a subscriber requested pay-per-view cable service, personnel were dispatched to the site to remove the appropriate filter. The high labor cost for service changes and the inability to provide pay-per-view services led to the use of addressable converters to control service access. These addressable converters are basically enhanced tuners that descramble the signals of premium channels when authorized by the network. In this way, the premium channels are blocked or opened for the subscriber.

CATV networks currently use analog technology and coax cable in the backbone. One of the main reasons for using analog technology is that today's TVs are analog and receive only analog signals. It is not possible to change the analog TVs to digital because there are more than 200,000,000 TVs in the United States alone. To save additional equipment costs for converting the signals, the CATV providers will keep the analog format all across the network until TVs are equipped with digital-to-analog converters.

18.2.2 Switched-star architecture

Switched-star architecture was developed in Europe for delivering video services. A typical switched-star system is shown in Figure 18.2. Switching systems minimize the complexity and cost of transmission systems. The switched-star system was developed to minimize the cost and complexity of any headend switching matrix and to reduce the cost of the final subscriber

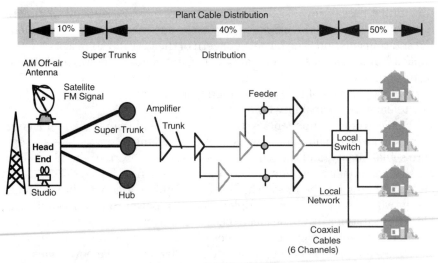

Figure 18.2 Switched-star architecture.

loop by allowing the use of a simple, low-cost coaxial cable. In this network, the switching is moved closer to the subscriber using small local switches, and the subscribers communicate with the switch via their set-top box to choose the TV channel to be sent to their home. All channels (typically 60 or so) are sent to the local switches, and the selected channels are delivered to the subscriber's home from the local switch.

Flexibility has been added to the switched-star systems by using higher-performance coaxial cable for the subscriber loop (local loop), enabling the transmission of up to six channels (of the 60 available at the local switch) to each subscriber. Usually, four or five of these might not be switched at all, similar to the broadcast channels, and are reserved as pay-per-view channels. One or two can be used for selectable programming by the subscriber via the set-top box to the local switch.

18.2.3 Tree-and-bush architecture

Figure 18.3 shows a typical tree-and-bush architecture. The latest tree-and-branch systems actually have a star distribution node and have been designed to enable individual broadband drop to each home. This architecture has been termed *tree-and-bush* and is being adopted in the United Kingdom. It caters to the use of high technology at the headend switching to achieve total network flexibility. With the addition of the headend switching, it will provide a full broadband switchable network and will be able to handle all telecommunications services, including video.

Constant improvements in fiber have made the tree-and-bush approach much more effective in the UK. Network builders are using far more fiber nowadays than in the past few years. Although differences between franchises exist in the development of fiber, fiber is generally run from a headend through a hierarchy of nodes serving fewer and fewer premises until the last mile connecting the subscriber or customers. For heavy-use business customers, the provision of fiber directly into the building for that last mile is increasing. Typically, fiber goes all the way to nodes that serve up to 600 homes. Some cable providers are planning to go farther, providing fiber to nodes serving 60 homes.

Currently, the telephony network is in overlay; that is, it occupies the same duct but is a separate network. In the future, it will be possible to have all services (including telephony) provided from a single network with a headend or central office without the need for an overlay.

There is no theoretical reason why these systems should be functionally different. Broadband systems can accomplish the same tasks as switched systems by using techniques such as time or frequency division multiplexing used in today's system. These approaches can provide dedicated distribution channels to every subscriber in the system. The real difference lies in cost. Multiplexing signals distributed to every subscriber is expensive in terms of bandwidth provided to each subscriber. It is important to note that by building a broadband tree-and-bush distribution network with optical fiber from the start, the upgrade path can be simplified by deploying equipment in existing locations and thus avoiding major changes within the subscriber premises.

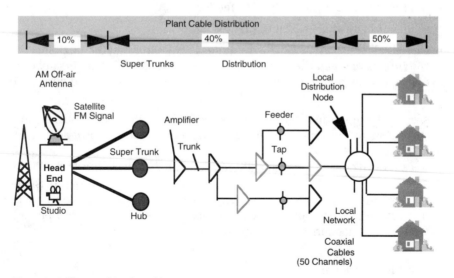

Figure 18.3 Tree-and-bush architecture.

In the architecture, the headend switch costs can be significantly reduced by focusing on the data services that can be efficiently switched, along with broadcast video. As these services share the same transmission medium, the switch cost can be distributed. When digitized, video becomes practical. Multiplexing of switched television signals can become more realistic for videoconferencing, video telephony, and other services, thus offering additional potential for generating revenue. These services could be offered using a multiplexed transmission channel, with several links time-sharing a single channel. Multiplexing is always transparent to the end user because of its high speed. The video signals are converted into ATM cells, which are then switched across the network.

18.3   Features and Benefits of CATV Architectures

Some of the features and benefits of the architectures discussed in the earlier section are summarized in the following subsections:

18.3.1   Tree-and-branch or tree-and-bush

These two architectures are similar in general. So they are grouped together as having the same features and benefits:

- lower up-front costs
- transparent service to end user
- fiber-optic cable from headend to feeder
- multisource (switched-star is single-sourced)
- carries all the bandwidth to the home, providing greater flexibility in the future, and can then be used for the expansion of channel capacity and new transmission formats (such as D-MAC, HDTV and digital hi-fi audio).
- switched-star's prime selling point of offering interactive services (and the higher revenue streams that go with it) is not as good as it sounds.

Recent developments in tree-and-branch technology have resulted in little distinction between tree-and-branch and switched-star in terms of provisioning of low-level interactive services, such as pay-per-view, and high-level interactive services, such as home-banking, database access, and video on demand (VOD). VOD is not currently available.

18.3.2   Switched-star

Switched-star is similar to the telephone architecture used around the world:

- proven and reliable now (in Europe)
- more secure; less signal theft because only the switched channel currently being viewed reaches the home, while the other channels are left at the switching point in the network
- less expensive equipment in the home because most of the electronics are in the street, causing replacement cost of home equipment to be less
- pay-per-view is standard; all requests by the subscribers are sent through the cable remote and stored at the local switch before being polled by the headend computer, thus providing better management of requests
- ease of upgrade; as new products and services are introduced, upgrade costs are lower because the changes can be made to the switch without visiting the home

18.4   CATV Environment

In this section, we address the CATV environment, including technology, CATV economics, and operating economics of different technologies related to CATV. Each is discussed in the following subsections.

18.4.1   Technology

Three basic technologies are considered in CATV system:

- AM (amplitude modulation)
- FM (frequency modulation)
- digital

Table 18.1 compares the bandwidth, cable length, distortion, video signal-to-noise ratio (SNR), and audio dynamic range of the three systems. Since the current CATV system with its coaxial analog technology cannot provide any of the additional channels or bandwidth required to provide residential broadband service, adoption of new technology is required by CATV providers. The AM and FM systems are familiar modulation schemes to CATV network providers, but these providers are not familiar with digital technology. A general tendency within CATV network designers exists to continue to maintain or improve AM and FM technology for their network upgrades. The CATV network equipment providers like this mindset of CATV providers because it prevents major digital transmission equipment providers from entering their marketplace. Some CATV companies are against deploying digital technology in the CATV network because the TVs are analog. Additional equipment must therefore be deployed for digital-to-analog conversion, which is expensive.

TABLE 18.1 CATV Technologies

AM system	FM system	Digital system
Cable length: 30 km	Cable length: 40 km	Cable length: 40 km
Bandwidth: 40 chs (420 MHz)	Bandwidth/fiber: 16 chs (700 MHz)	Bandwidth/fiber: 8–12 (600 MHz–1.2 GHz)
Intermodulation and reflection distortions	Intermodulation noise	No intermodulation noise Noise from AD to DA conversion
Channel spacing design	Channel spacing design	Channel spacing design not required
Regeneration on amplifier generates distortion	Regeneration on amplifier generates distortions	Regeneration without distortions
Video SNR: 55 dB	Video SNR: 65 dB	Video SNR: 57–67 dB
Audio dynamic range: 65 dB	Audio dynamic range: 65 dB	Audio dynamic range: 65–85 dB

In the digital system, a CATV network will not require the expensive telemetering and monitoring systems typically used in telecommunications transmission systems. Thus, digital CATV networks can be built at much lower cost. Advantages of using digital technology in the CATV system are in its performance, which is far superior to the AM/FM system and ease of migration from a coaxial to a fiber-based system, quality of service (QOS), and long-term cost effectiveness. Digital technology also enables easy provisioning of additional services and channels. By compressing the digital signal, more services can be provided because the bandwidth per channel is decreased, depending on the type of compression used. One major disadvantage of using digital technology, as mentioned earlier, is that it must be converted back to analog technology, because input to the TV sets has to be in analog. Thus, the additional equipment that converts analog to digital signals and vice versa called (A/D and D/A) converters is required at some point in the network.

Fiber-optics technology in CATV networks is also important. No network will benefit more from fiber-optic technology than the CATV network. Its inherent characteristics eliminate most of the bottlenecks existing in today's network, such as increasing the channel capacity and reducing the number of amplifiers in the network. We mentioned earlier that in coax cable systems, there are repeaters every 1000 to 2000 feet to regenerate the signal, causing the deployment of numerous amplifiers in the network at additional capital, maintenance, and operations costs. By deploying fiber, a typical CATV network can reduce the number of amplifiers to four or six, and still get substantial gain in bandwidth and cost. Fiber can be used to carry both analog and digital signals, so fiber does not need to be replaced if the technology evolves. Only the fiber terminals at the end of the physical fiber cable need replacing. Fiber terminals are located every 40 km if single-mode fibers are used.

18.4.2 Technology economics

Figure 18.4 shows the economics using different technologies in a CATV network. AM technology on fiber is still a low-cost solution for CATV network design. Currently, the digital system is the most expensive on a per-channel basis. Further cost reductions in the digital system will make it economically attractive to CATV. In addition, technologies like gallium arsenide (GaAs), digital signal processors (DSP), and lasers will make digital systems more attractive to CATV providers. This graph shows the cost dynamics of reducing cascaded amplifiers in the CATV network and the relative improvement in quality that can be achieved.

The intersection of the curve and the bar in the graph shows that by reducing the number of amplifiers to four or six, an 80 percent improvement in quality is possible, and CATV providers can charge less than $50 per customer. Most of the CATV providers are using these dynamics to improve their networks by systematically reducing cascaded amplifiers and using available technologies.

18.4.3 Operation economics

Figure 18.5 shows a comparison of operation cost for the three different types of systems—AM, FM, and digital—under consideration to be used with fiber-optic cable.

The comparison is on the maintenance cost of AM, FM, and digital systems as a function of optical power gain. Because an AM system is more susceptible to noise, it requires advanced optics and laser systems to achieve a bet-

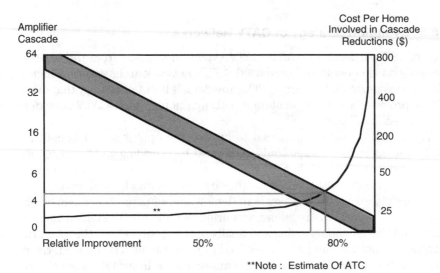

Figure 18.4 CATV economics.

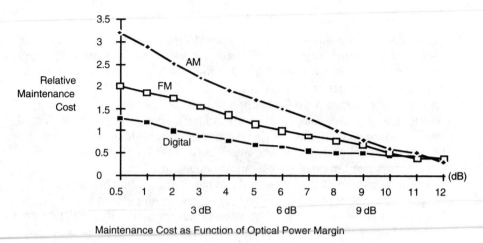

Maintenance Cost as Function of Optical Power Margin

Figure 18.5 CATV operations cost.

ter signal-to-noise ratio. The FM and digital systems outperform the AM system in low budget and margin requirements. The performance of the digital system is close to FM, but at a lower maintenance cost. For a digital system to interface with existing AM-VSB (vestigial sideband), it needs advanced DSP technology for the codecs, which incurs additional cost. The deployment of digital technology in the CATV network will depend on advances in DSP technology, which can provide bandwidth compression, and codecs for digital signal interface to AM-VSB TV sets. Initial reduction of the cost in DSP technology can make a digital solution more attractive to CATV networks.

18.5 Evolution Strategy of CATV Networks

Figure 18.6 suggests a possible CATV evolution strategy from the existing coaxial-based network to fiber-based BISDN network architecture, which is under consideration by many CATV providers. It is not necessary that every CATV provider adopt the evolution path mentioned here; CATV providers can bypass some stages.

This evolution path is designed to provide the optimum cost benefit for CATV providers. Each stage leads to the next by providing additional service to the customers with minimal capital investment, enabling CATV providers to recover the capital cost by charging the customers for additional services.

A possible evolution strategy is to deploy fibers first on the super trunks which are between the headend and primary hubs, then between the primary and the secondary hubs, and finally to the feeder. Using this plan, the number of cascaded amplifiers needed between the hubs is reduced. Every phase of the deployment plan brings improvement in overall network performance and channel capacity. By deploying fiber technology in the CATV network, CATV providers can build an open access system for residential broadband services. This system will support 150 channels without compression, or about 500 channels with compression. In the case of channels without compression, 50 channels will be used for broadcast services, and 100 channels can be shared between 150 and 200 households for asymmetric on-demand switched services (one-way switched services).

In Figure 18.6, the first stage is today's coax-based CATV network. The next stage is the fiber backbone stage. In this stage, the CATV backbone network, which is currently coax cable, is replaced by fiber-optic cable. This fiber-optic cable enables the CATV providers to reduce the number of amplifiers and provide additional channels to the customer because of the increased bandwidth. The next stage of the evolution is the deployment of a technology called subcarrier multiplexing (SCM). (The details of this technology and the architecture used are addressed in Section 18.5.1.) The next stage is the deployment of a technology called digital time-division multiplexing (TDM). (The details of this technology and the architecture used are addressed in Section 18.5.2.) The last stage, which is our target architecture, is BISDN, using ATM and SONET, which is discussed in Section 18.5.3. This evolution is an example of a real-world architecture.

18.5.1 SCM-based architecture

Figure 18.7 shows the SCM architecture from the headend to the interface node between the fiber and coaxial cable. Between the headend and the optical network interface (ONI), SCM is used, where the signal is mapped to a subcarrier and carried on the network until it is converted back into its original format for the subscribers. To create subcarriers, digital or analog in-

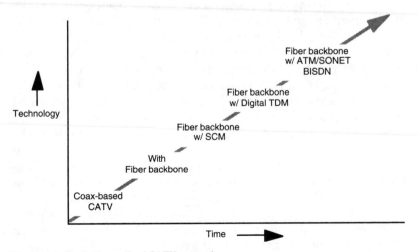

Figure 18.6 Evolution path of CATV networks.

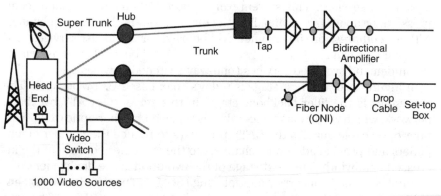

Figure 18.7 SCM fiber backbone.

formation is upconverted to a narrowband channel at a high frequency. For digital signals, the data can be added directly to the tuning voltage of a voltage-controlled oscillator, resulting in a frequency shift key (FSK) channel with modulation index determined by the magnitude of the applied data signal. Analog signals can be upconverted using any number of conventional modulation techniques, such as AM-VSB or FM-VSB, which are used in today's CATV systems. These subcarrier signals can be combined in a microwave power combiner, and then added to the bias of a laser diode, resulting in intensity modulation. Because laser output power is ideally a linear function of the input current and the detected current is linearly proportional to the received power, this linear link provides a convenient low-loss medium of microwave transmission. Thus, SCM technology enables one to multiplex several sources, each modulated with one or more subcarrier signals, that can be combined optically in a CATV network.

The basic operation of SCM-based CATV systems is as follows. The distributive video and switched video signals arrive at the headend at the compressed STS-1 rate (4.5 Mbps). A broadband switching matrix receives the on-demand video signals to be switched. A total of 54 signals are modulated over the frequency spectrum between 1.7 and 7 GHz, assuming 100-MHz spacing and combination using a subcarrier multiplexing scheme. The ONI connects the fiber backbone to the coaxial cable distribution plant. At the ONI, microwave receivers simultaneously provide reception for each channel, after which the signals are frequency shifted to the appropriate channel and fed to an amplifier for transmission over the coaxial cable. In addition to carrying conventional analog, video and voice can be carried digitally. These CATV architectures are also capable of handling wireless traffic. Wireless communication can be achieved by placing an antenna next to the ONI, which is used as a cell site to convert the signal from the medium of air to the fiber backbone. As mentioned earlier, the study of wireless communications is quite interesting, but it is beyond the scope of this book. The final segment of coaxial cable requires two-way amplifiers for the bidirectional signals on the cable. When a channel request is indicated by a subscriber, a microprocessor at the ONI polls the addressable converter on the subscriber premises and directs the video requests upstream to the headend.

18.5.2 Digital TDM-based architecture

The working of digital TDM-based architecture is similar to the telephone network, except that video signals dominate. Figure 18.8 shows the digital TDM architecture.

For this architecture, the video signal arrives at the headend at an STS-1 rate. The distributed video channels combine with the switched video channels using a SONET multiplexer to form data signals over the fiber backbone at an STS-12 rate (622 Mbps) or STS-48 rate (2.4 Gbps). With a backbone speed of STS-12 corresponding to 12 video channels per fiber, each channel serves one subscriber connection, and the network architecture is essentially a logical star. At the ONI, a SONET demultiplexer separates the channels, which are then decompressed, converted to analog NTSC format, and modulated for transmission over the coaxial cable to the subscriber. Downstream communications are handled in the same way as the SCM architecture's upstream communications.

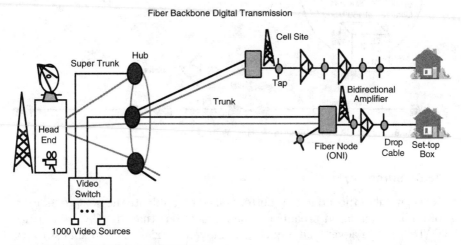

Figure 18.8 Digital TDM-based architecture.

18.5.3 ATM-based target architecture

Figure 18.9 shows the target architecture proposed by Time Warner to provide a full range of services. The network is based on ATM and SONET for its switching and transmission system. Here, the ATM switches are deployed at the headend with hub sites using a digital controller. Video is connected to the headend via an OC3 or OC12 interface. The interface from the headend to the hub (digital controller site) is at DS3 speed (45 Mbps). This digital controller is connected to the ONU via OC1 (optical equivalent to STS-1) or 45-Mbps interfaces. Each ONU serves 50 homes. This architecture is capable of providing voice (via both wired and wireless) and data, all in the same network, although it is still in its initial stage.

The next stage is to connect this headend with LEC and IEC networks (telecommunication networks) in a ring, thus providing survivability in the backbone. The server then can be located anywhere on the ring and accessed by any network (LEC or IEC).

The above-mentioned speeds are transmitted without compression. Once compression is achieved, about 500 channels can be provided to each home.

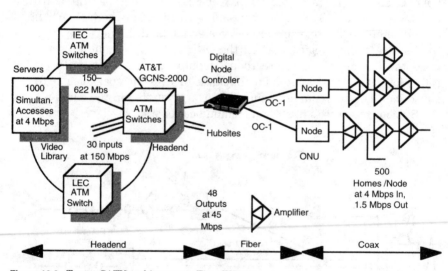

Figure 18.9 Target CATV architecture—Time Warner example.

18.6 Summary

No segment of the industry is better poised to dominate the next generation of multimedia-based broadband communications than the CATV industry. CATV has the advantage of having deployed coaxial cable to almost every home in the United States in the 1980s. It is thus in the dominant position to carry and provide next-generation services. Table 18.2 shows the high-level capability of fiber-based systems to provide future services.

This table does not imply that just by deploying fiber,one could provide all the services mentioned. It means that the basic technology to provide all the future services is fiber-based. One still needs to implement the different technologies mentioned in this book.

We have addressed the evolution strategy taken by CATV providers to evolve their network to next-generation broadband ATM/SONET-based networks, which can position them to provide cost-effective multimedia services. Here, CATV networks are in a tremendously advantageous position for providing the existing services such as plain old telephone system (POTS), and future wireless services such as personal communications services (PCS). PCS is the technology that CATV providers are planning to provide as telephone services, along with the video services they currently provide.

TABLE 18.2 Comparison of CATV Network Services

Network alternatives	Narrowband switched services	Network services				
		Broadcast video	Switched video	Video on Demand (VOD)	Interactive video	Data services
Coaxial cable (Today's system)	-	✓				
Fiber backbone	✓*	✓	✓	✓	✓	✓

*Need additional switching infrastructure.

Broadband Network Design

So far, we have looked at different technologies and different environments applicable to ATM. We have also looked into ATM switching and transmission networks. Having seen all these pieces, we need to know how these technologies all fit together in a network. We now look into how to design a broadband network or how to upgrade an existing telephone network to handle integrated broadband traffic. We discuss the issues related to broadband network design.

In theory, the access network is completed before the backbone network. In the real world, however, the backbone network is built or upgraded first, because it is the most cost-effective portion of the network shared by many users. The access network is then designed in phases, spread out over time, based on justification and traffic demand.

Remember that broadband network design is completely different from conventional telecommunications or computer network design because broadband must handle various types of traffic from narrowband to integrated broadband traffic (voice, video, and data) each with different characteristics, such as holding time, bit rate, peakedness, and average bit rate. This part addresses the access network design and backbone network design at a very high level without doing detailed performance analyses.

Broadband Access Network Design

19.1 Overview

In this chapter, the design of the access portion of the public network is covered. Figure 19.1 shows the reference access network in the United States, which is used as an access to the public backbone network. The access network is where all traffic is originated and terminated. Without the access portion, no traffic could exist in the backbone. This network is so critical to backbone network providers that they pay up to 40 percent of their revenue to the providers of local access networks.

This chapter addresses the design issues of estimating traffic, link (transmission system) sizing, and node (switching system) sizing in the access network. The different components in a network are the following:

- CPE, which means they are owned by the public carrier located at the customer premises
- local loop—the link connecting the CPE and the central office (CO), which is the first interface to the public network
- trunk links—the links connecting different COs
- access tandem—the second level of switching providing traffic consolidation.

Each of these components must be considered in designing the access network.

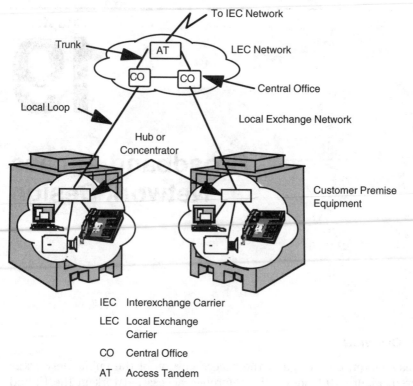

To IEC Network

Trunk

AT

LEC Network

CO   CO

Central Office

Local Loop

Local Exchange Network

Hub or
Concentrator

Customer Premise
Equipment

IEC   Interexchange Carrier

LEC   Local Exchange
       Carrier

CO   Central Office

AT   Access Tandem

Figure 19.1 Reference access network.

19.2 Broadband Access Network Definition

Before listing the requirements, let's define an access network. An access
network can be categorized into two possible access levels: the user or application level and the public network level. Figure 19.2 depicts these two
levels.

The user level defines the access required to the public network resources. The link between the customer's hub to the central office is the
access interface to the public network. The user level provides the greatest diversity in interfaces, protocols, architecture, technologies, and standards compared to any other level in the network hierarchy. Some of these
were described in chapter 15. User network design is based on the amount
of traffic generated from different sources, such as voice from PBX, data
from LANs, and video from desktop or a videoconferencing center within
the user's premises. All this traffic must be identified during the design at
the user/application level. This information is usually available within the
company.

The next level is the access interface to the backbone network, which is usually via a router, bridge, PBX, switch, customer service unit/data service unit (CSU/DSU) or any other device that has a standard interface related to the public networks and capable of performing protocol conversion at the customer premises. These access devices are sometimes called *hubs* or *access nodes*. In the broadband environment, this device should be able to handle all types of traffic (voice, video, and data). For simplicity, assume one interface exists, through which all traffic leaves the user's premises. This traffic can be connected directly to the backbone network, bypassing the local exchange network; however, it is connected via the LEC. Thus, the network from the customer's premises interface point to the backbone network forms the access network.

19.3 Network Design Requirements for Broadband Access

In a broadband environment, design of the access network is primarily based on existing traffic projections for each customer or user. The most difficult part here is identifying the traffic requirements of a customer. Customers can design their own private network; they have all the traffic and usage pat-

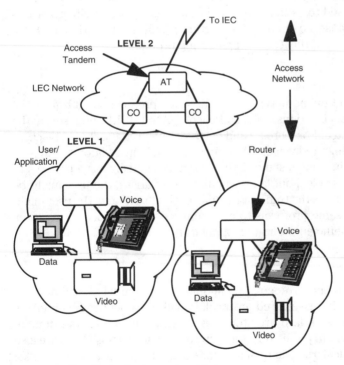

Figure 19.2 Two levels of access network.

tern information. For a public network provider, however, it is difficult to project the exact traffic that could be generated from the many user applications, especially if the traffic is nonexistent, such as broadband traffic. Network requirements are set in such a way as to meet these uncertainties, allowing for expansion if the projections are not close to the actual traffic. It is thus difficult to design a public network. In this section, we divide the public access network requirements into the following general categories:

- traffic
- protocol
- architecture
- services, features, and functions
- access route diversity (redundancy)

19.3.1 Traffic

One of the most important aspects of the network design is to estimate traffic. In a broadband network, traffic from the user comes from various applications, each with varying characteristics and usage patterns. All these parameters are used to estimate traffic, but it is difficult to estimate broadband traffic, because no such traffic currently exists. In Section 19.4, we look into the details of this traffic engineering problem.

19.3.2 Protocol

To interface to the public network, the interface protocol should meet the public network specifications, such as Bellcore, ANSI, IEEE, etc., set by the standards committee. For a broadband network, the protocol should be capable of handling integrated voice, video, and data traffic in addition to the published standards. In some instances, the user might initially use a protocol to handle only data or voice, but later, when the need arises, the user might be required to integrate all the traffic at the hub level. At that time, the user might prefer to use the same protocols to avoid unnecessary protocol conversion, which results in inefficient utilization of public network bandwidth.

19.3.3 Architecture

Having decided on the protocol, one should determine what architecture should interact with the selected protocol. Usually, protocol and technology go together. Many standard architectures exist, such as star and double-star, which are used in public telecommunications networks. There are also logical-ring, physical ring, and bus architecture for local area networks. These architectures vary, depending on the technologies. The logical-ring architecture is typically used with fiber technology. In the design of access networks, the selection of architecture should be in such a way to provide both short-term and long-term benefits in terms of handling current and future traffic and user growth. For example, in a star topology, as shown in Figure 19.3, a dedicated line and port must be added to add a new user.

In the case of the ring topology, shown in Figure 19.4, no additional link or port needs to be added to the hub. Here, the link capacities are already in place for existing users who share the facilities. In this topology, adding a new user is easy, but new users reduce the bandwidth available to the other users because of sharing. Thus, a limitation on the number of users that can be added must be set if reasonable performance is to be maintained. In a ring topology, the transmission rate is inversely proportional to the capacity of the ring. It is therefore important to select the right architecture for broadband networks, and each architecture has its own advantages and disadvantages.

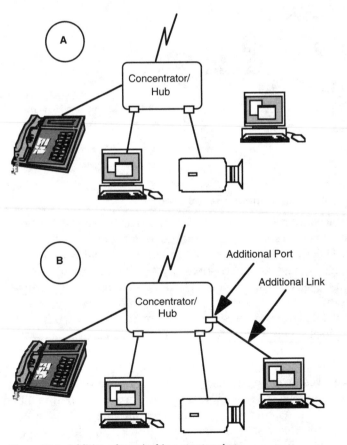

Figure 19.3 Addition of terminal in a star topology.

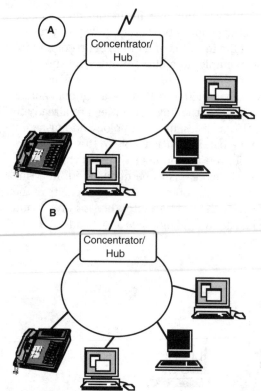

Figure 19.4 Addition of terminal in a ring topology.

19.3.4 Services, features, and functions required

When a user interfaces to the public network through an access device, certain features, functions, and services are required for standard interface and protocol support. The services, features, and functions vary with the protocol used. For example, the virtual path identifier (VPI) and virtual channel identifier (VCI)-based services are dependent on the ATM protocol. Because these services are protocol dependent, if the ATM protocol is changed, the services also change. It is thus necessary to decide on the protocol, based on the services it provides. The services, features, and functions provided with the protocols must be considered in the design of the access network. Today, service providers can provide service transparency, i.e., the user service characteristics are preserved regardless of the access protocol, which is achieved by service mapping. The disadvantage with service mapping is that the overhead involved in each protocol consumes additional bandwidth, which inefficiently uses the system resources. The user should therefore consider all services offered when deciding on a particular protocol.

19.3.5 Access route diversity

Integrated broadband networks are unique in that all traffic going outside the premises exits via a single very high-speed interface. Thus, the customer's only communications interface to the outside world is via one interface. Any damage or failure to this interface makes the customer vulnerable to being cut off from outside communication. It is therefore necessary to provide a completely diverse, redundant access route to each interface point in the network, as shown in Figure 19.5.

19.4 Broadband Access Network Design

In this section, we look at a high-level process for designing a broadband access network and explain the steps involved in the design of the access network. Like other networks, there are many ways to approach the design of a broadband network. Let us assume a bottom-up approach, as illustrated in Figure 19.6.

As mentioned earlier, designing a broadband network is a lot more complicated than designing a conventional circuit-switched telephone network or a packet-switched computer network because these networks are designed for specific applications. For example, the telephone network is designed to carry voice and is optimized for voice traffic. Thus, although data can be carried, it is not transmitted efficiently. Likewise, a computer network is designed to transfer data, whose requirements are quite different, and full-duplex communication is necessary. Voice transmission is not possible at all in a computer network.

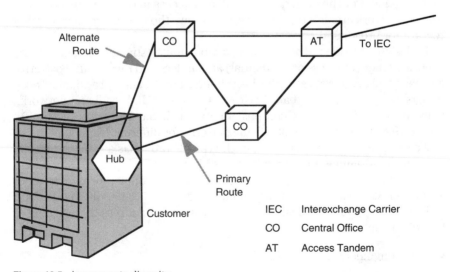

Figure 19.5 Access route diversity.

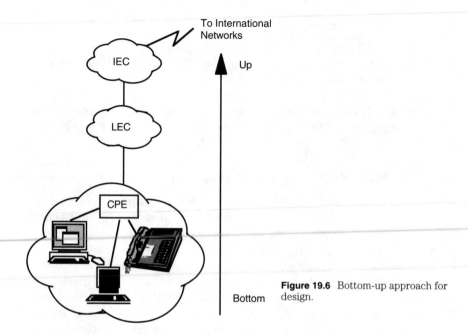

Figure 19.6 Bottom-up approach for design.

In broadband networks, integrating the different types of traffic onto a single network so that its resources are used efficiently is a formidable task.

We mentioned in chapter 1 that a typical network consists of nodes (switching centers) and links (transmission lines). Broadband networks are no exception. For the design of a broadband access network, we need a reference network architecture. Figure 19.7 shows the reference access architecture for broadband.

To design a broadband access network, one must estimate the traffic from each customer (business or residential) and understand these traffic patterns and usage characteristics in the busy hours. For example, residential customers' telephone usage peaks between 4 and 6 PM. In broadband network, video traffic is included, which peaks between 7 and 10 PM (prime time). Figure 19.8 shows approximate traffic patterns for different traffic types.

We can see from the chart that each type of traffic has a different usage pattern. In designing a network, the busy hour for each pattern should be considered when designing the network. For example, peak traffic is around 11 AM. Thus, the network should be designed to handle that traffic.

The following are the design issues that we must address in designing a broadband access network:

- traffic engineering
- access technology selections

- broadband network link design
- broadband network node design

Each is described in the following subsections.

19.4.1 Traffic engineering

The most important component when designing a network is estimating the network traffic. Traffic becomes difficult to estimate when traffic sources such as switched video, personal communication service (PCS), etc., do not yet exist. While some data are available from existing traffic, new traffic sources have made traffic projection more guesswork than concrete engineering. Here, we propose a simplified way to estimate different types of traffic.

In traffic analysis, the basic assumption is that all traffic can be categorized into voice continuous bit rate (CBR) or data variable bit rate (VBR).

Assume that video traffic belongs to one of the two categories. If video traffic is packetized, it is VBR; if not, it is CBR of high bandwidth and large holding time (compared to voice traffic holding time). Voice traffic can also

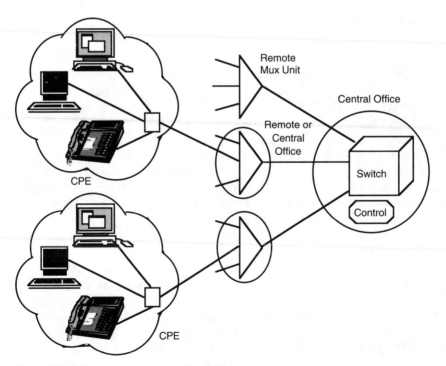

Figure 19.7 Reference access network architecture.

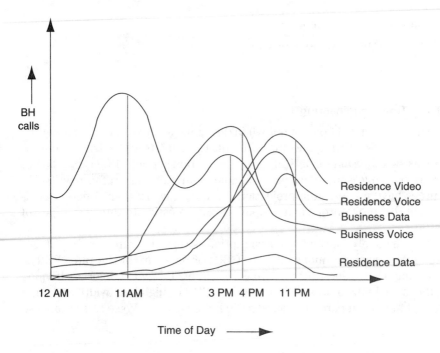

Figure 19.8 Example of traffic patterns for different traffic on a busy day.

be packetized to become VBR traffic. At this time, packetized voice traffic on a broadband network is far from a reality, so we can assume that most of the voice traffic is CBR.

Usually, each traffic type has its own units. Voice traffic is measured in CCS (hundred call seconds, where the first C represents the Roman hundred symbol) or Erlangs, depending on the country of origin. Data traffic is usually measured in packets, messages, frames, or cells. For simplicity, assume all types of traffic are converted to bits per second. Data traffic units are usually expressed in packets per second or bits per second. Video traffic units are usually bits per second.

The first step is to convert all the traffic into bits per second. Usually, the available information on traffic is in bytes per day, but the networks are designed based on busy hour (BH) traffic. It is customary to assume BH traffic to be 25 percent of the day's traffic, depending on the country. Figure 19.9 shows a typical traffic source for a business customer. Figure 19.10 shows typical traffic sources for a residential customer.

From Figures 19.9 and 19.10, the following can be derived. Note that the busy hours for voice, video, and data might be different.

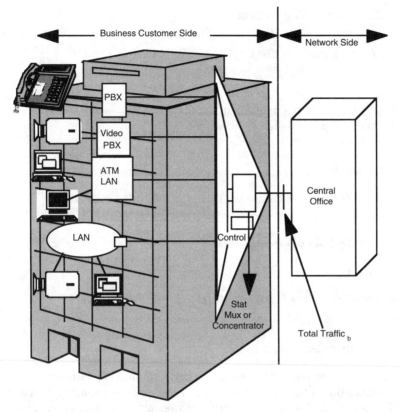

Figure 19.9 Business access traffic.

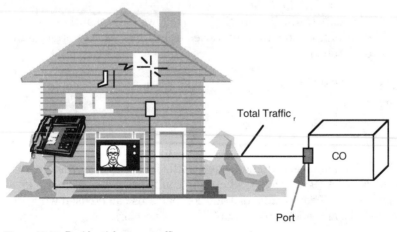

Figure 19.10 Residential access traffic.

Total traffic offered to the network =

$$\sum_{i=1}^{n} \text{(Total residential traffic)}_i \, l_{BH} + \sum_{i=1}^{m} \text{(Total business traffic)}_i \, l_{BH}$$

where
　　n = Total number of residential customers
　　m = Total number of business customers
　　BH = Busy hour

Residential traffic l_{BH} =

　　Σ voice traffic $l_{res\,BH}$ + Σ data traffic $l_{res\,BH}$ + Σ video traffic $l_{res\,BH}$

where
　　res.BH = residential busy hour.

Business traffic l_{BH} =

　　Σ voice traffic $l_{bus\,BH}$ + Σ data traffic $l_{bus\,BH}$ + Σ video traffic $l_{bus\,BH}$

where
　　bus BH = business busy hour

Once traffic is estimated, all the traffic in BH is summed to estimate total offered traffic to the network. This estimate is based on extremes, such as peak traffic for data. The offered traffic varies with the usage pattern. Because the effective traffic changes with time, it is important to keep the user traffic information up to date.

Typical voice/data traffic distribution is as follows. Note that these are approximate distributions; distribution varies from application to application, customer to customer.

- offered traffic =100%
- intra-premises = 70%
- extra-premises = 30%

Of the extra-premises traffic, 80 percent is intra-LATA and 20 percent is inter-LATA. For instance, switched video traffic for residential customers is usually intra-LATA (93 percent), where the traffic originates from the local video store server. These traffic distribution patterns can be used while designing the links and nodes in the public access network.

19.4.1.1 Access technology selection. Having estimated the traffic, the customer must decide on the best technology suitable for access to the public network. Many factors exist in selecting the right technology. For the sake of simplicity, assume the selection is based on technology capability. Table 19.1 shows the appropriate technology based on customer traffic. The appropriate technology can then be selected from Table 19.1.

19.4.2 Link or transmission network design

As mentioned in chapter 1, one of the components of the network is the physical link connection between the different switches or nodes in the network. In a broadband network, these links operate at very high speed and carry huge amounts of information. These links are part of the transmission system of a network. The transmission system carries the traffic from point A to point B error-free, as shown in Figure 19.11.

It is easy to estimate the traffic on the link if the traffic originates at A and terminates at B. Because points A and B are usually several hundred miles apart in an access network, care must be taken that the link is reliable and error-free. In a real-world network, not all nodes are fully connected as direct links. Usually, each node is connected to two or three adjacent nodes. To reach other nodes, the traffic must traverse the intermediate nodes that

TABLE 19.1   Access Technology for Different Traffic Rates

Traffic type	Effective BH traffic in bps in the access line*	Access speed required	Appropriate technology
Voice only	≤ 1.544 Mbps	T1 (PRI)	ISDN/POTS
Voice + data	≤ 1.544 Mbps	T1 (PRI)	ISDN
Data	≤ 1.544 Mbps	T1	FR
Data	> 1.5 Mbps ≤ 45 Mbps	FT1 or T3	SMDS/ATM
Voice + data	> 1.544 Mbps	FT1, T3, FT3, OC3	ATM
Voice + data + video	> 1.544 Mbps	FT1, T3, FT3, OC3	ATM

*Traffic including allocation for uncertainties

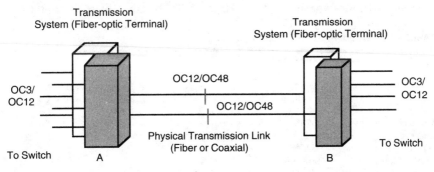

Figure 19.11   Typical transmission network.

connect the source and destination node. Thus, to size the intermediate link, all traffic passing through the link in the BH must be taken into consideration. (The link design between the switches or nodes in the access networks is discussed in chapter 20). The most critical portion of the access network is the link between the customer and the network. As this portion is more relevant to the backbone network design, it is also covered in chapter 20.

Because only one link exists from the customer premises carrying all the traffic to the network, the capacity of the link should be designed based on the effective sum of all the traffic. Care must be taken to consider the utilization of the link caused by each type of traffic. Figures 19.9 and 19.10 showed typical sources of traffic from a business and residential customer, respectively, in a broadband network environment. Table 19.2 shows the typical traffic characterization required in designing the access network.

Assume each device at the customer premises generates x_1 to x_n Mbps l_{BH} traffic, which is calculated based on the traffic characteristics in Table 19.2.

Therefore, total offered traffic is

$$\sum_{i=1}^{n} x_i \text{ Mbps } l_{BH} = K$$

where
  n = total number of traffic sources
  i = traffic source
  K = total offered traffic

Of these, a certain percentage of the traffic usually remains within the premises. The total traffic leaving the premises is calculated as shown below. Let A be the percentage of traffic leaving the premises.

Total traffic leaving the premises = $K \times A = Z$ Mbps l_{BH}

Using the traffic value with Table 19.1, one can find the appropriate technology suitable for access to the offered traffic. It is necessary to predict the customer's traffic over the next 5 to 10 years. Then, using the traffic numbers,

TABLE 19.2 Broadband Traffic Characteristics

Service	Peak Rate (bps)	Duration (hours)	Burstiness	BH call attempts
NTSC Video	45 Mbps	0.25	1.70	4
HDTV	150 Mbps	0.25	1.35	4
Voice	64 Kbps	0.05	1	2
Video conference	1.5 Mbps	0.67	5	2
Imaging	1 Mbps	0.25	15.60	3
Bulk data	1.5 Mbps	0.0056	15	6
Transaction data	100 Kbps	0.0083	200	20

BISDN Resource management by J. Burgin, North Holland, 1990

the required access technology can be determined from the table. These calculations enable the carrier/customer to select the appropriate access technology and speed required by the traffic at the end of the study period. The speed is the link access speed required from the customer premises to the network node.

To design the rest of the links in the public access network, the effective utilization of each access link (customer link) must be calculated. Summing the traffic of all customers provides the effective trunk-side link size. Usually, the trunk-side links are connected to more than one node, usually two or three. In that case, the traffic must be distributed to reach its destination via the shortest path, which is based on either distance or link utilization. Figure 19.12 shows the relation between access side and trunk side and its traffic distribution.

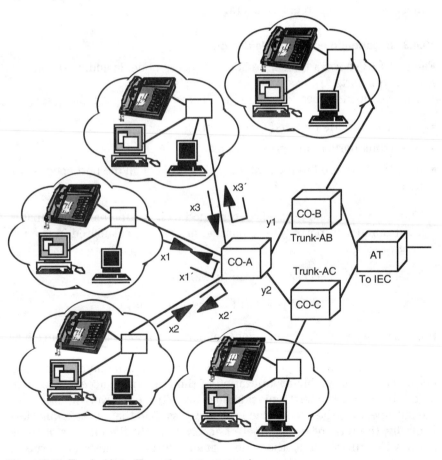

Figure 19.12 Trunk-side traffic on the access network.

Figure 19.12 shows how the trunk traffic is calculated in the network.

$$\text{Total traffic} = x_1 + x_2 + x_3 = K \text{ Mbps.}$$

Of these, $x_1' + x_2' + x_3'$ Mbps = M terminate within the CO-A switch.

Total trunk traffic on node

$$\text{CO-A} = y_1 + y_2 = N \text{ traffic units or CO-A} = K\text{–}M = N \text{ traffic units}$$

The calculation of y_1 and y_2 trunk traffic is similar to the calculation of traffic between nodes in the backbone network, which is addressed in chapter 20.

On the trunk side, the effective utilization of the link is usually 70 to 80 percent of link capacity including overhead. On the access side, the utilization is less than 10 percent on average.

19.4.3 Broadband access node design

Figure 19.13 shows a typical broadband access node, including the access side, access ports, trunk side, and trunk ports.

The broadband access node design is based on three main factors:

- total access ports required (access links)
- total trunk ports required (trunk links)
- total traffic offered to the switch, i.e., the sum of all the port speeds (total ports)

Thus, the access node size is determined by its traffic-handling capability and number of ports. These parameters can be calculated in the following way:

Total broadband access node size (in Mbps) =

$$\Sigma \text{ traffic per port (access and trunk) } l_{BH}$$

Broadband access node size in number of ports = Σ access link + Σ trunk links

The number of access ports is usually higher than the number of trunk ports because of the higher utilization of the link on the access side compared to the trunk side. Trunk-side links are more effectively utilized, and the link speed on the access side is usually less than that of the trunk side. Typically, the ratio of the number of access ports to the number of trunk ports is 10:1. In designing the node, plans for future expansion in terms of switch capacity and additional ports must be considered.

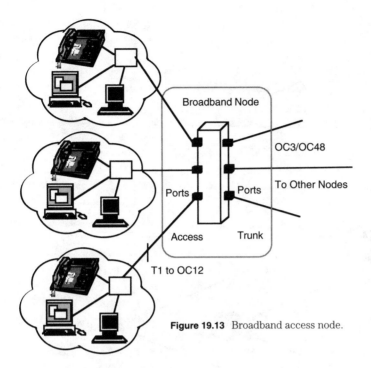

Figure 19.13 Broadband access node.

19.5 Broadband Access Network Topologies

Typical topology for a telecommunications or computer network is a star or double star, as shown in Figure 19.14.

Because of the requirements set forth in the design of broadband communications, along with the availability of new technologies such as fiber for high capacity, numerous topologies are under consideration for access networks. Some of these topologies are the following:

- physical star/logical star
- physical ring topology
- logical star/physical ring
- physical ring/logical ring
- bus/star and bus/bus

We describe the most popular architectures in the following subsections.

19.5.1 Physical star/logical star topology

Figure 19.15 illustrates physical star/logical star topology. This topology offers maximum potential bandwidth to each customer and hence provides maxi-

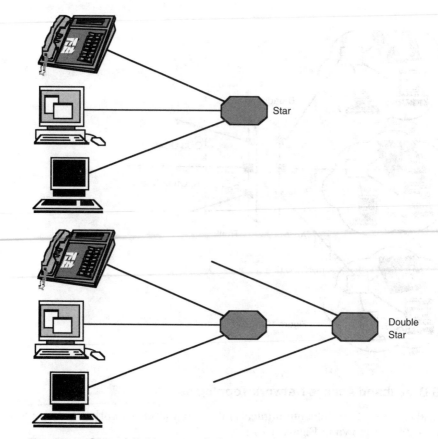

Figure 19.14 Star and double star topologies.

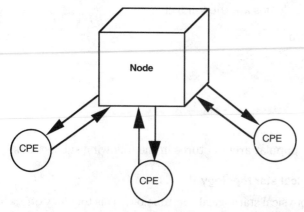

Figure 19.15 Example of physical star/ logical star topology.

mum flexibility in service provisioning and future upgrading. This architecture simplifies the administration of bandwidth requirements for individual customers and the diagnosis and partitioning of faults, so the lifecycle operational costs are low. The dedicated facilities of this architecture provide a high degree of security and limit unauthorized access, but can result in a higher initial cost. This topology should be considered when potential demand exists for a full range of services, when high reliability/security is required, or when the distances are short and hence initial cost is not a sensitive parameter.

19.5.2　Physical ring/logical star

This topology is physically interconnected as a ring, but a wavelength or time slot is allocated for each node. The initial cost for this architecture might be lower than that for a physical star because of the sharing of fiber, but the electronics cost can be higher because of the higher bandwidth requirements. This architecture takes advantage of the logical-star topology by providing security and privacy.

19.5.3　Physical star/logical ring

This topology has the same physical layout as the physical star/logical star, as shown in Figure 19.16. At the central node, however, the receive fiber from each node is connected to the transmit fiber of another node. Because information passes through all nodes and links on the ring, bandwidth is more difficult to administer, higher-speed interfaces are required, and fault partitioning is more difficult.

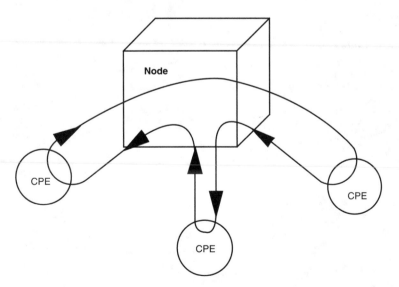

Figure 19.16　Example of physical star/ logical ring topology.

In this topology, the number of switch ports can be reduced. For example, the nodes at either end are connected to the switch directly, and the other nodes are connected to a patch panel (cheap manual connector board) and looped back. In the case of broadcast video type, the signal can be broadcast to one CPE and can flow to other CPEs without utilizing switch resources.

19.6 Summary

Design of the broadband access network poses interesting design issues mainly because all types of traffic are carried next to each other without differentiation. Because each type of traffic has its own characteristics, it must be treated separately. To meet these requirements, the issues of traffic, link, and node must also be considered separately and designed with care. In addition, the appropriate access topology must be selected to meet both short- and long-term requirements.

20

Broadband Backbone
Network Design

20.1 Overview

Once the access design is complete for the broadband network, one must decide how to connect it. It is not possible to connect all the access nodes together because of the network's complexity. For example, there are about 25,000 central office switches at the local exchange level in the United States; to connect them together, the local exchange carriers have a second level called the access tandem, which does traffic aggregation. There are about 1200 access tandem nodes. Thus, on average, a ratio of 10:1 central office locations to access tandem locations exist (more than one CO can exist in a location). These central office and access tandem nodes form the switching portion of the local access network. These networks were part of AT&T until 1984, and the nodes were part of the hierarchical network. Today, local exchange carriers have two levels, the central office switches and access tandem switches, as shown in Figure 20.1.

This network is called the *access network* for the backbone network. The next level above this access network is called the *backbone network* or interexchange carrier network, which, according to a U.S. Modified Final Judgment (MFJ) court ruling, can carry only inter-LATA and interstate traffic. Thus, this network forms the backbone, and the local exchange carrier becomes the access network for the interexchange carrier, also called the long-distance network. Figure 20.2 shows a typical backbone network with access networks subtending these backbone network switches.

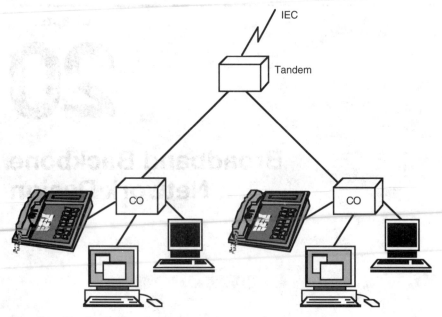

Figure 20.1 Two levels of local exchange carrier network.

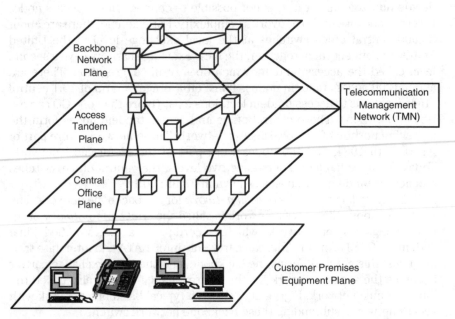

Figure 20.2 Backbone, access, and CPE plane.

Having defined the backbone network, let's design one. One of the main purposes of this network is to transport the traffic across the geographical borders, while the access network originates and terminates a connection.

20.2 Broadband Backbone Network Design Requirements

In broadband communication networks, the backbone network should be 100 percent perfect in all ways, i.e., the network should have zero downtime, zero percent blocking, and unlimited bandwidth to handle all the traffic all the time. Users expect these requirements from the backbone network. Of course, it can be costly to design such a perfect network. So, the objective of the network designers is to find a balance between cost and perfection. When compared to access networks, backbone networks are far more efficient and cost-effective because of the amount of sharing done by many users. The backbone network can therefore provide more services than the access network. Some of these services provided include the following:

- intelligent network services
- dynamic bandwidth allocation
- distributed network management
- redundant network for protection
- efficient utilization of resources
- efficient economics of scale
- advanced technology-based services

For a customer with multiple locations connected via a private network, the public network itself usually acts as a virtual backbone network that connects the different sites. Figure 20.3 illustrates a typical public network, and Figure 20.4 shows a private network.

In most cases, customers are connected via the local public network, but sometimes can be directly connected to the public backbone network and bypass the local access network in a private network. Sometimes, private network customers use their own facilities for the access portion of the network, enabling customers to avoid paying access charges to the local access provider. Sometimes, customers bypass the public network completely and use private facilities managed end-to-end. This trend is no longer popular with large customers because of changing technology and the cost involved in managing a network, especially if their core business in not in telecommunications. Thus, most customers prefer to bypass the local access network to get to the backbone network because of the advanced services provided by backbone network providers.

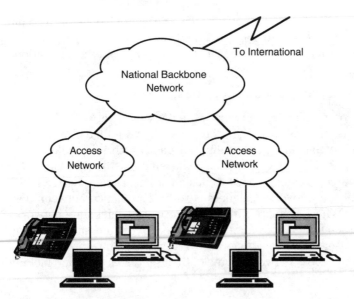

Figure 20.3 Public switched network.

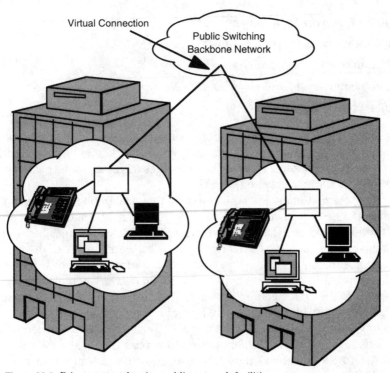

Figure 20.4 Private network using public network facilities.

In chapter 19, we defined network requirements under four categories:

- interfaces
- protocols
- architectures
- features, functions, and services

In addition to these requirements, other requirements for these categories are described in the following sections.

20.1.1 Interfaces

Interfaces to public networks always follow the standards. Certain interfaces are required for voice and data traffic, with varying speeds depending on the traffic requirements of the customer or on the serving access node in the local exchange network. In a broadband network, a common interface exists for transporting voice, data, and video. The typical access interfaces to a broadband backbone are shown in Table 20.1.

A variety of protocols can be mapped to DS1 and DS3. Protocols such as ISDN PRI, 24 DS0, or 1.544 Mbps can be mapped to DS1. About 40 Mbps of SMDS or ATM cells or 28 DS1 cells can be mapped to a DS3 protocol. The traffic in the network is transported at DS1 or DS3 speed, however the traffic is mapped and transported. It becomes the function of the switching equipment and termination equipment in the transmission network to identify the appropriate protocol for the traffic to be converted.

20.2.2 Protocol

Although access to the backbone network consists of various protocols, such as frame relay, SMDS, X.25, and ATM, all must be mapped to a common backbone protocol. The current trend in the public network is to use SONET/SDH for transmission. For data traffic switching, frame relay, SMDS, or ATM is used. Voice traffic is still handled by conventional voice switches. Although ATM can handle voice, it will be a while before voice traffic goes on the ATM backbone. In the case of a data network, the current protocol is transmission convergence protocol/internet protocol (TCP/IP). Most of today's video traffic is carried via CATV in a broadcast mode using traditional AM/FM analog technology.

TABLE 20.1 Broadband Interfaces

Interface terminology	Interface speed	Domain type
T1	1.544 Mbps	Electrical
DS3	45.76 Mbps	Electrical
OC3	155 Mbps	Optical
OC12	622 Mbps	Optical
OC48	2.4 Gbps	Optical

20.2.3 Architecture

Backbone architecture is always ahead of access architecture in terms of technology, features, and services provided to the end user because of the requirements set forth by end users and regulatory bodies. Only service providers who provide effective service operate the backbone networks because of the presence of competition. Because broadband networks should provide numerous advantages to the end user at minimal cost, the backbone architecture must always improve and look for better cost-cutting alternatives. The usual backbone architecture is almost fully meshed and is expected to be in the form of rings.

20.2.4 Features, functions, and services

The features, functions, and services in a backbone network vary with the type of technology being supported. If the backbone network is SMDS, it can support DQDB or SMDS interface. If ATM is used, it can support all types of interfaces, ranging from X.25 to ATM. Thus, the type of technology selected dictates the services, features, and functions, which vary from technology to technology.

20.3 Broadband Backbone Network Design

Three major components exist in the design of a broadband backbone network, similar to the ones in the access network:

- traffic engineering
- link design
- node design

Before addressing these areas, let's look at the backbone network in the United States. Currently, the United States has both a public telephone network and computer network, each designed and optimized for its specific application, the public telephone network for voice traffic, and the public computer network for data. The traffic pattern and characteristics of the applications for each network are different and therefore must be engineered differently, which in turn affects the design of the nodes and links in the networks. In a broadband communications network, the backbone network needs to handle all the traffic, where the network optimization is transparent to each user who requests the service. The user can be a local access carrier who provides the access function of originating and terminating the traffic or an end user who is a business or residential customer.

20.3.1 Traffic engineering

Traffic engineering in the backbone network is as complicated as in the access network. In the backbone network, the access speeds are very high; a failure in the link can cause damage to many customers. Thus, as a part of design, traffic is usually split between two backbone nodes to prevent any traffic loss in the case of a link failure. Before we go into traffic segmentation, let's look at the traffic source. For a backbone network, most of the traffic comes from the access networks, as shown in Figure 20.5. In some cases, it bypasses the local access network.

The traffic is generated from the access network via the access node, which provides all traffic to the backbone network. The problem is the difficulty in identifying the destination of the traffic. Voice, video, and data each have their own pattern of traffic distribution, which again can vary depending on industry and business. To simplify the network design, engineers use the *gravity method.* The principle behind the gravity method is that each node receives the same amount of traffic it offered to

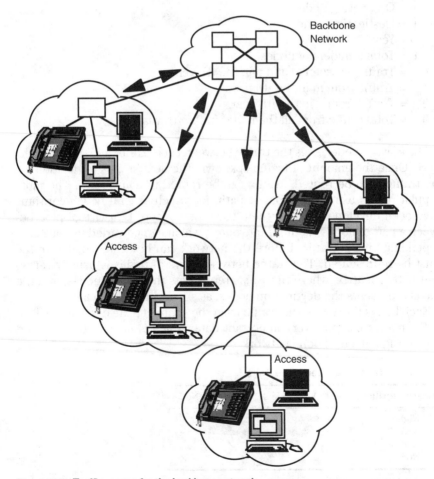

Figure 20.5 Traffic source for the backbone network.

the network. This theory might not be an accurate measurement, but it is sufficient to design the network. The formula for calculating the traffic to different destination nodes is given below.

$$t_{m,n} = (T_{m,i} \times T_{n,i})/T_{L,i}$$

$$t_{m,i} = \sum_{j=1}^{n} t^j_{l,i}$$

$$T_{L,i} = \sum_{l=1}^{L} \sum_{j=1}^{n} t^j_{l,i}$$

where

m = Originating node
n = Destination node
i = Year
L = Total numbers of nodes
$t_{m,n}$ = Traffic between m and n in the ith year.
$T_{m,i}$ = Traffic from m in the ith year.
$T_{n,i}$ = Traffic from n in the ith year.
$T_{L,i}$ = Total traffic from all the nodes in the ith year

The gravity model gives the traffic between each pair of nodes in the network. Once the amount of traffic is known, the next step is to decide how the traffic is to be routed. Traffic can be routed in many ways. The most popular one is using the shortest path, for which several new algorithms have been developed; one of them uses the concept of a virtual pipe between a pair of nodes via different routes, which can be varied in real-time, depending on the traffic. Before the network can be designed, the traffic must be represented. The traffic between pairs of nodes is usually represented by a matrix, where the x axis represents the source nodes, and the y axis represents the destination nodes, as shown in Table 20.2.

Each box is filled with the traffic from the source to the destination. The traffic matrix contains very important data, which is used in the link and node design of a backbone network.

TABLE 20.2 Traffic Between Node Pairs

Source/destination	Node 1	Node 2	Node 3	Node 4
Node 1	✓			
Node 2		✓		
Node 3			✓	
Node 4				✓

✓ Traffic terminates within the same node.

20.3.2　Link or transmission network design

Once the traffic is known, especially the traffic between each node pair, one can estimate the traffic on the link between the pair. This calculation is simple if a direct link exists between the two nodes because the traffic that passes through the link is the same as the traffic between the node pair. In the real world, however, no network is fully connected, even in the broadband backbone network. Thus, the traffic must pass through many nodes via links to reach its destination. For example, Figure 20.6 shows a typical four-node network. Assume that the traffic from A to D is given, and the network is connected as shown in the figure.

The network consists of four nodes: A, B, C, and D and three links l_1, l_2, and l_3. For the traffic from node A to reach node D and vice versa, it must pass through nodes B and C via links l_1, l_2, and l_3 before reaching destination node D. The traffic of link l_1 is

$$\text{Traffic in one direction} = \text{traffic (AD)}$$

$$\text{Traffic in other direction} = \text{traffic (DA)} + \text{traffic (BA)} + \text{traffic (CA)}$$

Thus, the size of link l_1 should be based on the maximum of two directions. Similarly, for links l_2 and l_3, the total traffic in both directions of each link must be estimated, and the link size should be the maximum of the traffic in two directions. When designing the link, all the intermediate traffic must be taken into consideration. There should be sufficient room for uncertainties—some extra capacity must be allocated for each link. This variable is simple if the network is small, but in today's complex broadband network, link sizes must be optimized without incurring additional cost and while taking every possible uncertainty into consideration. Link size depends on how much traffic is routed via that link. Routing depends on the

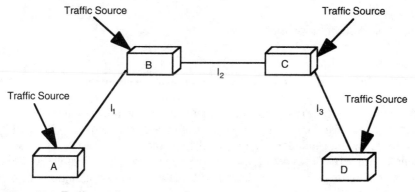

Figure 20.6　Traffic matrix.

type of algorithm used. Many algorithms are based on distance, cost, or a combination of both. For broadband networks, the routing algorithm must be dynamic and should be adjustable, depending on the situation of the traffic at the time of routing.

In most cases, carriers use existing routing methods unless carriers are convinced that a new algorithm is easy to use and cost-effective. Broadband networks require dynamic routing algorithms because of the bandwidth-on-demand type of traffic. In dynamic routing algorithms, routing decisions reflect changes in the traffic pattern. Because we do not know the exact broadband traffic between each pair of nodes, it is crucial to estimate the traffic correctly. Without knowing the traffic between the nodes, it is difficult to perform routing. In the case of broadband networks, because of the uncertain traffic patterns, the best algorithm to use is the distributed routing algorithm, which uses dynamic routing. In distributed routing, each node periodically exchanges explicit routing information with each of its neighbors.

Once the traffic between the node pair is available, we can find the route taken by the traffic under various conditions using any routing algorithm. It is possible to estimate the link size from the results of the routing algorithm. The link is sized considering the loading factors during the busy hour. Usually, it is assumed that the busy-hour traffic is 80 percent of the link capacity. In other words, almost 80 percent of the link is utilized at any given time. Thus, the link bandwidth in one direction (we can assume the same bandwidth in the opposite direction) can be given as

$$\text{link bandwidth } (l_{ij}) = \sum_{j=1;\ i=1}^{j=m;\ i=n} \text{point-to-point traffic from } i \text{ to } j +$$

$$\text{transit traffic between } i \text{ and } j + 20\% \text{ loading factor}$$

where
  i, j = Node on the network
  l_{ij} = link between nodes i and j

Total link capacities are obtained by summing all the individual link capacities. Link capacity does not mean the total traffic traversing the network. Sufficient bandwidth is reserved in the link to handle the uncertainties so that almost all traffic can be rerouted without bringing down the network. In addition to designing the link capacity to existing traffic, it should be designed to handle future traffic whose characteristics are not very predictable. All the traffic calculations are currently based on trials and the extrapolation of existing traffic by the carriers.

Thus, to summarize the steps in designing the broadband backbone link:

1. Calculate the total access traffic on a per-node basis.
2. If node-to-node traffic is not available, use the gravity method to estimate node-to-node traffic.
3. Use a routing algorithm to route the traffic in the given network topology.
4. Sum all the traffic on a link-by-link basis to calculate link traffic.
5. Add "fudge factors" for uncertainties, such as 20 percent of additional
   bandwidth to provide an estimated link capacity.

20.3.3 Node design

After calculating traffic and link capacity, we can now design the size of the
backbone node. Typically, four parameters define the design and size of a
broadband node:

- Total traffic handling capacity of the node, which is typically the size of
  the backbone switching matrix.
- Total access ports and each of their capacities
- Total trunk ports and each of their capacities
- Total ports on the node [access + trunk].

Figure 20.7 shows these parameters.

If these parameters are determined, one could come up with a good estimate on the size of a broadband node. The size (speed) of each port and the
number of ports (access and trunk) depend on the traffic arriving and leav-

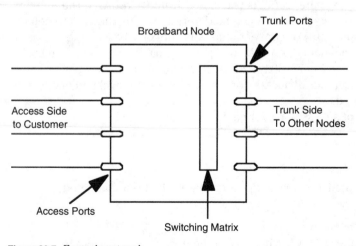

Figure 20.7 Example network.

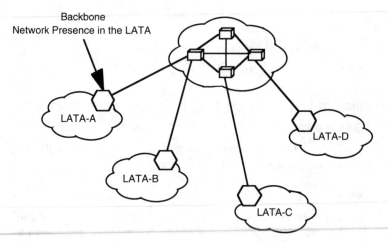

Figure 20.8 Typical node.

ing the node and the number of traffic sources. In the U.S. backbone network, the traffic is offered from different LATAs, as shown in Figure 20.8. The carrier that owns the backbone network should have a presence in each LATA. All traffic is collected at a point in the LATA and offered to the backbone network on a single port to the backbone node. If the traffic exceeds the capacity of the port, the carrier decides whether to add another port or put a concentrator at the interface point in the LATA. This decision depends on many factors, such as distance between the LATA and backbone node, cost of that distance in dollar per circuit, and cost of the port in the backbone node. These factors are illustrated in Figure 20.9.

Once the approximate traffic is calculated from each LATA, the number of ports required from the LATA can be determined. Justification to deploy an ATM port (stat mux) or conventional port (mux) is based on the cost of the port to the backhaul cost of the traffic to the backbone node. The backhaul is illustrated in Figure 20.9. Based on the number of ports per LATA to each node, the number of access ports required for the backbone node can then be calculated.

The number of access ports for each broadband node is

Number of access ports $1_{node} =$

$$\sum_{i=1}^{n} [(\text{traffic from LATA } i)/\text{port speed on the broadband node}]$$

where
n = total number of LATAs homing to the node

Note that port speed is the effective speed multiplied by 80 percent. For example, if effective speed is 155 Mbps, port speed is 155 × 0.8, or 124 Mbps.

$$\text{Total number of access ports} = \sum_{j=1}^{m} \text{Number of access ports for node } j$$

where

m = total number of nodes

Having calculated the total number of access ports, let's address the size of the switching matrix. The size is basically twice the effective traffic offered, plus sufficient capacity (an extra 20%) to handle traffic growth for the next 5 to 10 years.

Total switch size =

(Total traffic offered × 2) × 1.2 + Projected traffic for the next ten years

The next step is to calculate the number of trunk ports. Usually, the number of trunk ports in a node is the same as the number of adjacent nodes to that node. If the traffic to a node exceeds port capacity, an additional port to the same node can be added. In either case, one still needs to estimate the traffic on the trunk at each node. In addition to handling the traffic at each node, the trunk should have the capacity to handle almost all the traffic being carried on the other trunks. This type of backup is required because of the amount of traffic being carried and the reliability the carriers have guaranteed to their customers.

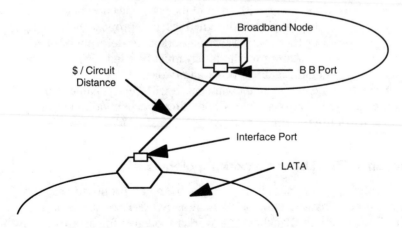

Figure 20.9 Backbone network interface at LATAs network.

To calculate the number of trunk ports and their speeds, we use the gravity method to distribute the traffic between the nodes and route them using one of the routing algorithms such as shortest-path algorithm. This algorithm gives the traffic at each node on the trunk side and the speed of the trunk ports from which an estimated number of trunk ports can be found. The number of trunk-side ports used as a backup to provide for redundancy are then added to the traffic number. For simplicity, assume the ports have a one-to-one backup. The number of trunk ports is then exactly twice the trunks required to handle the traffic.

To summarize:

$$\text{Total trunk ports required} = \sum_{i=1}^{n} (\text{traffic from/to adjacent node i})/\text{port speed}$$

where
  n = number of adjacent nodes

Note that we use the maximum of the two directions of traffic between a node pair.

$$\text{Total trunk ports with backup added} =$$
$$\sum_{j=1}^{m} (\text{number of trunk ports for traffic}$$
$$+ \text{number of trunk ports for backup})j$$

where
  m = number of broadband nodes

The network design proposed here is for a typical real-world public network, where the network, along with the number of nodes and nodal locations, already exists. This network design does not propose how many locations are required or the location for the nodes. To determine this information, a completely different methodology must be used. Usually, nodal location is decided when designing a private network.

The network design can be made complicated with dual homing and the distribution of traffic to various nodes depending on the conditions set by the carrier to distribute the traffic pattern.

20.4 Broadband Backbone Network Topologies

For a broadband network, certain topologies are being proposed that can effectively use broadband technologies such as SONET transmission systems and ATM switching. Backbone network topologies for a broadband backbone network are either a ring or meshed topology.

The topologies for access to the broadband backbone network are

- star topology
- ring topology
- dual-homing ring topology
- dual-homing star topology

The topologies for the backbone and access networks vary, depending on the type of technologies used in those networks. Figure 20.10 shows the ring/star topology combination.

Figure 20.11 shows a topology with multiple rings, which is the proposed topology for the broadband network environment. This architecture provides all the features, services, and functions required for broadband network.

The topology for a network is selected based on the ability to add links and nodes when necessary so that the traffic is shared evenly, and the existing links or nodes are not overloaded. Of course, a link or node is added only when no other alternative exists, such as rerouting the traffic on a fully utilized link.

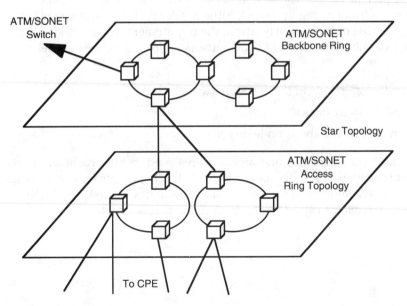

Figure 20.10 Backhaul cost.

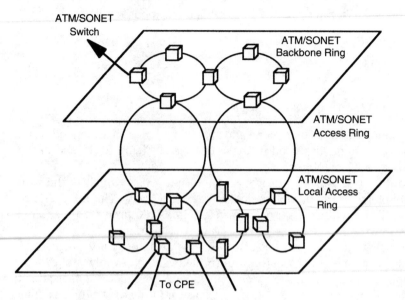

Figure 20.11   Combination of star and ring topologies.

20.5 Summary

This chapter presented the design issues related to the broadband backbone network and its requirements. Although some of the requirements are similar to those of the access network, the requirements are more stringent in terms of failures in the backbone network. Three categories are related to the design:

- the traffic offered from the access network
- the link capacity design
- the broadband backbone node design

The different topologies and architectures used in the broadband network environment were discussed. For a SONET/ATM environment, ring topology is recommended because of the advantages offered by a SONET ring network, such as an alternate route if one ring is cut off.

Miscellaneous

Here we address issues that do not fit in any of the previous sections of the book. In this part, there are four chapters. Chapter 21 discusses broadband/ATM deployment around the world. Chapter 22 discusses the telecommunications regulatory bodies around the world which are involved in deciding telecommunication policies. Chapter 23 discusses the different opportunities for ATM equipment vendors in the telecommunication network. Chapter 24 discusses the future of broadband and ATM switching.

Miscellaneous

21

ATM/Broadband
Around the World

21.1 Overview

In the previous chapters, we discussed the various broadband technologies and protocols. We then looked into the different environments where ATM/broadband is applicable. Finally, we discussed how to design a broadband network (public network). We mentioned that the most important element of design is to estimate customer traffic from the application used. The best way to determine an application's characteristics is to test it out with real customers running real traffic. This chapter gives a tour of different trial efforts underway around the world. These trial activities can offer service providers insight into traffic characteristics. Having seen the theoretical aspects and concepts behind broadband in the earlier chapters, the questions that immediately follow are

- Is ATM hype or real?
- Is ATM a universally accepted technology?
- What is the proof that ATM will deliver the advantages it promises?
- What applications are suitable for ATM/broadband?

A simple answer to these questions is that ATM is for real, it is a universal standard, and proof is given in terms of trials done and commitments made by service and equipment providers around the world. Some of these trials are mentioned in this chapter.

Regarding applications suitable for ATM/broadband, the answer is in chapter 2. Specific applications with respect to trials are mentioned in this chapter. Tests and trials answer these questions and help understand the real-world broadband communication environment.

In addition, suggestions are offered for countries that have no ATM/broadband trials, but want to jump onto the ATM/broadband bandwagon by conducting trials, testing, and deployment.

21.2 ATM/Broadband Trials in North America

No other country has more broadband trial activities and commitments than the United States. In all segments of the network—voice, data and video—some activity related to trial, testing or deployment of broadband is underway. When it comes to deployment of service, however, the United States is a bit cautious, especially in the public local exchange network areas, with some exceptions. Some of the ATM/broadband testbeds and trials are mentioned with the applications being tested on it. In general, the applications fall into one of the three major categories, voice, video, or data.

The Canadians are also testing broadband/ATM with their long-distance carriers. The Canadian efforts are joint ventures between business and universities.

21.2.1 United States

Contrary to the popular belief that the average researcher is a single professor with a group of students, most of the research in broadband networks is being conducted by large teams of senior researchers from large organizations. In some cases, the research involves multiple organizations. The best known broadband networking testbeds are called *gigabit networking testbeds,* because the transmission systems are in gigabits. They are funded by the National Science Foundation (NSF) and Advanced Research Projects Agency (ARPA). These gigabit testbeds are formed by funding from Government and industry. The initial work on testbeds was created in 1989 and coordinated by the Corporation for National Research Initiative (CNRI). The current gigabit networking testbeds are the following:

- Blanca testbed
- Aurora testbed
- Nectar testbed
- Project Zeus testbed
- CASA testbed
- VISTA testbed

- MAGIC testbed
- BAGNET testbed
- ACRON testbed
- BBN testbed
- COMDisco testbed

Figure 21.1 shows a gigabit testbed coordinated by CNRI.

The goal of the gigabit trials is to study and understand the network parameters by simulating throughput conditions and determining the effect of delay with different application mixes. In addition to these testbeds, the U.S. IECs have deployed their own nationwide ATM network to provide ATM-based services. Some of the carriers are Sprint, Wiltel, AT&T, MFS, and MCI. In the following subsections, we give an overview of some of the testbeds.

21.2.1.1 Aurora testbed. This testbed is sponsored by ARPA/NSF grants. Figure 21.2 shows an Aurora transmission network spanning from Boston (MIT) to Philadelphia (U. of Penn). The transmission networks are SONET OC48 systems from Northern Telecom; the ATM switches are from Bellcore and IBM. This network was deployed in three phases.

- Phase 1: White Plains to West Orange in 4Q92
- Phase 2: West Orange to Philadelphia (U. of Penn) in 2Q93
- Phase 3: White Plains to Boston (MIT) in 4Q93

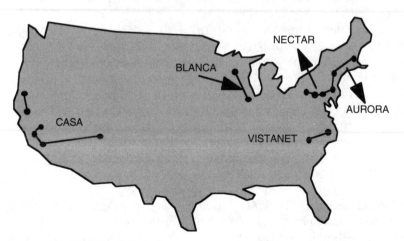

Figure 21.1   CNRI gigabit testbeds.

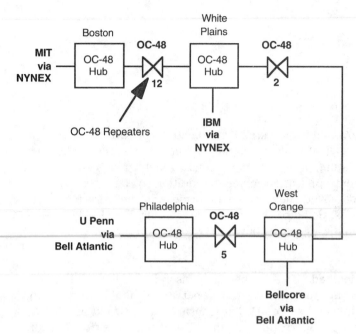

Figure 21.2 Aurora Network.

Based on the research conducted on this network, numerous papers have been published. The research was conducted in many areas, including switches, architecture, flow control, congestion control, and ATM protocol. In addition to these areas, this project is now used for conducting experiments in interworking of different protocols with ATM/SONET.

21.2.1.2 CASA network testbed. Figure 21.3 shows the CASA network topology. The network was designed to experiment on applications of distributed computing. This network is in California and funded by ARPA/NSF. The network was established in 1990 to link research centers such as JPL (Jet Propulsion Lab), Los Alamos Lab, and Caltech. The public carriers involved in the projects are Pacbell and MCI. The transmission network consists of OC48 terminals from Northern Telecom. The network consists of supercomputer centers in the above-mentioned labs. All the supercomputer centers are connected via high-performance parallel interface (HIPPI) to a SONET backbone. ATM switches will later be added to the network.

21.2.1.3 MAGIC testbed. The MAGIC (Multidimensional Applications and Gigabit Internetwork Consortium) research network is one of the newest testbeds, funded by ARPA in early 1992. It is a SONET/ATM-based network in the midwestern United States. The participants in this testbed are the following:

- Sprint
- Minnesota supercomputer center
- U.S. Army battle lab
- U.S. earth resources observation systems data center
- University of Kansas
- SRI International
- U.S. Army command battle lab
- DEC
- Southwestern Bell Telephone
- Northern Telecom

The purpose of MAGIC is to study two general topics (1) gigabit visualization application over a high-speed network, especially military problems involving terrain visualization and (2) simulation and integration of protocols such as TCP/IP into high-speed SONET/ATM environments.

Figure 21.4 shows the MAGIC network testbed. It consists of wide area BISDN network, a gigabit LAN, gigabit LAN/WAN interfaces, supercomputer access, gigabit workstation interface to LAN, ATM switching, and SONET transmission system with OC3 and OC12 as access and OC48 as backbone. This testbed will experiment on every part of the broadband network.

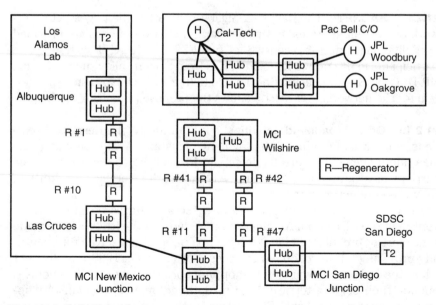

Figure 21.3 CASA Network.

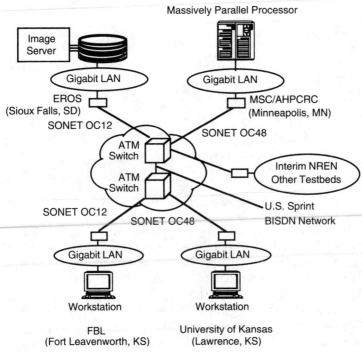

Figure 21.4 MAGIC network testbed.

21.2.1.4 BAGNET.   The Bay Area Gigabit Network is the largest ATM network in the United States. In this network, the ATM switches are provided by Fujitsu. This network consists of two FETEX-150 ATM switches. The participants in this network include firms from Silicon Valley such as Xerox, HP, DEC, NASA, and Pacific Bell. Pacific Bell provides the transmission system for the network. Figure 21.5 shows the proposed BAGNET testbed.

21.2.1.5 Other broadband activities.   In addition to the described testbeds, numerous broadband deployment plans and ATM/broadband service announcements were made public in 1993 by the carriers. These carriers have announced the first ATM-based T1 service. In addition, they are conducting international trials to provide ATM services across the Pacific and the Atlantic. Some of these are addressed in the intercontinental trial section. Among the LECs, most of the RBOCs have made some sort of broadband plans or have positioned themselves for eventually providing ATM/broadband services. For example, 1993 was the year when the Baby Bells went on a shopping spree to buy CATV franchises outside their serving territory and made huge investment plans for upgrading their network. One RBOC—Ameritech—announced a detailed plan to deploy an end-to-end broadband network for high-speed voice, data, and video services by the end of 1994.

Ameritech plans to use AT&T GCNS-2000 ATM switches in 14 cities in the Ameritech serving area (Chicago and Urbana-Champaign, Illinois; Indianapolis, Indiana; Ann Arbor, Detroit, Grand Rapids, and Pontiac, Michigan; Cleveland, Columbus, and Dayton, Ohio; Appleton, Eauclaire, Madison, and Milwaukee, Wisconsin). All these cities were planned for the first half of 1994. Prior to Ameritech's announcement, Pacbell announced a $16 billion broadband network enhancement where the ATM network will be used for public and private telecommunications. For more information, contact Ameritech data processing center at 1-800-TEAM-DATA.

In addition to the public carrier plans, the U.S. government is leading the efforts for setting up a National Information Infrastructure Testbed (NIIT). The customers in this network are distributed across various industries in the country. NIIT is led by a group of corporations, universities, and governments who will jointly conduct real-world applications (some of them are described in chapter 2). For more information on membership, contact 1-800-299-9973.

Because all activities have not been covered, more up-to-date information on ATM/broadband service can be obtained from magazines such as *Network World, Communication Week*, etc. A detailed list of magazines is provided in appendix C. This list gives an idea of how big ATM/broadband is.

21.2.2 Canada

An ATM/broadband testbed called OCRINET was announced for Canada in late 1993. This network connects Carleton University and BNR, an Ottawa research center, to conduct research on a variety of high-speed

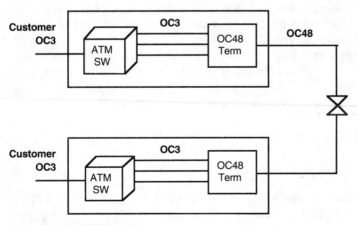

Figure 21.5 BAGNET testbed.

voice, video, and data services. As a part of the expansion, Canada's long-distance service provider, Stentor, joined the trial with the help of its U.S. partner for ATM trials sometime in 1994. The ATM switch used for the trial is Northern Telecom's Magellan gateway switch.

21.3 ATM/Broadband Trials in Europe

Before we discuss the different countries, we address the ATM/broadband trials on the European continent. It is more difficult to deploy a single uniform standard network in Europe than in any other part of the world. To realize a single network by putting the required infrastructure in place, service providers such as British Telecom, U.K., will not only have to invest a combined $270 billion, but will also have to enter cooperative ventures for the development of a new network architecture. To achieve such a goal, the European Commission (EC) did some background work, such as starting the Special Telecommunication Action for Regional Development (STAR), whose function is to help in the construction of a telecommunications infrastructure by using large, strong carriers. The EC has formulated the measures for telecommunication public policy in the EC Green Paper, 1987. The policy pursues the dual goal of providing Europe with high-quality telecommunications service while strengthening the European telecommunications industry for international competition.

To achieve such a network, European carriers must extensively adopt European standards such as ETSI as well as standards developed by the international standards bodies. In fact, some of the standards they have agreed upon are SDH for transmission and ATM for switching. Based on these standards, the European carriers have developed a network called Global European Network (GEN), consisting of carriers like German Telecom (DB Telecom), British Telecom, France Telecom, Telephonica of Spain, and STET of Italy, who have agreed to construct a fiber-optic transmission network. Figure 21.6 shows the GEN network.

GEN is a European-wide digital network. This network will evolve into METRAN (Managed European Transmission Network), which beginning in 1995, will provide fast and flexible provisioning of transparent transmission links up to 155 Mbps. METRAN is based on SDH standards, enabling it to support ATM switching that can be used for future BISDN services. Currently, 25 European carriers are involved in METRAN.

21.3.1 Germany

In Germany (formerly West Germany), the BISDN/ATM pilot project of DB Telecom, based on the internationally standardized ATM, went into service in 1994.

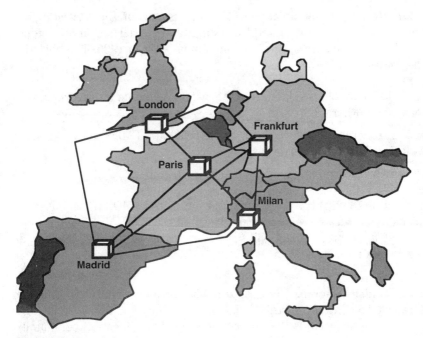

Figure 21.6 GEN network.

Figure 21.7 shows the German BISDN network configuration. The network consists of three BISDN exchanges in Berlin, Hamburg, and Cologne. The network will initially provide ATM-based permanent virtual circuit (PVC) by the end of 1994. At the beginning of 1995, ATM-based SVC services will be offered. The ATM switches are based on the Siemens ESWD platform, along with other Siemens Vision O.N.E.® products. In the network, each BISDN node consists of two remote ATM units with terminals, terminal adapters, and connectionless servers. Each exchange has 32 BISDN ports. Apart from testing technical features of BISDN, the following are the applications or services that are tested:

- data services via LAN connections
- multimedia services including videoconferencing
- incorporation of existing voice services

21.3.2 England

Similar to the United States, there are multiple long-distance carriers in the United Kingdom, and each one is conducting its own ATM trials. Two of the most popular carriers are Mercury Communication and British Telecom. In this section, we address the ATM trials planned by these carriers.

21.3.2.1 Mercury Communication. Mercury is one of the largest public network operators in the United Kingdom. Its ATM trials began in October 1993. Mercury is planning to offer commercial ATM in 1995. The objective of the trial is as follows:

- test ATM technology related to ATM switching and its interworking with router and other equipment
- look at practical interfaces with frame relay and SMDS networks along with the SDH transmission system
- identify broadband applications that help customers

In addition to ATM trials, Mercury has conducted SMDS trials, but it has no plans to offer frame relay service. Based on the trials, Mercury has identified that LAN interconnect will be the driving force for broadband wide area networking.

21.3.2.2 British Telecom. The largest public carrier in the United Kingdom was awarded a contract worth £18 million over four years to develop a high-speed fiber-optic network for the country's higher education community.

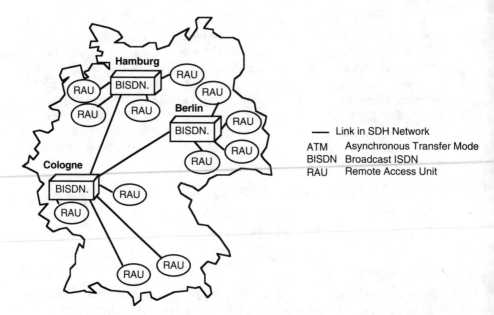

Figure 21.7 BISDN network connectivity in Germany.

Figure 21.8 Super Janet network in U.K.

The pilot network is called Super Janet (Joint Academic Network). The initial contract was awarded in November 1992. Figure 21.8 shows the Super Janet network configuration.

The core network will operate at plesiochronous digital hierarchy (PDH) speed ranging from 34 Mbps to 140 Mbps. The initial network consists of six sites, Cambridge; Edinburgh; the Manchester Universities; Imperial College of Science, Technology and Medicine; University College, London; and SERC Rutherford Appleton Lab. Six additional sites are to be added—the Universities of Wales, Birmingham, Nottingham, Newcastle, Glasgow, and Leeds. The advisory board is planning to add 40 more universities later using BT SMDS network.

Super Janet is a complement of existing Janet networks that constitute an X.25 packet network. The initial contract included the introduction of synchronous digital hierarchy (SDH) in the backbone by the end of 1993. Along with SDH, ATM services are to be introduced based on ITU-T recommendations.

Companies that manufacture ATM switches, such as Fore, Netcom, New-bridge, Stratacom, and Net/Adaptive, have been selected for Super Janet. The initial application of Super Janet will be data-oriented, and most applications are currently associated with it. The following are the new applications proposed for Super Janet:

- distance learning
- experimental electronic journal testbed
- information services, such as library document distribution
- high-quality medical imaging
- distributed group communications
- advanced data visualization

Janet Network team members selected ATM because it is a strategic networking technology and can handle all types of traffic.

21.3.3 Finland

Telecom Finland, the telecommunication side of Posti ja Tele (P & T) launched a pilot ATM network in May 1993, connecting sites in the Helsinki and Tampere metropolitan areas. The trial ran for one year. Full commercial ATM service is planned. Funding for the project comes mainly from the Technology Development Center of Finland.

The core network consists of four Netcom DV2 ATM switches, out of which two are based at the Tampere University of Technology. The other two switches are based at Telecom Finland headquarters in Helsinki. The two locations are connected using PDHs at 34 Mbps. The workstations used in the trial are fitted with adapter cards from the Fore switch, which implements ATM AAL 3/4 and 5 layers. The main objective of this trial is to provide Finnish industry with up-to-date knowledge and experience of the latest ATM technology.

21.3.4 France

The public carrier of France, France Telecom, announced that it will provide backbone ATM service by the end of 1994. The initial public service will be for LAN interconnect services based on the ETSI standard. Initial access speeds for ATM service are 2 Mbps, 10 Mbps, 16 Mbps, 25 Mbps, and 34 Mbps.

France Telecom is currently installing two international and four national ATM nodes with plans for 15 additional ones. These nodes are expected to be in place by the end of 1994. The ATM switches are provided by Alcatel, Siemens, and Matra. The service can be accessed via LAN interfaces such as

802.3, 802.5 and FDDI using Internet Protocol (IP) as well as via other interfaces, such as SMDS (SNI) and frame relay UNI (FRI).

France Telecom's strategy for ATM service, which is scheduled for commercial launch in 1995, is phased in the form of three experiments:

- BREHAT ATM project, 1st Quarter of 94
- BETEL (Broadband Exchange Trans European Links) project
- Successor to BETEL, European ATM pilot project of virtual path-based services.

Each is described in the subsections following.

21.3.4.1 BREHAT ATM project. The applications of the BREHAT project include audiovisual transport services, multimedia information retrieval services, and downloading of multimedia applications. The network, built in collaboration with Alcatel and TRT, consists of three ATM cross-connects located at Centre National d'Etudes des Telecommunications (CNET) Lab at Rennes in Brittany and Lannion and several sites in Paris. The application areas are categorized as

- LAN interconnect
- videoconferencing
- video transmission
- circuit emulation

21.3.4.2 BETEL project. The BETEL project is actually an ATM project between France and Switzerland. As two sites are in France, we listed this project under France instead of Switzerland.

This phase of the project includes two applications:

- sharing of supercomputers for scientific computing tasks
- distance learning using videoconferencing.

In this trial, two sites (one in France and another in Geneva, Switzerland) are connected via 34-Mbps links using the ATM switch provided by Alcatel. The user sites are equipped with FDDI LAN via Cisco AGS+ and Cisco 7000 routers. Figure 21.9 shows the BETEL testbed network.

21.3.4.3 Future BETEL ATM pilot projects. Eighteen public network operators are involved in this project. Each one has made a commitment to purchase and install at least ATM cross-connects conforming to the ITU-TS recommendation. These cross-connects are initially linked via a 34-Mbps

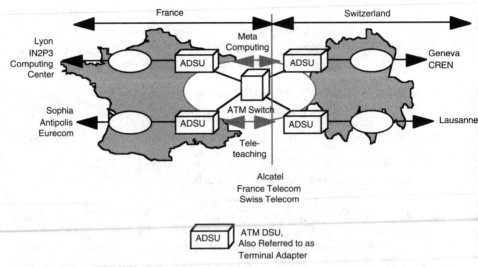

France   Switzerland

Lyon
IN2P3
Computing
Center

Meta
Computing

ADSU   ADSU

Geneva
CREN

Sophia
Antipolis
Eurecom

ADSU   ATM Switch   ADSU

Lausanne

Tele-
teaching

Alcatel
France Telecom
Swiss Telecom

ADSU   ATM DSU,
Also Referred to as
Terminal Adapter

Figure 21.9 BETEL testbed network.

link. Following interoperability tests, the first demonstration of the network is expected to be in 1994.

21.3.5 Switzerland

The largest ATM testbed in Europe has been completed in Basel, Switzerland. The R1022 ATM Technology Testbed (RATT) covers 800 square miles and is a culmination of six years' work under the EC backed by the Research and Development in Advanced Communications Technologies in Europe (RACE) program.

The key objective of RACE is "the introduction of Integrated Broadband Communications, considering the evolving ISDN and national introduction strategies, progressing toward community-wide services by 1995." Basel was selected because of its proximity to German and French networks and because it was the site where a 140-Mbps SDH pilot trial called BASKOM was held.

In the RATT network, three ATM switch prototypes were tested, each working at 155 Mbps at both user-network interface (UNI) and network-network interface (NNI). The two main exchanges were from Alcatel Bell Telephone in Belgium and Philips Kommunikations Industrie (PKI) from Germany. A remote exchange from AT&T was hooked up via a 622-Mbps SDH link. Other subsystems include five ATM terminal adapters, two NT-2s, and one NT-1. The applications tested in this testbed are AAL 1, AAL 3/4 and AAL 5.

The RATT network is intended to be hooked up with other broadband networks in Belgium, Germany, France, Holland, Spain, Portugal, and Denmark. Numerous other countries are also in the process of being linked.

21.4 ATM/Broadband Trials in Asia

Asia is the largest continent with the world's largest population. With the exception of certain countries such as Japan and Singapore, Asia lags far behind the rest of the world with regard to telephone systems. Broadband in Asia is thus far from reality. Some carriers are planning to pursue a broadband network, however, and we address two, China and Japan.

21.4.1 China

SCM/Book Telecommunication LP and China's Galaxy New Technology Co. have joined forces to design, build, and operate a commercial and civilian BISDN system in one of China's most affluent areas. The initial prototype is anticipated to be a $20 million project, which was deployed in January 1994. The network is being deployed at Gvangzhov and is expected to be operational by January 1995.

Gvangzhov is located less than 200 miles northwest of Hong Kong and has a population of 70 million, roughly equal to that of old Germany. The China America Telecom company is working with the Gvangzhov government to build and operate the network, which will provide foreign cable links for TV programming, as well as telephone, cellular, and data services.

21.4.2 Japan

As American and European companies continue to jump on the broadband bandwagon, companies in the Asia Pacific region are also moving as quickly as possible. NTT Corp., Japan's largest public carrier, is performing extensive evaluation of BISDN services, which incorporates voice, data and video transmission in an optical network.

In Japan, the first phase of the experiments was centered on the expansion of NTT-INS (integrated network services), new broadband corporate network services based on ATM technology, and an optical transmission system for subscribers.

In phase 2, experiments were aimed at incorporating ATM, intelligent network, and a radio access system into the BISDN trial.

21.4.3 Other Asian countries

Some countries, such as Singapore, Taiwan, and Korea, have some form of broadband trial under way. Other countries need to go a long way to build the basic infrastructure to provide any form of telecommunication service. These countries, with the opening of telecommunications to private industries, have the potential to get on the broadband bandwagon when the ATM/broadband standards are mature enough and if no political barriers exist to the provision of telecom services.

21.5 Intercontinental ATM/Broadband Trials

This section discusses two intercontinental ATM trials, one across the Pacific and the other across the Atlantic.

21.5.1 Transpacific ATM trial

KDD of Japan and AT&T of the United States have linked the two nations via the world's longest intercontinental ATM network, spanning 9000 miles. The ATM trial began operating in July 1993. The link was set up between Shinjuku, Japan, and Holmdel, New Jersey. This network examines the performance and quality levels of the ATM infrastructure, carrying traditional data and emerging multimedia traffic. Network operation issues are also examined.

The ATM trial network will evolve over three years in three phases. Phase 1 will include the characterization of performance of its DS3 facilities and the different types of cabling that exist in the 9000-mile span. Phase 2 is expected to begin in the second quarter of 1994 and will include testing of applications such as voice, data, and videoconferencing over the ATM network. Phase 3 will include a thorough evaluation of the operation and network management features and is expected to begin in 1995.

The ATM trial network consists of AT&T GCNS-2000 ATM switches at both Shinjuku and Holmdel. Initially, 45 Mbps will be used. Figure 21.10 shows the network between Japan and the United States.

21.5.2 Transatlantic ATM trial

France Telecom demonstrated the feasibility of ATM over intercontinental distances. The test used connectionless broadband data service technology over ATM virtual paths for high-speed interconnection of conventional Ethernet LANs located in New York and Paris.

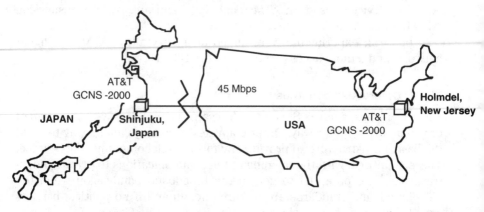

Figure 21.10 Transpacific ATM trial.

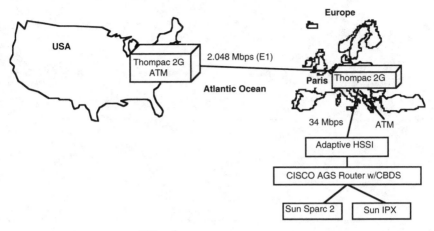

Figure 21.11 Transatlantic ATM trial.

The Thompac 2G ATM switches were supplied by Thompson, which does both VP cross-connections and constant bit rate data service (CBDS) switching. The test demonstrated how ATM technology can be implemented in the design of an international private enterprise network. The transatlantic lines used were the recently available TAT-11, which was provided by Sprint International, and the access line in New York, provided by Teleport Communications Group. Figure 21.11 shows the transatlantic connectivity for the ATM trial.

21.6 Summary

The objective of this chapter was to show that ATM/broadband is no longer a hype or a technology only for researchers at universities. It exists now, and by 1995 many of the trials mentioned will become commercial. In the United States, long-distance carriers have already announced a tariff for T1-based ATM service, which is comparable in cost with frame-relay and private-line T1. This tariff is another positive indication that ATM services will be comparable in price with existing services.

22

Telecommunications Standards

22.1 Overview

Standards are a set of specifications that groups of people from different countries and regions (continents) or from within a country agree to follow. Standards are especially important in the telecommunications industry because of the need for interworking among different groups of people. Telecommunications is a major industrial sector for two reasons:

- it is vast in terms of revenue generated per year
- it affects all aspects of commercial and domestic life

Standardization within telecommunications is intended to perform three basic functions:

- facilitate interconnection between different users
- facilitate the portability of equipment within different applications and regions
- ensure that equipment bought from one vendor can interface with that from another

In this chapter, we provide an overview of the standards-making process and of the different standards organizations at national and international levels around the world. Almost every organization has some sort of involvement in the broadband communications standards-making process. Study groups or working groups within the standards organizations address the needs and issues of broadband communications.

22.2 Standards-Making Process

The most difficult process in the standards arena is at the global level. Historically, each country has had its own postal, telephone, and telegraph (PTT) authority. Countries have set standards and monopolized local manufacturing and service provisioning. There has been little interest in standardization. Even when standards bodies were set up, they produced recommendations rather than requirements (hence not enforceable) containing many country-specific exceptions.

Producing standards is not an easy task. It is important to understand that standards formation does not hold back technological progress. At the same time, once a technology becomes established, and companies have invested in rival systems, each player is eager to promote its methodology as the international standard. Furthermore, leaders in any field are reluctant to slow down and agree to standards that can eventually help their rivals.

Standards-making is an expensive business. It is estimated that hundreds of millions of dollars per year are spent on worldwide telecommunication standardization activities. The standards-making process is one of cooperation at many levels, nationally and internationally. Cooperation exists between industrial concerns within a country, between these concerns and their national governments, and between nations at the international level. Users groups and trade organizations usually have members from several countries. Figure 22.1 shows an example of a standards-making process in the United States.

Many industrial organizations are multinational in operation and whose turnovers exceed the gross national product of some countries. Cooperation is important to obtain agreement on standards, but a danger exists that the

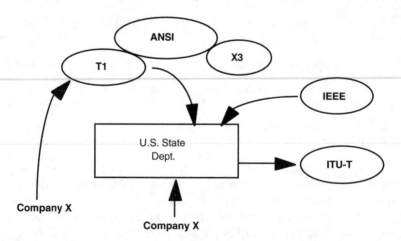

Figure 22.1 Example of standards-making process in U.S.

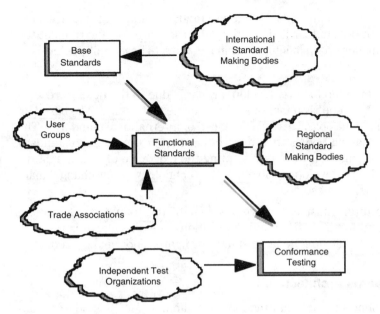

Figure 22.2 Standards-making process.

many separate groupings and interests can result in different standards be-ing prepared for the same item. This fact has resulted in the world being divided into two transmission standards, 2.048 Mbps (A-law codec) in Europe and 1.544 Mbps (μ-law codec) in the United States and Japan.

To accommodate the many conflicting interests, the international standards-making organizations often concentrate on producing *base standards*, which contain variants or alternative methods to allow flexibility for the implementor. Adopting the variants means simply that the implementor will be compliant with the standard, and there is no guarantee that equipment based on separate variants would interwork. It is the responsibility of the implementor to ensure that different equipment is interoperable.

The problem of interworking is being tackled by regional and national standards bodies, often consisting of trade organizations and users groups, with well-defined requirements. These groups might be manufacturers or users. They adopt internationally based standards as functional standards or profiles, which contain only a limited subset of the permissible variants. Agreed-upon test specifications and methods are also developed to ensure that equipment designed to the different variants permitted within the functional standards can interwork. Independent test houses such as Bellcore or Cable Labs then conduct conformance tests against the selected profiles and certify products that meet these requirements. The standards-making process can be categorized into three stages as shown in Figure

22.2. They are base standards, functional standards, and conformance standards. Each of the standards down the ladder is more specific to meet the needs of a smaller group, such as within a country. These multiple levels of standards have led to several problems:

- Standards take too long to develop, primarily due to the need to reach a consensus between rival factions. Thus, the de facto standards, based on proprietary solutions, are available ahead of international standards. The standards-making bodies then must either accept the de facto standard and abandoning their own, or accepting the existence of two standards. If standards are developed too early, the technology might change, making the standards obsolete.

- Standards often must cover every aspect of the intended application, usually resulting in overlap and duplication. The alternative is to avoid duplication by allowing de facto standards in areas not yet covered.

22.3 Standards Architecture

Figure 22.3 shows today's information/telecommunications standards-making architecture. This architecture gives an idea of the complexity of telecommunication standards-making. The arrows show the direction of the flow of information and interworking required for that organization to reach a consensus. One can see that there are different levels of standards-making, including national, regional, and global/international. In Figure 22.3, the central circles represent ITU-T and JTC1, which are international organizations. Other organizations deal with the issues at a regional level and contribute their methods or ideas to the international organization to influence the decision in their favor. These regional organizations are formed by user-interest groups or other specialized organizations, who hope to have their ideas and technology adopted as standards at a national level. These organizations are shown in the periphery.

Each region around the world is controlled by certain standards bodies. Some of the standards bodies and regions where they have jurisdiction are as follows:

T1	North America
ETSI	Europe
TTC	Japan
TTA	Korea
AOTC	Australia
CITEL	Latin America

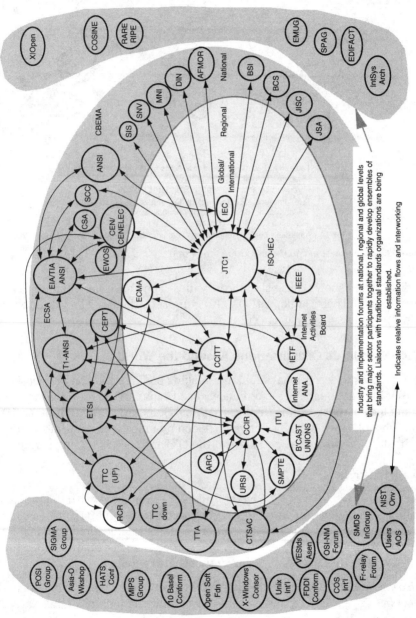

Figure 22.3 Global telecommunication standards architecture.

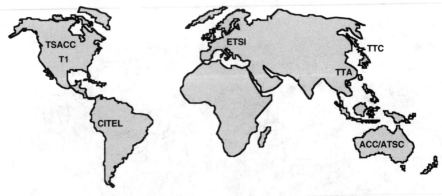

TSACC	Telecommunications Standards Advisory Council of Canada
CITEL	Telecommunication Commission
ETSI	European Telecommunication Standards Institute
TTC	Telecommunication Technology Committee
TTA	Telecommunication Technology Association
ACC	Australian CCITT Committee
ATSC	Australian Telecommunications Standardization Committee

Figure 22.4 Regional standards bodies around the world.

Figure 22.4 shows these standards bodies. They have the ultimate say in their region. These organizations also represent their regions in international organizations such as ITU and JTC.

22.4 Standards Organizations Around the World

The foremost international standards-making body for telecommunications is the International Telecommunications Union (ITU). For information technology standards, it is the International Standards Organization (ISO). With the blurring of lines between telecommunications and information technology, however, the activities of the ITU and the ISO often overlap.

22.4.1 The ITU

The ITU was founded in 1865 as the Union Telegraphique, with the primary goal of developing standards in telecommunications. In 1947, it became a specialized agency for United Nations telecommunications, under UN Charter Articles 57 to 63, when it was renamed the ITU. Figure 22.5 shows the new structure of ITU. ITU currently has three main functions:

- To encourage the interconnectivity of telecommunications equipment and services by promoting and establishing technical standards in these areas.

- To promote the best use of scarce telecommunications resources by the implementation of international regulations. This task is especially important in the use of the radio frequency spectrum, which the ITU controls via the ITU-T, ITU-R, and the International Frequency Registration Board (IFRB), and conferences such as WARC and ITC.

- To encourage the growth of telecommunications in developing countries via the Technical Cooperation Department (TCD) and the Telecommunications Development Bureau (TDB).

In addition, the ITU carries out other ancillary functions for its members, such as organizing telecommunications exhibitions to keep members informed of the latest advances in technology. The best known exhibition is the Telecom Exhibition, which is held every four years.

At the end of 1993, the ITU had 173 countries as members, each with equal voting rights. Recently formed countries such as Lithuania have also become members of ITU. Only administrative bodies, or government departments concerned with telecommunications, can be ITU members. The United Kingdom is thus represented by the Department of Trade and Industry. Most other countries are represented by state-owned PTTs. For the United States, representation is by a complex mix of government agencies and suppliers, but the interface is only through the State Department. In addition, other organizational groups are recognized that can attend meetings of the ITU and its organizations but cannot vote. Some of these organizations are the following:

- *Recognized Private Operating Agencies (RPOAs).*   Over 80 of these have been recognized by the ITU and consist of telecommunications operators. Examples of these are BT and Mercury in the United Kingdom and Nippon Telephone and Telegraph Corporation (NTT) in Japan.

- *Scientific and Industrial Organizations (SIOs).*   There are over 150 of these, including IBM, Northern Telecom, Siemens, etc.

- *International Organizations (IOs).*   There are over 50 IOs recognized so far. These are mainly trade organizations and user groups, such as the International Telecommunications User Group (INTUG).

RPOAs can be members of ITU-T and ITU-R and have voting rights, although SIOs and IOs can only be advisory members.

22.4.2 The CCITT or ITU-T

The Comite Consultatif Internationale Telephonique et Telegraphic (CCITT), recently changed to ITU-T, is the primary vehicle for developing technical standards in the telecommunications field. Throughout this book, the new terminology ITU-T has been used. In the 1989 Plenipotentiary, the ITU-T and ITU-R were directed to conduct their activities with due consideration for the work of national and regional standardization bodies, keeping in mind that

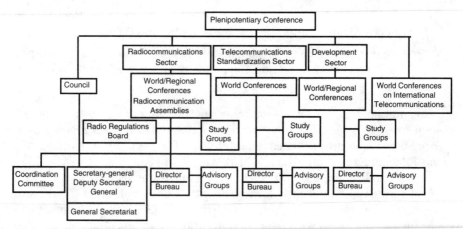

Figure 22.5 Structure of ITU.

the ITU must maintain its preeminent position in the field of worldwide standardization for telecommunications. The two ITUs (ITU-T and ITU-R) were given a mandate to coordinate global standards activities, with the ITU-R developing standards in the radio field, opposite to those of the ITU-T.

The International Frequency Registration Board (IFRB) comprises five specialists elected by the Plenipotentiary Conference. The IFRB acts as the custodian of an international public trust. Its main duty is to decide whether new frequencies assigned to radio users are in line with the conventional radio regulations. It also provides advice on improving the utilization of the radio spectrum and on the preparation and organization of the Administrative Radio Conferences.

The World Administrative Radio Conference (WARC), the World Administrative Telephone and Telegraph Conference (WATTC), and the Regional Administrative Radio Conference (RARC) are held at irregular intervals as required. These conferences approve changes to the regulations in their respective fields. The WATTC establishes principles relating to the operation of international telecommunications public services. The WARC revises regulations governing the use of radio frequency spectrum and geostationary satellite orbits. The RARC deals with specific radio communication questions of a regional nature.

The task of the ITU-T is vast and almost doubles with every Plenary. Because of the problems of producing standards in a timely fashion, the ITU-T, at its 1988 IX Plenary held in Melbourne, Australia, agreed to an accelerated procedure for issuing standards. Recommendations are now adopted as soon as they have been stabilized, rather than needing full ratification by the Plenary Assembly. Thus, standards could be ready within one year of starting work.

The ITU-T has also adopted more of a coordinating role between the various national and regional standards-making authorities. Key among these are the US T1 committees, the Japanese TTC, and the European ETSI. The ideas and work from these organizations, in addition to those from R&D organizations, manufacturers, users, and service providers, flow in and out of the ITU-T until they are stable and become accepted as standards.

An area of contention in ITU-T standards-making activities is the work carried out by Study Group III, which produces recommendations covering pricing and supply conditions, especially for international leased circuits. The European community has stated in a Green Paper that this work conflicts with the communities' competitive laws.

Table 22.1 shows the ITU-T study groups, where each study group is responsible for certain area. For example, Study Group XVIII is responsible for broadband-related activities. The table summarizes the function performed by each study group.

22.4.3 The ISO

Based in Geneva, the ISO is a non-treaty organization and body of the United Nations. The ISO's primary aim is to promote the development of international standardization to facilitate international trade in goods and services. The ISO mainly works in the information technology area, while its sister organization, the International Electrotechnical Commission (IEC), is involved in standards for electrical and electronic engineering. All standards developed by ISO are published as international standards. It is the responsibility of the individual national standards organizations to promote and distribute these standards within their own countries.

TABLE 22.1 ITU-T Study Groups

Study group	Services
Study Group I	MHS, Fax, Dir. Services, UPT Serv. Desc. GVNS, PCS/UPT, ICCN
Study Group II	Network operation, numbering/routing country codes, fax quality
Study Group III	Tariff and accounting principles, accounting rates, data accounting
Study Group IV	Maintenance
Study Group V	Protection against electromagnetic effects
Study Group VI	Outside plant
Study Group VII	Data communications networks MHS, Data Comm. X.25 for ICCN
Study Group VIII	Terminals for telematic services
Study Group IX	Telegraph networks and telegraph terminal equipment
Study Group X	Languages for telecommunication applications
Study Group XI	Switching and signaling; SS7, UPT, IN, ICCN, Network Outage; ISDN Services
Study Group XII	Transmission performance of telephone networks and terminals
Study Group XV	Transmission systems and equipment
Study Group XVII	Data transmission over the telephone network
Study Group XVIII *	ISDN, Broadband, CBDS, UPT Network Terminology

* Most of the broadband-related standards come out of this study group (SG XVIII)

The ISO's members are primarily national standards-making bodies, such as ANSI (U.S.), BSI (U.K.), and DIN (Germany). Members can be active or participating, designated as P members, or they can be corresponding members or observers, designated as O members. P members lead technical committees or subcommittees.

The ISO and the ITU-T work closely together in areas of common interest. For example, all ISDN activities within ISO are conducted in Technical Committee TC97, which is responsible for information processing systems. This committee has two subcommittees. SC6 is involved with telecommunications and information exchange between systems. This subcommittee is working with the ITU-T on common channel signaling (CCS) and the relationship of ISDN to the open system interconnect (OSI) model. SC21 is responsible for developing the seven-layer OSI model on which the ITU-T has modeled ISDN.

The ISO and the IEC work together in many areas through their joint technical programming committee (JTPC). This committee ensures that the two bodies avoid working on overlapping items. The ISO and the IEC have also set up a Joint Technical Committee on Information Technology, called JTC1, to develop generic information technology standards. This committee incorporates ISO TC97 and IEC TC93. JTC1 is responsible for producing international standards profiles (ISPs). Other organizations conduct conformance testing, such as the standards promotion and application Group (SPAG) in Europe and the corporation for open systems (COS) in the United States. All these organizations cooperate in the Feeders Forum, set up in 1987, which unifies the technical work and provides a forum for liaison with ISO and IEC.

22.4.4 International Trade and User Groups (INTUG)

INTUG was formed in 1974 to represent the telecommunications user organizations from several countries, including the United States, United Kingdom, Australia, and Japan. It is active in promoting the interests of its members and lobbying associations, such as the ITU, CEPT, PTTs, and ITU-T. Any person or group can join except PTTs and manufacturers, who need to be represented directly as individual members.

22.4.5 European Standards

The largest grouping of countries is in Europe, and the one that shows the greatest integration is the European Community (EC). Within the EC, Directorate General XIII (DGXIII) is responsible for telecommunications, information industry, and innovation. Its aims are the following:

- to assist in the development of the general economy by building a sound telecommunications infrastructure throughout Europe
- to foster growth of the telecommunications service sector so that it is effective and economically viable
- to develop the telecommunications industry within Europe so that it can compete effectively on the world stage.

DGXIII has six directorates and contacts with the senior officials group on telecommunications (SOGT), which comprises ministers of telecommunications and industry. They meet every six weeks under the chairmanship of the Director General. A subcommittee of SOGT is the analysis and forecasting group (GAP), which studies industrial developments in selected areas such as ISDN, broadband, cellular, etc. GAP organizes meetings that member countries attend and provide opinions. The output is the recommendations that can be made mandatory within the EC.

One of the major aims of the single European market is to ensure the free movement of goods and services, which requires the coordination of standards activities. A 1990 Green Paper proposed setting up the European Standardization Organization (ESO) to oversee the activities of the European standards-making bodies like CEN, CENELEC, and ETSI.

22.4.5.1 CEPT. The Conference European des Administrations des Postes et des Telecommunications (CEPT) was formed in 1958 by the PTTs to harmonize standards. It presently consists of 31 members, covering all the countries of the European Community and the European Free Trade Association (EFTA), plus PTTs from other European countries.

CEPT is a sister organization to CEN/CENELEC and participates in many of its work programs. This organization is administered by a member nation for two years, and meetings of the plenary body are held every two years. The CEPT presidency moved to Greece from the United Kingdom in October 1990.

Historically, CEPT has been noted for its restrictive bureaucratic policies rather than its commercial outlook, an image it is anxious to change. In the past, most of the PTTs paid only lip service to standardization. Because the output from CEPT was recommendations not enforceable and containing many country-specific exceptions, the standardization effort was largely ineffectual.

A 1987 European Commission Green Paper on competition in Europe within telecommunication markets clearly defined the PTTs as commercial undertakings (rather than monopolistic telecommunications administrations) that were subject to competition and with separate regulatory and operational activities. This paper resulted in CEPT setting up an independent body, in January of 1988, called the European Telecommunication Standard Institute (ETSI), to carry out all the standards activities on its behalf. CEPT still maintained the Technical Recommendations Application Committee (TRAC), formed in 1986, to approve standards for connection of equipment to public networks.

22.4.5.2 ETSI. ETSI is an independent organization funded by its members, who decide on its work program. The EC and EFTA, however, can fund ETSI to produce specific standards of interest to the community. ETSI's main interest is in telecommunications, although it also has interests in information technology, for which it cooperates with CEN/CENELEC, and in broadcasting, where it works with the European Broadcasting Union (EBU). The following are ETSI's main aims:

- to complete worldwide standards, in line with Europe's needs, and to choose a single option where many are allowed in international standards
- to anticipate the worldwide standards scene by adopting European standards and proposing these to international standards-making bodies
- to prepare a common European position for input to worldwide standards bodies such as ITU-T, ITU-R, IEC, and ISO and support these bodies in their work

ETSI membership is open to a wide spectrum of organizations of which there are six types:

- administrations that are part of the administration of a country
- public network operators
- manufacturers
- users and user organizations
- private service providers
- research organizations

It is this diverse range of membership that is the prime strength of ETSI, as it ensures a healthy interchange of views. In April 1991, there were 269 members of ETSI, of which manufacturers accounted for 63 percent; public network operators, 14 percent; national administrations, 10 percent; and users/service providers, 13 percent. Twenty three countries were represented from the EC and EFTA, Turkey, Malta, Cyprus, Czechoslovakia, and Poland.

There are currently 12 Technical Committees within ETSI:

- Radio, Equipment and Systems (RES)
- Groupe Speciale Mobile (GSM)

- Paging Systems
- Satellite Earth Stations
- Network Aspects
- Business Telecommunications
- Signaling Protocols and Switching
- Transmission and Multiplexing
- Terminal Equipment
- Equipment Engineering
- Advanced Testing Methods
- Human Factors

Figure 22.6 shows the relationship of ETSI with international organizations such as ITU.

Figure 22.7 shows the ETSI relationship with European standards organizations such as CEN.

22.4.6 American Standards Organization

In the United States, the T1 committee addresses telecommunications standards issues. Committee T1 currently has approximately 125 member companies, agencies, and other participants. To carry out its work program of 140 projects, Committee T1 has established six primarily function-oriented, technical subcommittees with subtending working groups and sub-working groups, as illustrated in Figure 22.8.

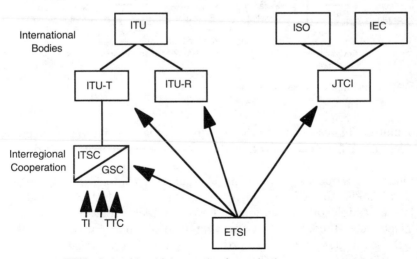

Figure 22.6 ETSI relationship with international organizations.

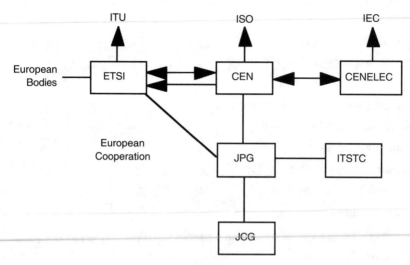

Figure 22.7 ETSI relationship with European organizations.

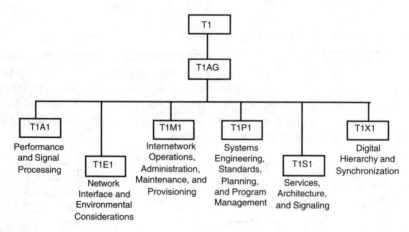

Figure 22.8 Committee T1 structure.

Committee T1 also has an advisory group (T1AG) comprising elected representatives from each of the four interest groups to carry out T1 directives and develop proposals for consideration by the T1 membership. Committee T1 is open to participants all around the world. The T1 committee comes under the ANSI standards organization, which deals with standards such as telecommunications, information technology, and electrical and electronics standards. It provides input to the Telecom Group in the State Department, which in turn provides input to international standards organizations such as ITU-T, ITU-R, etc.

22.5 ITU-T Related BISDN Standards

ITU-T began the standards activities on BISDN in 1985 in its study group 18. THe focus was on the next generation of network concepts, architectures, technologies, and services for high-speed and flexible communications. The basic Recommendation was completed in 1990. Since then, the BISDN Recommendations have evolved to meet the immediate needs of the telecommunications industry.

The detailed protocols specification are being released in three steps. Release 1 Recommedations were for the ATM adaptation layer, which does the adaptation for various data rate types such as frame relay, TCP/IP, etc. These recommendations are for a minimum set of functions that support point-to-point, constant bit rate services. The first set of signaling recommendations corresponding to release 1 was released in 1994. This set is compatible with existing Q.931 (NISDN-UNI signaling) and ISUP (NISDN-NNI signaling). The recommendation for release 2 (1994–1995) will support VBR capability for voice and video, point-to-multipoint connection, and QOS. The release 3 Recommendation (1996+) will cover enhanced connection configuration for multimedia. The recommendations related to the current stage of BISDN are shown in Figure 22.9 on page 438.

2.6 Summary

In this chapter, we gave an overview of standards bodies involved in the telecommunications industry. We covered the process involved in standards-making. Many standards organizations are present both at national and international levels. Every organization has some broadband-related activity as a part of its subgroup. Among the international standards-making bodies for telecommunications, the foremost is the International Telecommunication Union (ITU), which has specialized organizations to address different areas of standards. With the advent of broadband and other technologies, the lines between the telecommunications and information technology activities often overlap, forcing different organizations work very closely with each other.

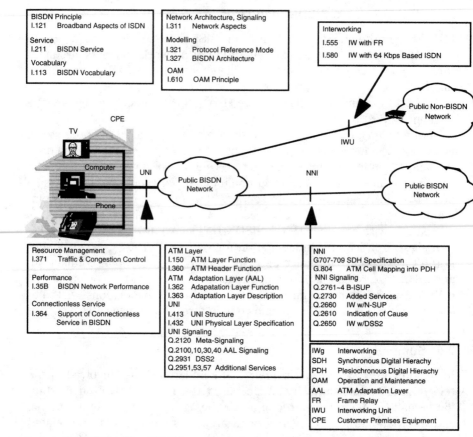

BISDN Principle
I.121   Broadband Aspects of ISDN

Service
I.211   BISDN Service

Vocabulary
I.113   BISDN Vocabulary

Network Architecture, Signaling
I.311   Network Aspects

Modelling
I.321   Protocol Reference Mode
I.327   BISDN Architecture

OAM
I.610   OAM Principle

Interworking
I.555   IW with FR
I.580   IW with 64 Kbps Based ISDN

Public Non-BISDN Network

IWU

TV
CPE

Computer
UNI
Public BISDN Network
NNI
Public BISDN Network

Phone

Resource Management
I.371   Traffic & Congestion Control

Performance
I.35B   BISDN Network Performance

Connectionless Service
I.364   Support of Connectionless Service in BISDN

ATM Layer
I.150   ATM Layer Function
I.360   ATM Header Function
ATM   Adaptation Layer (AAL)
I.362   Adapatation Layer Function
I.363   Adaptation Layer Description
UNI
I.413   UNI Structure
I.432   UNI Physical Layer Specification
UNI Signaling
Q.2120 Meta-Signaling
Q.2100,10,30,40 AAL Signaling
Q.2931 DSS2
Q.2951,53,57 Additional Services

NNI
G707-709 SDH Specification
G.804   ATM Cell Mapping into PDH
NNI Signaling
Q.2761~4 B-ISUP
Q.2730   Added Services
Q.2660   IW w/N-SUP
Q.2610   Indication of Cause
Q.2650   IW w/DSS2

IWg   Interworking
SDH   Synchronous Digital Hierachy
PDH   Plesiochronous Digital Hierachy
OAM   Operation and Maintenance
AAL   ATM Adaptation Layer
FR   Frame Relay
IWU   Interworking Unit
CPE   Customer Premises Equipment

Figure 22.9 BISDN-related ITU-T recommendations.

23

ATM Equipment Vendors

23.1 Overview

As mentioned earlier, ATM is the switching technology for future broadband communications, and it is a technology expected to span all the different segments of the network around the world. In this chapter, we see some of the equipment vendors who provide ATM-based equipment.

As far as ATM equipment vendors are concerned, it is not difficult for anyone to guess who is developing ATM-based products. In fact, it is difficult to tell which vendor has not made plans for ATM products. Everyone, from chip manufacturers such as TI to telecommunications switch manufacturers such as AT&T, have put forward their ATM plans in their respective fields of expertise.

This chapter reveals the commitments made by different types of equipment manufacturers related to their ATM-based products. Before we describe the different vendors in detail, however, we need to understand the different portions of the network where equipment vendors are positioning their ATM products. Figure 23.1 shows a generic network that identifies the opportunities for ATM products.

Based on the figure, opportunities for ATM equipment manufacturers span the network. Some of the areas identified are the following:

- chips
- adapter interfaces
- LAN switches
- hub or campus backbone switches

- access switches
- ATM backbone
- ATM test equipment to test the equipment and interfaces for compliance of standards

Although ATM is the common technology throughout the network, the environment and requirements for each of the ATM products are different. An equipment vendor can design a product to be positioned in more than one market by packaging the product accordingly, which is how equipment vendors differentiate their products and position themselves for a particular market. For example, backbone ATM switches must meet the government and regulatory bodies' standards requirements in terms of downtime, redundancy, etc., whereas an ATM LAN switch has to satisfy the customer whose needs might include the ability to connect the existing LAN networks to an ATM LAN switch.

In this chapter, we look into an overview of the products of some of the vendors in each segment of the market. Again, this discussion is only to

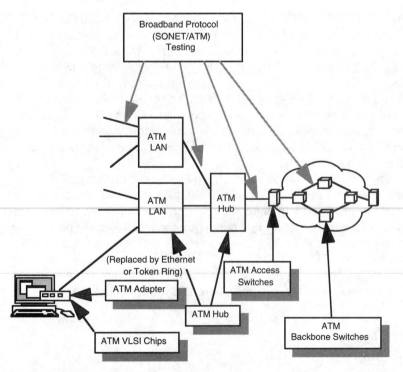

Figure 23.1 ATM opportunities in generic network.

show the commitment made by different vendors with respect to ATM products at the time of this writing.

23.2   ATM Equipment Vendors for LAN Products

No other area provides more opportunity for equipment vendors than LANs. Equipment manufacturers of all segments have developed some sort of ATM equipment, including

- chips
- adapters
- LAN switches

Each is discussed in the following subsections.

23.2.1   ATM chips

All the semiconductor manufacturers have positioned themselves to provide ATM chips, and most of them are already ATM forum members. Some of the manufacturers are Applied Micro Circuit Corp., Base2 systems, National Semiconductor, QPSX, Saturn Synoptics Chip Development, Texas Instruments, Transwitch and Vitesse G-Taxi chip.

The ATM chip does the function of the broadband protocol layer, including mapping the cells to the SONET OC3 interface payload; the DS3 PLCP ATM adaptation layer functions; and other protocol functions of varying interface speeds. These chips are used in the development of ATM products such as ATM adapters, ATM access switches, backbone switches, etc. These chips are used in all ATM equipment that perform standard functions.

23.2.2   ATM adapters

ATM adapters are the interface cards plugged into a computer typically in a workstation or PC bus. The information from the computer is transferred to external devices via the ATM interface card. The function of the ATM adapter is to adapt the data to the ATM format (cells) for routing through the ATM network without any additional protocol conversion delay. This interface card puts in appropriate virtual path identifier (VPI) and virtual channel identifier (VCI) values in the ATM cell.

Currently, workstation adapters are available from Adaptive and Fore Systems for many of the popular workstations, including SUN, Silicon Graphics, DEC, Next and HP. These adapters operate at 100 Mbps, using the AAL 3/4 or AAL 5 ATM adaptation layer.

Other manufacturers have developed lower-speed adapters. For instance, Newbridge has developed a 51-Mbps adapter for UTP (unshielded twisted pair) category 3 adapters for SBUS, NUBUS, Microchannel, ISA, EISA, VME and Futurebus-based systems. It is expected that by 1995, a wide range of cost-effective adapters for various systems will become widely available from various vendors.

23.2.3 Local ATM switches

Local ATM switches connect computers or workstations in a star topology as shown in Figure 23.2. These switches have interfaces to workstations at 100 Mbps or 140 Mbps and Ethernet interfaces at 16 Mbps. Currently, 100-Mbps (TAXI), 45-Mbps (DS3 ATM UNI with PLCP) and 155-Mbps (SONET-OC3c) interfaces for the workstations are available.

Some of the vendors with ATM products are Adaptive and Fore Systems. Adaptive's ATMX has 15 card slots with a total capacity of 1.2 Gbps whereas Fore's ASX has 8 to 16 ports each at 100 to 140 Mbps, with a total capacity of 2.4 Gbps. These switches can be used as hubs, depending on the type of interface available for the switch. In fact, these vendors manufacture ATM-based hubs.

23.3 ATM Hub or Campus Backbone

Hubs connect many LANs; usually there is a LAN for each department in a building. These shared LANs use the broadcast method, where the end stations in the LAN accept all the frames transmitted, discarding the ones that are not addressed specifically to them. The conventional technique for an over-subscribed shared LAN is to split the LAN into two and connect the segments with a bridge or router. This type of optimization leads to numerous LANs with very few users, which become difficult to manage. To help users set up small LANs, smart hubs came into existence. These hubs allow

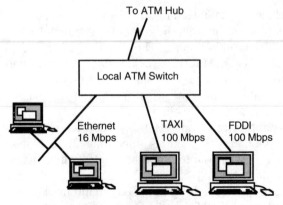

Figure 23.2 Typical local ATM switch environment.

any equipment (bridge, router, or LAN) to reside in any segment. Sometimes they are called switching hubs, because certain small groups of ports are assigned to a LAN segment for switching between the ports. Currently, most of the smart hubs are connected via fiber-optic cables, enabling the hubs to be linked to high-speed LANs such as fiber distributed data interface (FDDI). As the interwork became more and more complex, these smart hubs started directly connecting to desktop PCs and a single hub to connect multiple hubs back-to-back. These hubs will be upgraded to ATM modules and continue to have Ethernet and token ring interfaces while switching is performed on the ATM backbone between them. These ATM hubs will have more ATM ports and fewer number of other LAN interface ports such as Ethernet, FDDI, etc.

Currently, over a dozen vendors are in the hub market. These ATM hubs have aggregate bandwidths ranging from 2 to 10 Gbps, with interface ports of speed ranging from 1.5 Mbps to 155 Mbps. The switch architecture might vary from vendor to vendor, but the basic cell switching is done in the hardware (VLSI chips perform the switching) in all of them. In simple terms, the function of these hubs is to convert LAN frames into ATM cells from different interfaces. This equipment is sometimes called the *cell slicer.*

ATM has clearly hit its stride, and products will be rolling out en masse throughout the year. In addition to local-area and workgroup switches, a raft of component products—from ATM chipsets, ATM interfaces, and ATM LAN analyzers—will enable users to deploy enterprise-wide ATM-based networks.

Vendors who provide switches are Newbridge, VIVID, 3COM, DEC, Fibermux, Bytex, Cabletron, Fibercom, Synoptics, Ungermann-Bass, Adaptive/Net, Northern Telecom, and Fore Systems. The switches provided by some of the above-mentioned vendors are described in the following subsections.

23.3.1 3Com Corporation

3Com Corp.'s ATM strategy is aimed at helping users build collapsed backbones that consist of its ATM-capable LinkBuilder hubs and NetBuilder 11 routers and, eventually, a full-blown ATM switch. Dubbed *CellBuilder*, these modules are used in the hub and router to convert token ring and Ethernet data into ATM cells and support 155-Mbps ATM links to hubs on different floors. 3Com plans to unveil an ATM switch that can link to the router and hub. Like Cabletron, 3Com is also partnering with Fore to obtain ATM technology.

23.3.2 Adaptive/Net

Adaptive's ATM hub is called ATMX. It has 16 slots with a 1- to 2-Gbps backbone bus. The interface includes 100-Mbps fiber or high-speed serial interface (HSSI), FDDI.

23.3.3 Bytex

Bytex's product is based on Series 7700, a 3-Gbps nonblocking switch. Bytex plans to add an ATM interface card in 1994.

23.3.4 Cabletron

Integrated ATM in the MMAC hub is based on Fore Systems' ATM technology. It includes 100-Mbps multimode fiber interfaces. The switch is a 2-Gbps bus-based architecture.

23.3.5 Chipcom Corporation

Chipcom Corp.'s move into the ATM arena and its enviable alliance with IBM might be just the clout the hub vendor needs to make its mark in the next generation of networking. Chipcom's new ONcore hub is a 17-slot device that supports not only ATM and traditional LAN traffic, but also Ethernet switching and 100-Mbps Ethernet technology as well. The ATM switch module of the ONcore hub is based on IBM's Prizma chip, which offers a throughput capacity of more than 8 Gbps.

Chipcom second-generation ONline hubs, will come equipped with three 10-Mbps Ethernet buses, seven token ring buses, and four FDDI rings. Chipcom will provide users of its existing ONLine systems concentrator hubs a migration path to ATM via a passive ATM backplane on the ONcore that will be used to link LANs to the ATM switch module.

23.3.6 Cray Communications, Inc.

Cray's ATM product line is referred to by its high speed switching and networking product line that consists of five new ATM products, two of which are focused on providing wide-area ATM service, while the remaining three provide local area ATM functionality. The unnamed ATM LAN switch, expected to be available by the end of 1994, will scale in capacity from 1 Gbps to 20 Gbps. On the LAN side, there will be an ATM interface for the existing hub-based router, as well as a combination Ethernet/mini-ATM switch for collapsed backbone topologies that will come equipped with multiple Ethernet and ATM ports. It can be used to interconnect Ethernet segments in the System Center hub or provide ATM conversion. The ATM ports can be used to tie ATM devices together to a server outfitted with an ATM interface.

23.3.7 Digital Equipment Corporation

In Oct. 1993, Digital announced plans to roll out an array of ATM products in 1994—scaling the work group to the wide area—to enable its enormous base of LAN users to migrate to the world of ATM. The product line will include a 3.2-Gbps and 10-Gbps ATM switch for work group and departmental networks, a line of workstation and server adapter cards, and ATM interfaces for its existing line of routers and hubs.

The 13-slot, 10-Gbps premises-based switch can be positioned as a departmental hub of hubs, a campus backbone, or a backbone node for a private wide-area ATM network. The second ATM switch—a four-slot 3.2-Gbps device—is a scaled-down version of the former targeted at smaller ATM workgroups.

23.3.8 Fibermux, Inc.

Fibermux's entry into ATM is called Project ATMosphere, and is expected to be out in 1994. It will essentially add ATM switching functions to its existing family of Crossbow LAN hubs. Fibermux will provide a new Ethernet-to-ATM conversion module for the Crossbow hub that will transport the ATM cells across a 100-Mbps ATM backbone to a new cross-connect module. The cross-connect module is a 16-port ATM cell switch that will be installed in a centrally located hub to provide ATM switching for up to 16 Ethernet segments.

23.3.9 Fore Systems, Inc.

Fore Systems is an early entrant into the ATM game. This vendor is leading the market in actual ATM switches shipped to more than 250 customers as of mid-1994. The vendor's flagship ForeRunner ASX-100 ATM switch is equipped with a 2.5-Gbps non-blocking switching matrix that supports both permanent and switched virtual circuits. A maximum of 64 switches can be connected locally at speeds up to 155 Mbps for an aggregate capacity of 10 Gbps or over the wide area via T3 or SONET OC3 for a total of 1024 workstations.

23.3.10 General Datacom, Inc.

General DataComm's contribution to the ATM LAN arena includes an ATM switch that supplies both pure cell switching and adaptation switching interfaces for non-ATM-based traffic. The APEX switch can be configured as a LAN ATM hub, a high-speed backbone router, or a WAN switch. It comes in 3.2-Gbps and 6.4-Gbps versions and features a cross-point switching matrix that supports a total of 64 ports. General DataComm also provides a high speed serial interface module that allows its ATM switches to support speeds as low as T1 for connecting to devices such as routers and mainframe channel extenders.

Hub vendor LANNET Data Communications, Inc., formed an alliance with General DataComm to provide the LAN-to-ATM routers that will enable users to access ATM-based servers, backbones, and public ATM services.

23.3.11   LightStream Corporation

LightStream Corp., a joint venture formed last year by Bolt Beranek and Newman, Inc., and Ungermann-Bass, plans to develop and market a family of ATM-based products, and will roll out its first ATM switch in fall 1994. The 2-Gbps LightStream 2010 ATM switch is designed to support the functions of both a wide area bandwidth manager and a multiprotocol router. The initial release of the LightStream 2010 is intended for use primarily in wide area and metropolitan area networks, but future releases of the switch will support native LAN interfaces and integrated routing, thereby allowing users to interconnect LANs over an ATM network without adding additional routers or special ATM interfaces.

23.3.12 Network Equipment Technologies, Inc.

NET's Adaptive Division was one of the first to bring a local ATM switch to market. The ATM work group hub, dubbed the ATMX, is a local area net switch equipped with a 1.2-Gbps backplane that supports as many as 15 ATM interfaces, each of which has a maximum capacity of 100 Mbps shared across six ports.

NET plans to release additional interfaces on the ATMX hub in 1994— including Ethernet, token ring, T3, and synchronous optical network adapters—all of which will enable users to incorporate other LAN traffic into their ATM networks. When used in conjunction with a router outfitted with an ATM interface, the ATMX will have the ability to anchor collapsed backbone networks and, eventually, ATM WANs. NET plans its ATM to be deployed as a backbone premises technology that will supersede shared-media LANs such as FDDI.

23.3.13   Newbridge Networks, Inc.

In 1993, Newbridge announced an ATM switching hub as part of a comprehensive ATM LAN product line called VIVID, an acronym for video, voice, image, and data. The VIVID ATM hub is an eight-slot device that has an overall switching capacity of 2.4 Gbps. Each slot can support Ethernet, token ring FDDI, T3, and ATM modules, with the ATM port supporting local-and-wide-area speeds of up to 140 Mbps. The VIVID hub can be connected to Newbridge's 36150 MainStreet wide-area ATM switch via the 140-Mbps ATM link or over a T3 line.

Intelligent hub vendors like Optical Data Systems, Inc., (ODS) and Xyplex, Inc. both announced plans to incorporate Newbridge's VIVID technology into their respective hubs in 1994. ODS will provide ATM switching interfaces for its line of Infinity hubs that utilize the VIVID ATM technology, while Xyplex will use a similar strategy for its Network9000 Routing Hub.

23.3.14 Sybernetics

The ATM hub is based on LANplex 5000, which is an FDDI hub. Here, the ATM hub is used as a backbone hub for linking multiple FDDI hubs.

23.3.15 SynOptics Communications, Inc.

Like Fore, NET, and Newbridge, hub giant SynOptics is another early entrant into the ATM game. SynOptics began developing its Lattis Cell ATM switch more than four years ago, along with its own ATM chip set, which the vendor jointly developed with Washington University in St. Louis.

To support legacy Ethernet traffic in an ATM backbone network, SynOptics introduced its EtherCell switch in early 1994, a standalone packet-to-cell conversion device that ties SynOptics' existing hubs with its Lattis Cell ATM switch.

23.3.16 Ungermann-Bass

This ATM hub is based on Dragon Switch Ethernet switching modules. It has 10 slots and a variety of interfaces.

There are many other vendors positioning their products in this segment. Some products cover a wide range of environments. As mentioned, these switches can be used to cover LANs scattered over a wide geographical area. In addition, vendors are also upgrading their routers to ATM-based ones so that LAN networks can be connected over WAN on ATM backbone. Router vendors such as Cisco System Inc., Network Systems Corp., Proteon, Retix, and Wellfleet have developed routers with ATM DXI interfaces based on standards specifically designed for this type of application.

23.3.17 IBM

IBM recently announced its ATM product plans covering a wide range. It included an ATM switch, ATM LAN hub, and ATM chip set. IBM's ATM switch is based on a variation of ATM called packet transfer mode (PTM). The ATM switch itself is called transport network node (TNN). This switch has an 8-Gbps backplane and the capability of adding another 8-Gbps backplane. It is a 16-port switch supporting interfaces such as T1, HDLC, ISDN, fiber channel, and frame relay. In addition to these standard interfaces, IBM's switch will support TCP/IP, SNA, 3745, 3174, 3172, LAN-host interfaces, and 6611 via frame relay.

IBM's ATM LAN hub is based on the Intelligent Hub family of 8250 (a version of Chipcom Corp's ONline System box). It has an 8-Gbps backplane, capable of supporting ATM-attached workstations and servers. For personal computers and UNIX-based systems, IBM plans to develop ATM adapters of 25 Mbps, 100 Mbps, and 155 Mbps, connected via shielded or unshielded twisted-pair cables.

23.4 ATM Equipment Vendors in WAN

In this section, we address the products of ATM vendors designed to connect remote sites. Figure 23.3 shows a typical ATM switch (ATM hub) used in WAN applications.

We described the requirements for a typical product in a WAN environment in chapter 16. This product can be an ATM switch or an ATM concentrator. Vendors targeting this market are T1/T3 multiplexer or mux and X.25 packet switch equipment vendors. Current T1/T3 mux vendors such as Ascom Timeplex, GDC, NET, Newbridge, Stratacom and X.25 vendors such as ADN and BBN are positioning their existing products to handle ATM. Table 23.1 shows the vendors, their products, and their technology source. Each vendor is further described in the following subsections.

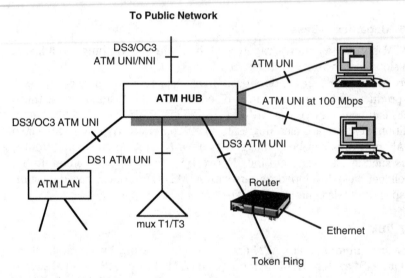

Figure 23.3 ATM switch as WAN interface.

TABLE 23.1

Vendors, Products, and Technology Sources

Company	Product name	Technology source
Ascom	Timeplex INP	Washington University
GDC	APEX	Netcom
Net	ATMX	Adaptive
Newbridge	36150	MPR
Stratacom	BPX	Internal
ADN	HSS	Alcatel
BBN	Emerald	Internal

23.4.1 Ascom Timeplex, Inc.

Timeplex is developing three switches that will scale in capacity from 1 Gbps to 19 Gbps. All three, which are expected to be available in 1995, will be equipped with signaling and traffic management capabilities and support existing LANs as well as traditional time-division multiplexing equipment.

The switches will employ an architecture that supports their existing line of multiplexers, routers, X.25 packet switches, and SONET switches. The work group ATM switch will have the capacity of up to 600 Mbps, while the departmental and enterprise switch will have a maximum capacity of 2.4 Gbps and 19.2 Gbps, respectively.

23.4.2 Cabletron Systems, Inc.

Cabletron's ATM switch is called the MMAC-PLUS hub, a 14-slot device that will support Ethernet, token ring, FDDI, and ATM traffic across a 10-Gbps switching backplane. The switch's distributed architecture will eliminate the need for dedicated management and internetworking modules by letting users interconnect cell-and-packet-based networks using its switch. Cabletron has also partnered with Fore Systems to use Fore's ATM technology in the MMAC-Plus.

23.4.3 BBN

BBN's ATM switch is called Emerald. It is a 10-port ATM switch, each capable of handling 155 Mbps. It has Ethernet, FDDI, T1, and T3 interfaces.

23.4.4 Cascade Communication

Its ATM product is based on B-STDX 9000. It is a 16-slot, 1.2-Gbps bus with DXI HSSI, V.35 with frame relay, and SMDS trunking interfaces to the public networks.

23.4.5 Gandalf

Its ATM switch is called the 2050 ATM enterprise switch. It is a 1.6-Gbps switch capable of integrating voice, data, and video traffic.

23.4.6 GDL

This ATM switch is based on the APEX 6.4-Gbps ATM switch. It has 16 ports, each at 200 Mbps. It has frame relay, HDLC, SNA, Ethernet, and ATM interfaces for access. Its WAN interface varies from T3 to SONET OC12 interface.

23.4.7 GTE

Its ATM switch is called SPAnet. Its basic fabric is an 8-×-8 non-blocking switching matrix. Its interfaces include T1, T3, OC3, OC12, and TAXI. This switch is designed for military applications.

23.4.8 NewBridge

Its ATM switch is called the 36150 Mainstreet ATMnet. Its entry level switch is a 4-port, 160-Mbps non-blocking switch matrix. This switch can be expanded to 32 ports—5 Gbps scalable to 10 Gbps. It has a variety of interfaces, such as Ethernet, FDDI, TAXI, DS3, T1, E3, HSSI, OC3, and OC12, along with frame relay (FR) and SMDS. It has redundancy as an option. Figure 23.4 shows the NewBridge ATM switch interwork with existing New Bridge products such as T1 mux. Here, the ATM switch provides the backbone connectivity for the existing equipment.

23.4.9 Stratacom

Stratacom pioneered cell relay with its IPX (fast packet), which is currently used as a frame-relay switch. It evolved this ATM switch and named it BPX. It is a 9.6-Gbps crosspoint ATM switch. It has an automatic bandwidth allocation with twelve 800-Mbits slots and is compatible with Stratacom's IPX (frame-relay switch). Its interfaces include ATM UNI, frame relay interworking, SMDS, FDDI, and 2.048 Mbps (E1). In addition to these, it is fully redundant (100 percent), with features such as backup for addressing, interface, cooling, switch, processor, backplane, and hot card replacement.

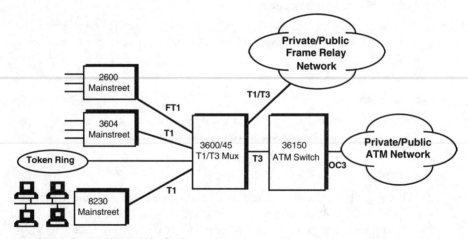

Figure 23.4 Interworking of NewBridge equipment.

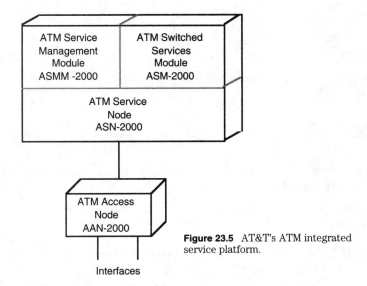

Figure 23.5 AT&T's ATM integrated service platform.

23.5 ATM Vendors' Strategy in Public Networks

This market segment is the one where only a very few large players exist, operating on a global basis. They are telecommunication switch and transmission equipment manufacturers whose annual revenue exceed $1 billion. As mentioned in chapter 17, the requirements for equipment in a public network are more stringent than any other segment. In this section, we will address some of these vendors. The largest market for ATM switches in the public network is those that are used either as a central office or an end-office switch.

23.5.1 AT&T's ATM strategy

AT&T's ATM switch called GCNS-2000 is based on the evolution of the BNS-2000 switch. The ATM switch developed by AT&T is a backbone switch for long-distance carriers. It is based on shared memory switching technology with a backplane speed up to 20 Gbps. Its interfaces include ATM UNI-based 45-Mbps (DS3), 155-Mbps (OC3), Circuit Emulation-based 1.544-Mbps (T1), 45-Mbps(DS3), SMDS SNI-based 1.544-Mbps, 2.048-Mbps (E1), 34-Mbps (E3), 45-Mbps (DS3), Frame Relay-based n × 64 Kbps, 1.544-Mbps (T1), 2.048-Mbps (E1). These interfaces are both on the trunk and access sides.

AT&T developed an ATM integrated service platform that consists of four building blocks: ATM Service Management Module (ASMM), ATM Switched Services Module (ASM), ATM Service Node (ASN), and ATM Access Node

(AAN). Figure 23.5 illustrates these modules. This building block approach ensures a smooth progression to ATM as service needs evolve. The functions of these different modules are as follows:

- *Service Node (ASN 2000).* It provides core switching and service functions. It can be configured with a capacity ranging from 2.4 Gbps to more than 100 Gbps.
- *Switched Service Module (ASM 200).* It performs call processing and signaling.
- *Access Node (AAN 2000).* It provides the required access, concentration, and multiplexer functions.
- *Service Management module (ASMM 2000).* It performs the management of the modules.

The AT&T GCNS-2000, which is AT&T's first product based on this platform, utilizes a shared memory fabric that enables it to carry voice, data, and video transmissions simultaneously. As such, it can handle mixed bursty and constant bit rate traffic. In addition, the GCNS-2000's ATM access interfaces support DS1/E1, DS3/E3, and SONET/SDH levels, while trunk interfaces support DS3/E3 and SONET/SDH, allowing for an easy fit into the existing service provider networks, as well as the emerging broadband fiber network.

The GCNS-2000 allows network providers to access the operations systems (software) to add new services or remedy potential problems. It also offers a network management system and will be complemented with a full line of router and hub products, including those from AT&T Network Systems, NCR, and AT&T Paradyne.

23.5.2 DSC Communications

The MegaHub iBSS is designed around two key components. The first is the broadband switching module (BSM), which is a switching module capable of providing a variety of services on a variety of communication link types. BSMs are managed by the administration module (AM), which is a general computing resource with the ability to administer and maintain switch operations. One MegaHub iBSS system can include several BSMs depending on the number and type of interfaces required. The BSM is the key building block of the MegaHub iBSS. Each system can contain from 1 to 64 BSMs.

The MegaHub iBSS supports basic ATM interfaces in its first commercial release. Future releases will extend this support. The basic ATM interfaces supported by the system include the following:

- user-network interface (UNI)
  —DS3, E3, OC3c/STM-1 physical interfaces
  —DS3/E3: up to 4,096 PVCs
  —OC3c: up to 16,384 PVCs
  —integrated local management interface (ILMI) support
- broadband intra-switch interface (B-ISI)
  —DS3, E3, OC3c/STM-1 physical interfaces
  —used for intraswitch (inter-BSM) traffic
  —supports multiple service types per trunk

The switching architecture of the MegaHub iBSS is distributed and centered around a non-blocking "cellbus" within each BSM. Each BSM can switch up to 900,000 packets per second and up to 6,840,000 cells per second (2.9 Gbps). When future OC12 intraswitch interfaces are included, the total capacity will exceed 4 Gbps per BSM.

23.5.3 Fujitsu network switching

Fujitsu was one of the first to introduce an ATM switch in the United States with Fetex 150, an equivalent central office switch. The Fetex switch is a 15-port module at 622 Mbps. This switch is expandable to a 255 port module with aggregate capacity of 160 Gbps. It has frame relay and SMDS interface capability. There are numerous trials being conducted in the U.S. using Fetex 150. Its narrowband switch is sold in over 20 countries.

Fujitsu recently introduced an ATM access vehicle called SMX-6000. It is expected to provide access to public ATM networks that do not contain built-in ATM interfaces. The SMX-6000 ATM service multiplexer is one of the first commercially available products to give access to public ATM networks at rates as high as 155 Mbps. The SMX-6000 multiplexer has ports for six interface connections. The interfaces have variable speeds for both customer premise interfaces and network interfaces. On the network side, the product has ports for ATM OC3 speeds (155 Mbps) and DS3 coaxial (45 Mbps) and SMDS interfaces.

23.5.4 TRW

TRW's ATM switch is called BAS 2010C. Its switch architecture is similar to DSC's megahub. It is a non-blocking switching matrix with 16 OC3 ports per system. The switch has a throughput of 3.2 Gbps. Its interfaces include DS3, OC3, and OC12. The switch is expandable in 0.8-Gbps increments up to 12.8 Gbps.

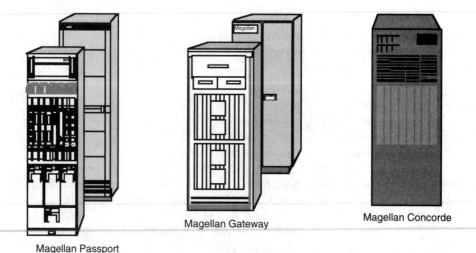

Magellan Gateway

Magellan Concorde

Magellan Passport

Figure 23.6 Northern Telecom Magellan ATM product (*Northern Telecom*).

23.5.5 Northern Telecom

Northern Telecom announced a series of ATM products for different environments under a family called Magellan products in 1993. In this family, NT announced the following products:

- Magellan Passport Switch
- Magellan Gateway Switch
- Magellan Concorde Switch

All Magellan family products share a common architecture and multipriority system (MPS), which allows carriers and customers to create classes or grades of service. Figure 23.6 shows these products. Each is described in the following subsections.

23.5.5.1 Magellan Passport Switch. The Magellan Passport Switch is the smallest of the family used for enterprise networks. It is a dual-bus 1.6-Gbps ATM switch with 16 slots. It has a variety of interfaces, such as V.35, T1, T3, FDDI, frame relay UNI, and NNI. Its modular design and scalable multiprocessor architecture create a flexible platform for private and public WANs. The passport switch supports voice as well as data via frame relay interfaces. Its features and benefits are the following:

Features:

- high-performance ATM architecture
  —1.6-Gbps aggregate shelf capacity

—scalable multiprocessor architecture
—object-oriented software environment
- network services
—frame and cell switching
—frame relay
—frame relay to ATM interworking
—DPN-100 interworking
- wide range of interfaces and protocols
—V.11, V.35, T1, E1, OC3, T3, E3
—token ring, FDDI, Ethernet
- sophisticated network capabilities
—Magellan MPS support
—end-to-end networking
—connectionless and connection-oriented routing

Benefits:

- flexibility
—integration of data, voice, video, and image
—frame switching to support today's data services
—cell switching to support ATM networking
- cost savings
—network consolidation
—bandwidth consolidation for service access and networking
- evolvability
—performance to match service needs
—high availability and reliability
—suitable for mission-critical environments
- network integrity
—dynamic route recovery for network connections
—easy-to-use sophistication, combined with Magellan Network Management System (NMS) to provide control of complex networking environments.

23.5.5.2 Magellan Gateway Switch. The Magellan Gateway Switch is a broadband multimedia switch targeting access to carrier ATM networks. The gateway was developed through an alliance with GTE Government Systems. This switch can be used for private networking. The gateway switch is based on the SPANet switch, which is a 1.2-Gbps 8-x-8 single stage ATM switching fabric. This switch is fully redundant in terms of 1:1 protection for the ports and power. This switch is designed as a public carrier access vehicle. The following are features and benefits offered by these switches.

Features:

- high-performance ATM access switch
  —initial 1.2-Gbps throughput
  —8-x-8, input-output buffered switch matrix
- network services
  —traffic consolidation and switching
  —ATM bearer, ATM private line
  —isochronous services
- network interfaces
  —T1 isochronous, DS3 and OC3 ATM UNI
  —DS3, OC3 ATM NNI
- Magellan MPS support
  —four hardware and four software priority levels
- sophisticated congestion management fully redundant packaging
  —redundant control and switching architecture
  —modular shelf design
  —EMI and NEBS compliant
- local and remote management control
  —remote provisioning, alarms, and test.

Benefits:

- flexibility
  —network and access capabilities to address application requirements
  —suitable for central office and customer premises applications
- evolvability
  —modular structure provides both port and network growth capacity
- high availability and reliability
  —fully redundant central office compatibility for non-stop reliable network operations
- manageability
  —remote management reduces operations costs
  —modular packaging and fiber management to ease craft interface
- network integrity
  —comprehensive network traffic management capability ensures service and network resource availability
- supports open interfaces

23.5.5.3 Magellan Concorde Switch.   The Magellan Concorde Switch is the latest and largest ATM switch of the Magellan family. This switch is scalable from 10 Gbps to 80 Gbps. Unlike other switches, it will primarily support SONET/SDH-compatible trunk-level interfaces packaged to optimize a fiber-optic transmission in a carrier core backbone network. This system is a fully redundant, multipriority system. The features and benefits of this switch are described as follows.

Features:

- high-capacity switch
  —shared memory fabric
  —10 to 80 Gbps throughput
- ATM-based network services
  —ATM bearer service
  —isochronous adaptation
  —PVC and SVC services
- network services
  —Magellan MPS support point-to-point, point-to-multipoint broadcast, multicast
  —dynamic network routing
- SONET/SDH compatible interfaces
  —OC3, OC12, OC48, STM-1, STM-4 UNI and NNI
  —packaging optimized for fiber transmission
- open network management system
  —HP Openview, Q3 (CMIP) interfaces
- scalability
  —down to 10 Gbps
  —up to 80 Gbps
- modular design
  —hot card and shelf changes
  —ESD/EMI compliant
  —global footprint (cabinet size)

Benefits:

- flexibility
  —layered architecture for flexible growth
  —switch elements can evolve independently
  —smooth real-time growth, with minimal processor start-up cost
- evolvability
  —robust traffic handling capabilities for evolvable services
  —bursty and continuous traffic segregated for assured performance
  —dynamically shared buffering to provide multiple priorities
- availability and reliability
  —end-to-end virtual path management for improved network availability
  —automatically monitored network performance
- network integrity
  —consolidated billing, quality-of-service analysis
  —security to protect privacy and minimize provisioning, configuration errors

- designed for customer specifications
  —architectural flexibility
  —operations and maintenance simplicity
  —throughput evolvability
- protects existing investments through server concept

23.5.6 Alcatel

Alcatel, currently the world's largest telecommunications equipment manufacturer, has been one of the pioneers of ATM in terms of design and concept. Its ATM product is based on the A1010 switch. It can be configured as a VC mux and VP cross-connect. The ATM product has interworking capability with frame relay and SMDS switches.

There are also other vendors, such as Ericsson, Siemens, Thompson, etc., who have announced ATM-based products addressing specific markets around the world.

23.6 Equipment Vendors in ATM/Broadband Testing

It is interesting that so many equipment manufacturers are addressing the different ATM environments. When the customer needs to know the features and how they operate or needs to verify their compliance with standards requirements, however, testing equipment is needed. Currently, very few vendors have come up with ATM/broadband testing equipment. They are HP, ADTech, Network General, Wandel, and Goltermann.

23.6.1 HP

HP has three broadband testing products:

- HP 75000 series 90 SONET/SDH analyzer

- HP 37704A SONET test set

- HP broadband series ATM/SONET/SDH analyzer

These products are used for testing the broadband equipment functionality for SONET and ATM. In this equipment, HP has taken a structured layer-by-layer approach, i.e., testing each layer of the BISDN protocol. Figure 23.7 shows the BISDN protocol used by HP to test different layers.

23.6.2 ADTech

ADTech approached testing from a different angle. It designed a box capable of generating multiple DS3/OC3 traffic, enabling the user to readily load

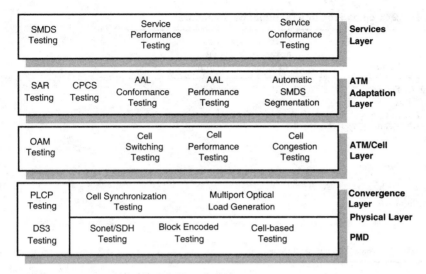

Figure 23.7   HP BISDN protocol stack for testing.

the switch and test it under full load conditions. It is obvious that ADTech understood it is not possible to generate gigabits of traffic in a lab environment. In addition, AdTech is designing new products capable of receiving multiple DS3/OC3s and conducting performance studies.

23.7 Summary

In this chapter, we discussed different vendors in each of the environments addressed in Part 5. The different environments are ATM chips, ATM adapters, ATM LANs, ATM hubs, ATM access switches, and ATM backbone switches. Some vendors package their products in such a way that they can address more than one market. It is interesting to note that among different vendors, Northern Telecom has positioned itself in such a way that it can address most of the ATM switch market. In fact, no equipment vendors have ATM products across the board that can span a wide range of environments in a network. Table 23.2 on page 460 shows the switches of some of the vendors.

TABLE 23.2 Comparison of ATM Equipment Vendors

Vendor	Carrier backbone switch	Carrier access switch	Enterprise network switch
Northern Telecom	✓	✓	✓
AT&T/NCR	✓		✓
Digital Switch Corp.	✓		
Siemens/Strombert Carlson	✓		
TRW	✓		
Fujitsu	✓	✓	
NEC	✓		
Alcatel	✓		✓
Newbridge Networks		✓	✓
StrataCom		✓	✓
N.E.T./Adaptive		✓	✓
IBM			✓

✓—market addressed by the vendor

Yankee group, August 1993

24

The Future of
Broadband Communications

24.1 Overview

The previous 23 chapters covered broadband communications topics with an emphasis on ATM. Broadband communications with ATM is still in its infancy; numerous issues are still pending. The focus of this chapter is to go further in the evolution of broadband communications and see what the future for broadband communications will be. Here, the next stages, such as optical add/drop and optical regenerators, are addressed. Currently, we have SONET systems up to 2.4 Gbps that require repeaters every 40 km. In these systems, the signal regeneration, add/drop, etc., is accomplished in the electrical domain, which obviously becomes a bottleneck for speeds greater than 100 Gbps.

To achieve the transmission system that can transmit faster than 100 Gbps and carry information without regenerator for 100 km or greater, the electrical bottleneck must be replaced with optical-level regeneration and add/drop capability. Optical amplifiers are already available on a limited basis.

24.2 Photonic ATM Switching Architecture

With the current growing demand not only for traditional voice, data, and still-picture services, but also for high-speed and broadband communications services such as video phone, videoconferencing, high-definition TV (HDTV) distribution, and high-speed file transfer services, the need for broadband communications networks has been increasing. ATM switching systems are flexible enough to handle a wide range of communication services. To put broadband services nationwide, a switching system should be capable of accommodating 100 to 10,000 user network interfaces (UNIs), i.e., more than 150 Mbps per UNI. Thus, a 1-Mbps ATM switch, which can handle the traffic of hundreds or thousands of input and output ports with several gigabits-per-second bit rates, will be required.

The bottleneck with existing ATM switches handling such high speeds is the electrical switching fabric or matrix. The only way to achieve these higher speeds is to migrate to a completely optical domain. A switch that uses an optical switching fabric is called a *photonic ATM switch*. Currently, several types of photonic ATM switches have been developed. In this chapter, we address two architectures of photonic switches.

24.2.1 Architecture 1 (first generation)

Figure 24.1 shows the configuration of a typical, conventional photonic ATM switch. Although most of the present research has focused on the efficient utilization of the available bandwidth in the optical matrix switch, electronic circuits are still used to control the optical matrix switch. Thus, it is necessary to convert the optical header to an electrical header and to generate an electronic control signal for the optical matrix switch by translating the electronic header. As a result, the switching speed is limited by the electronic control circuit operating speed. Thus, for a N×N optical

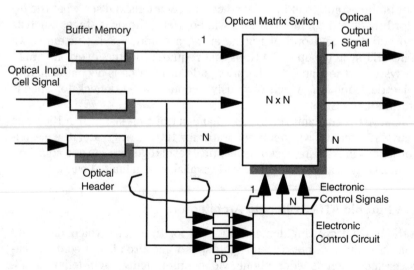

Figure 24.1 First generation photonic ATM switch architecture (*IEEE Communications Apr. '93, pg. 63 reprinted with permission*).

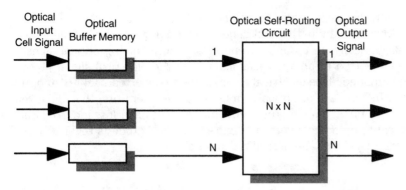

Figure 24.2 Second generation of photonic ATM switch architecture (*IEEE Communications Apr.93, pg. 64 reprinted with permission*).

switching matrix, N^2 electronic control signals are required. With the increase in the number of input and output ports, the electronic control circuit becomes quite complicated because of the rapidly growing number of high-speed electronic circuits.

An electronic buffer memory is used for the queue in an electronic environment. In the case of an optical environment for an optical buffer, a fiber delay line is usually used in the conventional photonic ATM switch. In the case of an electronic buffer memory, the switching speed in the photonic ATM switch is limited by the electronic operating speed of buffer memory. As for the optical fiber delay line, it can store extremely high-speed optical signals. To increase cell signal speed, strict precision in the delay line length is required. Furthermore, the fiber delay line memory is not suitable for such integration.

24.2.2 Architecture 2 (second generation)

Figure 24.2 shows the next-generation photonic ATM switch architecture. Here, a photonic input buffer ATM switch is depicted. This switch consists of optical first-in first-out (FIFO) buffer memories and an optical N-x-N self-routing circuit. When the cells bound for the same output are sent to the self-routing circuit from different FIFO buffer memories at the same time, the cell with the highest priority can be self-routed, while the other cell signal is rejected to prevent cell contention. The high-speed optical buffer memory is expected to be constructed by converting serial cell signals into parallel signals and by using massively parallel optical interconnections.

24.3 Broadband Transmission Systems

In the 1990s, commercial broadband transmission rates are 2.4 Gbps, and regenerators are placed every 40 to 80 km along the cable path. The next-generation transmission rates can go beyond 10 Gbps and only require repeaters every 100 km. In the next generation, the repeaters will regenerate the signal optically. In today's regenerators, the incoming optical signals must be converted to an electrical signal, and the electrical signal power must be amplified. This electrical signal is then converted back to an optical signal. In the case of optical amplifiers or regenerators, the signal remains in the optical domain, thus reducing the delay and loss incurred due to electrical-to-optical conversion. Figure 24.3 shows today's regenerators, and Figure 24.4 shows the optical regenerator.

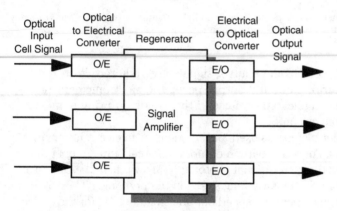

Figure 24.3 Today's electrical repeater or amplifier.

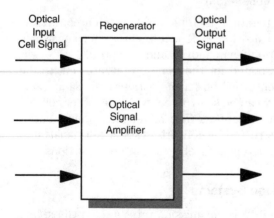

Figure 24.4 Optical repeater or amplifier.

24.4   Areas of Broadband Not Covered

Most of the chapters in this book addressed the issues related to ATM and broadband in the near future and the related technologies. Certain areas of broadband, such as residential broadband and bringing fiber to the curb (FTTC)/fiber to the home (FTTH), are not addressed in this book. Residential architecture is related to LEC (local exchange carrier) in the United States for providing video-on-demand services to the home. Although it is an important area, in general, its widespread applicability in the near future is negligible. If the reader is interested in that area, the best book to read is *Residential Fiber Optic Network*, by David P. Reed. This text covers residential broadband aspects.

Another area not discussed is the implementation of true BISDN because true BISDN implementation is far from a reality. Full implementation is expected to take more than 20 years and over a trillion dollars of investment.

24.4   Final Thoughts

Over the past few years, interest in broadband communications has grown drastically. It has become the talk of the town, with everyone having some role to play in making its future a reality. Numerous issues still linger and this will always be so. It is up to everyone in society to take part and contribute to whatever is possible. For instance, even a child can contribute to the future by exposing the needs of how things ought to be. Every effort is being made to enable one to understand the basic technology behind a broadband communications world. In my opinion, broadband communications is a reality in the near future. It is only a matter of time before everything needed for broadband communications falls into place. As this book goes to press, several enhancements in these areas have probably been made, which is always a never-ending process as the technology keeps evolving.

Standards Sources

Alpha Graphics
10215 N. 35th Ave., Suite A&B
Phoenix, AZ 85051
Phone: 1-602-863-0999
(IEEE P802 Draft Standards)

American National Standards Institute
(ANSI)
Sales Department
1430 Broadway
New York, NY 10018
Phone: 1-212-642-4900
Fax: 1-212-302-1286
(ANSI and ISO standards)

Association Francaise de Normalisation
Tour Europe - Cedex 7
92080 Paris La Defense
France
Phone: 33-1-4-778-13-26
Telex: 611-974-AFNOR-F
Fax: 33-1-774-84-90

Bell Communications Research
Bellcore Customer Service
60 New England Ave., Room 1B252
Piscataway, NJ 08854-4196
Phone: 1-908-669-5800 or 1-800-521-
CORE (1-800-521-2673)

British Standards Institution
2 Park St.
London, WIA 2BS
England
Phone: 44-1-629-9000
Telex: 266933 BSI G
Fax: 44-1-629-0506

Canadian Standards Association
178 Rexdale Blvd.
Rexdale, ON M9W 1R9
Canada
Phone: 1-416-747-4363
Telex: 06-989344
Fax: 1-416-747-4149

Comite Europeen de Normalisation
Rue Brederode 2 Bte 5
1000 Brussels
Belgium
Phone: 32-2-513-79-30
Telex: 26257 B

Computer and Business Equipment
Manufacturers Association (CBEMA)
311 First St., NW, Suite 500
Washington, DC 20001-2178
Phone: 1-202-626-5740
Fax: 1-202-638-4299, 1-202-628-2829
(ANSI X3 secretariat)

Dansk Standardiseringsrad
Aurehojvej 12, Postboks 77
DK-2900 Hellerup
Denmark
Phone: 45-1-62-32-00
Telex: 15-615 DANSTA DK

DDN Network Information Center
SRI International
333 Ravenswood Ave.
Menlo Park, CA 94025
Phone: 1-415-859-3695 or
1-800-235-3155
E-mail: NIC@NIC.DDN.MIL

Deutsches Institut für Normung
Burggrafenstrasse 4-10, Postfach 1107
D-1000 Berlin 30
Germany
Phone: 49-30-26-01-1
Telex: 184-273-DIN D
Fax: 49-30-260-12-31

Electronics Industries Association
(EIA)
Standards Sales
2001 Eye St., NW
Washington DC 20036
Phone: 1-202-457-4966
Telex: 710-822-0148 EIA WSH
Fax: 1-202-457-4985

European Computer Manufacturers
Association (ECMA)
Rue du Rhone 114
CH-1204 Geneva
Switzerland
Phone: 41-22-735-36-34
Telex: 413237 ECMA CH
Fax: 41-22-786-52-31

European Conference of Postal and
Telecommunications Administrations
(CEPT)
CEPT Liaison Office
Seilerstrasse 22
CH-3008 Bern
Switzerland
Phone: 41-31-62-20-78
Telex: 911089 CEPT CH
Fax: 41-31-62-20-78

Exchange Carriers Standards
Association (ECSA)
5430 Grosvenor Ln.
Bethesda, MD 20814-2122
Phone: 1-301-564-4505
(ANSI T1 secretariat)

Institute of Electrical and Electronics
Engineers (IEEE)
Standards Office/Service Center
445 Hoes Ln.
Piscataway, NJ 08855-1331
Phone: 1-908-564-3834
Fax: 1-908-562-1571
(IEEE standards)

International Organization for
Standardization
1 Rue de Varembe, Case Postale 56
CH-1211 Geneva 20
Switzerland
Phone: 41-22-734-1240
Telex: 23-88-1 ISO CH
Fax: 41-22-733-3430

International Telecommunications
Union
General Secretariat—Sales Service
Place de Nation
CH 1211 Geneva 20
Switzerland
Phone: 41-22-730-5860
Telex: 421000 UIT CH
Fax: 41-22-730-5853
(CCITT and other ITU
recommendations)

Japanese Industrial Standards
Committee
Standards Department, Agency of
Industrial Science & Technology
Ministry of International Trade and
Industry
1-3-1, Kasumigaseki, Chiyoda-ku
Tokyo 100
Japan
Phone: 81-3-501-9295/6
Fax: 81-3-680-1418

National Institute of Standards and
Technology
Technology Building 225

Gaithersburg, MD 20899
Phone: 1-301-975-2000
Fax: 1-301-948-1784

National Standards Authority of
  Ireland
Ballymum Road
Dublin 9
Ireland
Phone: 353-1-370101
Telex: 32501 IIRS EI
Fax: 353-1-379620

Nederlands Normalisatie-Instituut
Kalfjeslaan 2
P.O. Box 5059
2600 GB Delft
Netherlands
Phone: 31-15-61-10-61

Omnicom, Inc.
115 Park St., SE
Vienna, VA 22180-4607
Phone: 1-703-281-1135
Telex: 279678 OMNI UR
Fax: 703-281-1505

Omnicom International, Ltd.
1st Floor, Forum Chambers
The Forum
Sevenage, Herts SG1 1EL
United Kingdom
Phone: 44-438-742424
Telex: 826903 OMNICM G
Fax: 44-438-740154

Saudi Arabia Standards Organization
P.O. Box 3437
Riyadh 11471
Saudi Arabia
Phone: 9-661-4793332
Telex: 201610 SASO

SRI International
333 Ravenswood Ave., Room EJ291
Menlo Park, CA 94025
Phone: 1-800-235-3155
(Internet Protocol RFCs)

Standardiseringskommissionen i
  Sverige
Tegnergatan 11, Box 3 295
S-103 66 Stockholm

Sweden
Phone: 468-230400
Telex: 17453 SIS S

Standards Association of Australia
Standards House
80-86 Arthur St.
North Sydney N.S.W. 2060
Australia
Phone: 61-2-963-41-11
Telex: 2-65-14 ASTAN AA

Suomen Standardisoimisliitto
P.O. Box 205
SF-00121 Helsinki 12
Finland
Phone: 358-0-645-601
Telex: 122303 STAND SF

U.S. Department of Commerce
National Technical Information Service
5285 Port Royal Rd.
Springfield, VA 22161
Phone: 1-703-487-4650
(CCITT Recommendations, U.S.
  Government and Military Standards)

United Nations Bookstore
United Nations General Assembly
  Building
Room GA 32B
New York, NY 10017
Phone: 1-212-963-7680
(CCITT Recommendations)

B

Broadband
Equipment Manufacturers

Adaptive/Net Corp.
800 Saginaw Dr.
Redwood City, CA 94065
Phone: 1-415-366-4400

Alcatel ATFH
55 rue Greffulhe BP 302
Levallois-Perret
France F92301
Phone: 33-1-47-58-3000
Fax: 33-1-47-58-3596

Alcatel Bell Telephone
Francis Wellesplein 1
Antwerp Belgium B-2018
Phone: 03-240-40-11
Fax: 03-240-99-99

Alcatel Canada Wire
22 Commercial Rd.
Toronto, Ontario
Canada M4G 3W1
Phone: 1-416-421-7400
Fax: 1-416-421-5983

Alcatel Network Systems
1225 N. Alma Rd.
Richardson TX 75081

Phone: 1-214-996-5000 or
800-ALCATEL
Fax: 1-214-996-5409

Alcatel STR AG
Friesenbergstrasse 75
Zurich, Switzerland CH-8055
Phone: 41-1-465-2111
Fax: 41-1-462-5141

Ascom Timeplex Inc.
400 Chestnut Ridge Rd.
Woodcliff Lake, NJ 07675
Phone: 1-201-391-1111 or
1-800-669-2298
Fax: 1-201-573-6470

AT&T Global Business Communication
Systems
55 Corporate Dr.
Bridgewater, NJ 07054
Phone: 1-800-247-7000 in NJ

Bytex Corp.
4 Technology Dr.
Westborough, MA 01581
Phone: 1-508-366-8000

Cabletron System Inc.
35 Industrial Way
Rochester, NH 03867
Phone: 1-603-332-9400

Chipcom Corp.
7900 International Dr.
Bloomington, MN 55425
Phone: 1-612-854-7097

Digital Equipment Corp.
Telecomm Div.
550 King St.
Littleton, MA 01460
Phone: 1-508-474-2000

DSC Communications Corp.
1000 Coit Rd.
Plano, TX 75075
Phone: 1-214-519-3000

Fibermux
9310 Topanga Canyon Blvd.
Chatsworth, CA 92631
Phone: 1-818-709-6000

Fore Systems
1000 Gamma Dr.
Pittsburgh, PA 15238
Phone: 1-412-967-4040
Fax: 1-412-967-4044

Fujitsu Network Switching of America
  Inc.
4403 Bland Rd.
Somerset Park, Raleigh, NC 27609
Phone: 1-919-790-2211
Fax: 1-919-790-8376

Gandalf Technologies, Inc.
1051 Perimeter Dr., 6th floor
Schaumburg, IL 60173
Phone: 1-708-517-3600 or
  1-800-354-4224
Fax: 1-708-517-3627

GTE Government Systems Corp.
77 A St., MS20-25
Needham Heights, MA 02194-2892
Phone: 1-617-455-2858 or
  1-800-982-0381
Fax: 1-617-455-2858

IBM Corp.
53 Knightsbridge Rd.
PO Box 1299
Piscataway, NJ 08855
Phone: 1-201-885-3500

Newbridge Networks Inc.
593 Herndon Pkwy.
Herndon, VA 22070-5421
Phone: 1-800-343-3600

Northern Telecom
200 Athens Way
Nashville, TN 37228-1803
Phone: 1-615-734-4000 or 1-800-
  NORTHERN
Fax: 1-615-734-5191

Siemens Private Communication
  System
5500 Broken Sound Blvd.
Boca Raton, FL 33487
Phone: 1-407-997-9999
Fax: 1-407-997-3567

Sybernetics Inc.
85 Rangeway Rd.
North Billerica, MA 01862
Phone: 1-508-670-9009

Synoptics Communications, Inc.
4401 Great America Pkwy.
Santa Clara, CA 95054
Phone: 1-408-764-5000

3Com Corp.
5400 Bayfront Plaza
Santa Clara, CA 95052
Phone: 1-510-490-8000

UB Networks
3990 Freedom Cir.
Santa Clara, CA 95054
Phone: 1-408-496-0111

Telecommunications Journals and Magazines

Access (monthly)
Telecommunications Research
P.O. Box 12038
Washington, DC 20005

ACM—Transactions on Information Systems (quarterly)
Association for Computing Machinery
11 West 42nd St.
New York, NY 10036

Advances in Telematics (irregular)
Ablex Publishing
355 Chestnut St.
Norwood, NJ 07648

AES—Journal of the Audio Engineering Society (monthly)
AES
60 East 42nd St.
New York, NY 101165-0075

At the LATA level (weekly)
CCMI/McGraw Hill
50 S. Franklin Tpk.
Ramsey, NJ 07446

AT&T Technical Journal (bimonthly)
AT&T

550 Madison Ave.
New York, NY 10022

AT&T Technology (quarterly)
Richard A. O'Donnell
550 Madison Ave.
New York, NY 10022

British Telecom Journal (quarterly)
British Telecom
81 Newgate St. Fl.
A2 London EC1A 7AJ
England

Budavox Telecommunications Review (quarterly)
Telecommunication Foreign Trading Co. Ltd.
Budafoki tu 79, H-1392
Budapest X1, Hungary

Bulletin Signaletique des Telecommunications (monthly)
Centre National d'Etudes des Tell
 Service of Documentation
Interministerielle
38-40 Rue de General LeClerk
92131 Issy-Les-Moulineaux
France

Business Communications Review
(monthly)
BCR Enterprises, Inc.
950 York Rd.
Hinsdale, IL 60521

COMSAT Technical Review
(semiannually)
COMSAT Corp.
22300 COMSAT Dr.
Clarksburg, MD 20871

Cable TV Law and Finance
(monthly)
New York Publishing Co.
111 Eight Ave.
New York, NY 10011

Canadian Communications
(semimonthly)
MACLEAN-Hunter Ltd.
Business Publication Div.
MACLEAN-Hunter Bldg.
777 Bay St.
M5W 1A7 Toronto, Ontario
Canada

Communication (weekly)
NTIS
5285 Port Royal Rd.
Springfield, VA 22161

Communications Consultant
(monthly)
Jobson Publishing Corp.
352 Park Ave. S., 16th Floor
New York, NY 10010

Communications News (monthly)
Edgell Communications
7500 Old Oak Blvd.
Cleveland, Ohio 44130

Communications Week (weekly)
CMP Publications
600 Community Dr.
Manhasset, NY 11030

*Congressional Report on
Communications* (semimonthly)
New Media Publishing
1117 N. 19th, #200
Arlington, VA 22209

Data Communications (monthly)
McGraw-Hill
1221 Ave. of the Americas
New York, NY 10017

Data Communications Management
(bimonthly)
Auerbach Publishers
1 Penn Plaza
New York, NY 10119

Datamation (monthly)
Reed Publishing Co.
44 Cook St.
Denver, CO 80206

*Datapro Reports on
Telecommunications* (monthly)
Datapro Research Corp.
600 Delran Pkwy.
Delran, NJ 08075

European Telecommunications
(semimonthly)
Probe Research Inc.
3 Wing, #240
Cedar Knolls, NJ 07927

FCC Rulemaking Reports (biweekly)
Commerce Clearing House
4025 W. Peterson Ave.
Chicago, IL 60601

FCC Week (weekly)
Capital Publications
1101 King St., #444
Alexandria, VA 22314

Fiber Optic Sensor (monthly)
Information Gatekeepers Inc.
214 Harvard Ave.
Boston, MA 02134

Fiber Optics (monthly)
Taylor and Francis
3 East 44th St.
New York, NY 10017

*Fiber Optics and Communication
News* (weekly)
Information Gatekeepers Inc.
214 Harvard Ave.
Boston, MA 02134

Fiber Optics News (weekly)
Phillips Publishing Inc.
RD #2, Box 486, Saw Mill Rd.
Red Hook, NY 12571

Fiber to Home (biweekly)
Information Gatekeepers Inc.
214 Harvard Ave.
Boston, MA 02134

Fiber and Integrated Optics
  (quarterly)
Taylor and Francis, Inc.
79 Madison Ave. #1110
New York, NY 10016

Focus on Communications
  (monthly)
Business Communications
3190 Miraloma Ave.
Anaheim, CA 92806

Global Communications (monthly)
Cardiff Publishing Co.
6300 South Syracuse Way
Englewood, CO 80111

Globe Communications IEEE
  (annually)
IEEE
345 East 47th St.
New York, NY 10017

GTE Telenet Packet (monthly)
GTE Telenet
Communication Corp.
12490 Sunrise Valley
Reston, VA 22096

IEEE Communications Magazine
  (monthly)
Institute of Electrical and Electronics
Engineers
345 E. 47th St.
New York, NY 10017

*IEEE Journal on Selected Areas in
  Communications* (bimonthly)
Institute of Electrical and Electronics
Engineers
345 E. 47th St.
New York, NY 10017

IEEE Network (bimonthly)
Institute of Electrical and Electronics
  Engineers
345 E. 47th St.
New York, NY 10017

IEEE Spectrum (monthly)
Institute of Electrical and Electronics
  Engineers
345 E. 47th St.
New York, NY 10017

*IEEE Transactions on
  Communications* (monthly)
Institute of Electrical and Electronics
  Engineers
345 E. 47th St.
New York, NY 10017

*International Telecommunications
  Union Operational Bulletin*
  (monthly)
International Telecommunications
  Union
Place de Nation, CH1211
Geneva 20, Switzerland

ISDN Report (semimonthly)
Probe Research Inc.
3 Wing Dr. #240
Cedar Knolls, NJ 07927

ISDN User (bimonthly)
Information Gatekeepers Inc.
214 Harvard Ave.
Boston, MA 02134

Japan Telecommunication Review
  (quarterly)
Telecommunications Association
Tokyo, Japan

Japan Telecommunications
  (monthly)
Ciber Inc.
International Trade Commission Bldg.
500 E St., SW
Washington, DC 20024

Journal of Optical Communications
  (quarterly)
Fachverlac Schiele und Schoen
GMBH

Markgrafenstrasse 11PF
610280, D-1000
Berlin 61, Germany

Liberia Ministry of Posts and
Telecommunications Annual
Report (annually)
Ministry of Posts and
Telecommunications
Monrovia, Liberia

Lightwave (monthly)
Penn Well Publishing Co.
P.O. Box 987
1 Technology Park Dr.
Westford, MA 01886

List of International Telephone
Routes (annually)
International Telecommunications
Union
Place de Nation, CH1211
Geneva 20, Switzerland

Long Distance Letter (monthly)
Phillips Publishing Inc.
7811 Montrose Rd.
Potomac, MD 20854

Military Fiber Optics
Communications (biweekly)
Information Gatekeepers Inc.
214 Harvard Ave.
Boston, MA 02134

Mobile Communications Business
(monthly)
Phillips Publishing Inc.
7811 Montrose Rd.
Potomac, MD 20854

Mobile Phone News (biweekly)
Phillips Publishing Inc.
7811 Montrose Rd.
Potomac, MD 20854

Network World (weekly)
IDG Communications
161 Worcester Rd.
Framingham, MA 01701

Networking Management (monthly)
Penn Well Publishing Co.

P.O. Box 987
1 Technology Park Dr.
Westford, MA 01886

NTT Topics (quarterly)
Ruder, Finn, and Rotman
110 E. 59th St.
New York, NY 10022

Pay Phone News (monthly)
Telestrategies Publishing
1355 Beverly Dr., Box 1218
McLean, VA 22101

Perspective on AT&T and BCR
Products and Marketing (monthly)
BCR Enterprises, Inc.
950 York Rd.
Hinsdale, IL 60521

Planning Guide 1 Inter-LATA
Telecommunications Rates and
Services (monthly)
McGraw-Hill
50 S. Franklin Tpk.
Ramsey, NJ 07446

Planning Guide 2 Inter-LATA
Telecommunications Rates and
Services (monthly)
McGraw-Hill
50 S. Franklin Tpk.
Ramsey, NJ 07446

Planning Guide 3 Value Added
Networks and Data Private Line
Telecommunications Rates and
Services (monthly)
McGraw-Hill
50 S. Franklin Tpk.
Ramsey, NJ 07446

Postel (monthly)
National Press for the Department of
P&T
Pretoria, South Africa

Regulation News (monthly)
Interconnections Pub. Inc.
P.O. Box 128
Rhinebeck, NY 12572

Revue Francais des
Telecommunications (quarterly)

Ministere des PTT
Paris, France

Saskatchewan Telecommunications
Annual Report (annually)
Saskatchewan Telecommunications
Saskatchewan
Canada

Sasktel News (monthly)
Saskatchewan Telecommunications
Saskatchewan
Canada

Soviet Journal of Communications
Technology and Electronics (16 per
year)
John Wiley/Scripta Technica
7961 Eastern Ave.
Silver Spring, MD 20910

Swedish Telecom Annual Report
(annually)
Televerket
Farsta, Sweden

Telecom (English Edition)
(semiannually)
Televerket
Farsta, Sweden

Telecom Bulletin (semiannually)
Fleural Management Co.
Ottawa, Ontario
Canada

Telecom Australia Annual Report
(annually)
Telecom Australia
Melbourne, Australia

Telecom Insider (monthly)
International Data Corp.
Framingham, MA 01701

Telecom Today (monthly)
British Telecommunications
London, England

Telecommunication Journal
(monthly)
International Telecommunications
Union
Place de Nation, CH1211
Geneva 20, Switzerland

Telecommunications (monthly)
Horizon House—Microwave Inc.
685 Canton St.
Norwood, MA 02062

Telecommunications Abstracts
(10 per year)
R.R. Bowker, EIC
New York, NY

Telecommunications Alert (monthly)
Management Telecommunications Pub
New York, NY

Telecommunications Authority of
Singapore Telecoms Annual Report
(annually)
Telecommunications Authority of
Singapore
Singapore

Telecommunications Journal of
Australia (3 per year)
Telecommunications Society of
Australia
Melbourne, Australia

Telecommunications (Norwood)
(monthly)
Horizon House
Norwood, MA 01105

Telecommunications (Potomac)
(quarterly)
Phillips Publishing Inc.
7811 Montrose Rd.
Potomac, MD 20854

Telecommunications Policy
(quarterly)
Butterworth, Science
Guildford, England

Telecommunications Product plus
Technology (monthly)
Penn Well Publishing Co.
P.O. Box 987 1 Technology Park Dr.
Westford, MA 01886

Telecommunications Product
Review (monthly)
Marketing Programs and Services
Group
1350 Piccard Dr.
Rockville, MD 20850

Telecommunications Reports
(weekly)
Telecommunications Reports
1036 National Press Bldg.
Washington, DC 20045

*Telecommunications Technology
Dianxin Jishu* (monthly)
Guoji Shudian
Beijing, China

Telecommunications Sourcebook
(annually)
North American Telecommunications
Association
1036 National Press Bldg.
Washington, DC 20045

*Telecommunications Swedish
Edition* (quarterly)
Televerket
Farsta, Sweden

*Telecommunications Systems and
Services Directory* (irregular)
Gale Research Inc.
835 Penobscott Bldg.
Detroit, MI 48226

Telecommunications Week (weekly)
Telecommunications Reports
1036 National Press Bldg.
Washington, DC 20045

Telecommunicazioni (quarterly)
Societa Italiana
Telecommunicazioni
Milan, Italy

Telecoms International (monthly)
Computer World Communications
Neuilly Sur Seine cedex
France

Telecoms Technical Quarterly
(quarterly)
Directorate General of
Telecommunications
Taiwan, Republic of China

Teleconnect (monthly)
Telecom Library Inc.
12 West 21 St.
New York, NY 10010

Telektronikk (quarterly)
Elektrotechniek
Teledirektoratet
Oslo, Norway

Telematica (5 per year)
Etas Kompass S.P.A
Milan, Italy

Telematics and Informatics
(6 per year)
Pergamon Press Inc.
Journal Division
Maxwell House
Fairview Park
Elmsford, NY 10523

Telephone Engineer (monthly)
Edgell Communications
7500 Old Oak Blvd.
Cleveland, Ohio 44130

Telephony (weekly)
Intertec Publishing Co.
55 East Jackson Blvd.
Chicago, IL 60604

U.S. Telecom Digest (23 per year)
Capitol Publishers, Inc.
1101 King St. #444
Alexandria, VA 22314

*Year Book of Common Carrier
Telecommunications Statistics*
(once every 10 years)
International Telecommunication
Union (ITU)
Geneva, Switzerland

Technology Comparison

	Frame Relay	**FDDI**	**DQDB**	**SMDS**	**ATM**
Standard organization	ITU-T	ANSI	IEEE	Bellcore	ITU-T
Year of standard	1992	1987–1990	1991	1990	1992
Targeted services	Data	Data	Data, voice, video	Data	Broadband services
Main application	High speed data comm.	LAN interconnect	LAN interconnect	High speed data comm.	Broadband comm. network
Operation Public	Public network	Private network	Public network	Public network	Public
Network interwork	IEEE 802 LAN	IEEE 802 LAN	IEEE 802 BISDN	IEEE 802 BISDN	BISDN
Transmission rates	1.544, 2.048 Mbps	100 Mbps	150 Mbps	DS1, DS3, SONET	SDH
Distance (range)	No limits	100 km	No limits	No limits	No limits
Transmission medium	Copper wire	Optical fiber	Optical fiber	Copper, optical	Optical 4 fiber
Start of transmission	Contention	Token	Slot reserv., BW reserv.	Slot reserv.	Under study
Number of priorities	2	2	3	4	4

	Frame Relay	FDDI	DQDB	SMDS	ATM
End of transmission	Frame unit	Time out	Slot unit	Slot unit	Slot unit
Addressing	16-bit	16/48-bit	20-bit VCI or preassigned	Fixed (20 × 1)	8/12-bit VPI + 16-bit VCI
Error check	16-bit CRC	32-bit CRC	8-bit HCS 10-bit CRC	8-bit HCS 10-bit CRC	8-bit HCS 10-bit CRC
Data removal	Not required	Source	Not required	Not required	Under study

Bibliography

— "Gigabit Network Testbeds." *IEEE Computer*, Vol. 23, No. 9, IEEE, September 1990.

Aceveres, J. J. "A New Minimum Hop Routing Algorithm," INFOCOM '87, 1987.

Ahmadi, H., and W. E. Denzel. "A Survey of Modern High-Performance Switching Techniques," *IEEE Journal on Selected Areas of Communication,* Vol. 7, No. 87 Sep 1989.

Ahmadi, H., R. Guerin, and, K. Sohraby. "Analysis of Leaky Bucket Access Control Mechanism with Batch Arrival Process," GLOBECOM, 1990.

Akhtar, S. "Congestion Control in a Fast Packet Switching Network," Master's Thesis, Washington University, St. Louis, MO, December 1987.

Akinpelu, J. M. "The Overload Performance of Engineered Networks with Nonhierarchal and Hierarchal Routing," *BSTJ*, Vol. 63, No. 71, 1984.

Albanese, A., H. E. Bussey, S. B. Weinstein, and R. S. Wolf. "A Multi-Network Research Testbed for Multimedia Communication Services," ICC '91, 1991.

Anagnostou, M. E., et al. "Quality of Service Requirements in ATM Based B-ISDNs," *Computer Communications,* Vol. 14, No. 4, 1991.

Andrade, J., W. Burakowski, and M. Villen-Altamirano. "Characterization of Cell Traffic Generated by an ATM Source," ITC-13, 1991.

Ang, P.H., P. A. Ruetz, and D. Auld. "Video Compression Makes Big Gains," *IEEE Spectrum,* October 1991.

ANSI T1.105-1988, "American National Standard for Telecommunications—Digital Hierarchy—Optical Interface Rates and Formats Specifications," 1988.

ANSI T1.105-1991, "Digital Hierarchy—Optical Interface Rates and Formats Specifications (SONET)," 1991.

ANSI T1.105a-1991, "Supplement to T1.105," 1991.

ANSI T1.106-1988, "American National Standard for Telecommunications—Digital Hierarchy—Optical Interface Specifications (Single Mode)," 1988.

ANSI T1.117, "Digital Hierarchy—Optical Interface Specifications (SONET).

ANSI T1.606 add, "Addendum to T1.606," (T1X1/90-175), 1990.

ANSI T1.606-1990, "Telecommunication—Frame Relay Bearer Service—Architectural Framework and Service Description," 1990.

ANSI T1.6ca, "Core Aspects of Frame Protocol for Use With Frame Relay Bearer Service," (T1S1/ 90-214), 1990.

ANSI X3.139 (Draft), Fiber Distributed Data Interface (FDDI)—Media Access Control (MAC-2), Draft Maintenance Revision 4.0, October 1990.

ANSI X3.139-1987, Fiber Distributed Data Interface (FDDI)—Token Ring Media Access Control (MAC).

ANSI X3.148 (Draft), Fiber Distributed Data Interface (FDDI)—Physical Layer Protocol (PHY-2), Draft Maintenance Revision 4.0, October 1990.

ANSI X3.148-1988, Fiber Distributed Data Interface (FDDI)—Token Ring Physical Layer Protocol (PHY).

ANSI X3.166-1990, Fiber Distributed Data Interface (FDDI)—Physical Layer Medium Dependent (PMD).

ANSI X3.184-199X (Draft), Fiber Distributed Data Interface (FDDI)—Single Mode Fiber Physical Layer Medium Dependent (SMF-PMD), Rev. 4.2, May 1990.

ANSI X3.186-199X (Draft), Fiber Distributed Data Interface (FDDI)—Hybrid Ring Control (HRC), Rev. 6.2, May 1991.

ANSI X3T9.5/84-49, Fiber Distributed Data Interface (FDDI)—Station Management (SMT), Rev. 6.2, May 1990.

Ash, G. R., A. H. Kafker, and K. R. Khrishnan. "Servicing and Real Time Control of Networks with Dynamic Routing," *BSTJ,* Vol. 60, No. 8, 1981.

Ash, G. R., R. H. Cardwell, and R. P. Murray. "Design and Optimization of Networks with Dynamic Routing," *BSTJ,* Vol. 60, No. 8, 1981.

ATM Forum, "ATM User-Network Interface Specification," Version 2.0, 1992.

ATM Forum, "Network Compatible ATM for Local Network Applications," Phase 1, Version 1.0, 1992.

ATM forum, *ATM UNI specification* V. 22, July 1993.

Banwell, T. C., et al. "Physical Design Issues for Very Large ATM Switching Systems," *IEEE Journal on Selected Areas of Communication,* Vol. 9, No. 8, 1991.

Bar-Noy, A., and M. Gopal. "Topology Distribution Cost vs. Efficient Routing in Large Networks," *Proc.* SIGCOMM '90, 1990.

Bell, T., and K. Pawlikowski. "The Effect of Data Compression on Packet Sizes in Data Communication Systems," ITC-13, 1991.

Bellcore Framework Advisory, FA-TSV-001109, "BISDN Transport Network Framework Generic Criteria."

Bellcore Technical Advisory, TA-TSV-001059, "Generic Requirements for SMDS Networking," Issue 2, August 1992.

Bellcore Technical Advisory, TA-TSV-001061, "Operations Technology Network Element Generic Requirements in Support of Interswitch and Exchange Access SMDS."

Bellcore Technical Advisory, TA-TSV-001210, "Generic Requirements for High-Bit-Rate Digital Subscriber Lines."

Bellcore Technical Advisory, TA-TSV-001235, "SMDS Generic Criteria on Operations Interface—Information Model Supporting Intercarrier SMDS," Issue 1, April 1993.

Bellcore Technical Advisory, TA-TSV-001237, "SMDS Generic Requirements for Initial Operations Management Capabilities in Support of Exchange Access and Intercompany Serving Arrangements," Issue 1, June 1993.

Bellcore Technical Advisory, TA-TSV-001238, "SMDS Generic Requirements for SMDS on the 155.52 Mbps Multi-Services Broadband ISDN Intercarrier Interface (B-ICI)," Issue 1, December 1992.

Bellcore Technical Advisory, TA-TSV-001239 "Generic Requirements for Low Speed SMDS Access."

Bellcore Technical Advisory, TA-TSV-001240 "Generic Requirements for Frame Relay Access to SMDS."

Bellcore Technical Reference, TR-TSV-000772, "Generic System Requirement in Support of SMDS," Issue 1, 1991.

Bellcore Technical Reference, TR-TSV-000773, "Local Access System Generic Requirements, Objectives, and Interfaces in Support of Switched Multimegabit Data Service," Issue 1, June 1991; Rev., January 1993.

Bellcore Technical Reference, TR-TSV-000774, "SMDS Operations Technology Network Element Generic Requirement," Issue 1, March 1992; Sup.1, March 1993.

Bellcore Technical Reference, TR-TSV-000775, "Usage Measurement Generic Requirement in Support of Billing for Switched Multimegabit Data Service," Issue 1, June 1991.

Bellcore Technical Reference, TR-TSV-001060, "Switched Multimegabit Data Services Generic Requirements for Exchange Access and Intercompany Serving Arrangements (SMDS)," Issue 1, December 1991; Rev. 1, August 1992; Rev. 2, March 1993.

Bellcore Technical Reference, TR-TSV-001062, "Generic Requirements for Phase 1 SMDS Customer Network Management Service," Issue 1, March 1993.

Bellcore Technical Reference, TR-TSV-001063, "Operations Technology Generic Criteria in Support of Exchange Access SMDS and Intercompany Serving Arrangements," Issue 1, December 1992; Rev. 1, March 1993.

Bellcore Technical Reference, TR-TSV-001064, "SMDS Generic Criteria on Operations Interfaces—SMDS Information Model and Usage," Issue 1, December 1992.

Bellcore TR-NWT-000909, "Generic Requirements and Objectives for Fiber-in-the-Loop Systems," 1991.

Bellcore TR-NWT-001209, "Generic Requirements for Fiber Optic Branching Components," 1991.

Bellcore, SR-NWT-001756, "Automatic Protection Switching for SONET," Issue 1, 1990.

Bellcore, SR-NWT-002076 "Report on the Broadband ISDN Protocols for Providing SMDS and Exchange Access SMDS," Issue 1, September 1991.

Bellcore, SR-NWT-002224, "SONET Synchronization Planning Guidelines," Issue 1, 1992.

Bellcore, SR-TSV-002198 "Support of Intercarrier Aspects of SMDS in a BCC Multiswitch Network," Issue 1, March 1992.

Bellcore, SR-TSV-002395, "Switched Multimegabit Data Service First Phase for Exchange Access SMDS and Intercompany Serving Arrangements," Issue 1, July 1991.

Bellcore, SR-TSV-002422, "Phasing of Service Capabilities and Information for SMDS Customer Network Management Service," Issue 1, September 1992.

Bellcore, TA-NWT-001042, "Generic Requirements for Operations Interfaces Using OSI Tools: SONET Path Switched Ring Information Model," Issues 1 and 3, 1992.

Bellcore, TA-NWT-001250, "Generic Requirements for Synchronous Optical Network (SONET) File Transfer," Issue 2, 1992.

Bellcore, TR-NWT-000253, "Synchronous Optical Network (SONET) Transport Systems: Common Generic," Issue 2, 1991.

Bellcore, TR-NWT-001230, "SONET Bidirectional Line Switched Ring Equipment Generic Criteria," Issue 2, 1992.

Bellcore, TR-TSP-000496, "SONET Add/Drop Multiplex Equipment (SONET ADM) Generic Criteria," Issue 3, 1992.

Bellcore, TR-TSY-00023, "Wideband and Broadband Digital Cross-Connect Generic Requirements and Objectives," Issue 2, 1989.

Bellcore, TR-TSY-000303, "Integrated Digital Loop Carrier System Generic Requirements, Objectives, and Interface," Issue 1, Revision 3, 1990.

Bermejo, L., P. Parmentier, and G. H. Petit. "Service Characteristics and Traffic Models in a Broadband ISDN," *Electrical Communication*, Vol. 64-2/3, 1990.

Bertsekas, Dmitri, and Gallager, Robert. *Data networks*, 2nd Ed., Prentice-Hall, Inc., Englewood Cliffs, NJ, 1991.

Biocca, A., G. Freeschi, et. al. "Architectural Issues in the Interoperability between MANs and ATM Network," *Proc.* XIII ISS, Stockholm, 1990.

Boiocchi, G., L. Fratta, et. al. "ATM Connectionless Server: Performance Evaluation," *Proc. Modeling and Performance Evaluation of ATM Technology*, Perros, Pujolle, and Takahashi (eds.), North Holland, 1993.

Borelli, Vincent R. and Hermann Gysel. "Fiber-optic super trunking: A comparison of performance & topologies using analog and or digital technologies," NCTA technical papers, 1990.

Boyer, P. F. "Congestion Control for the ATM," 7th International Teletraffic Congress Seminar, Morristown NJ, October 9–11, 1990.

Bubenik, R., and J. Tuner. "Performance of Broadcast Switch," IEEE Trans. Comm., Vol. 37, No. 1, 1989.

Burgin, J., and D. Dorman. "Broadband ISDN Resource Management: The Role of Virtual Paths," *IEEE Comm. Magazine*, September 1991.

Catlett, C.E. "In Search of Gigabit Applications," *IEEE Communications Magazine*, Vol. 30, No. 5, April 1992.

CCIR Recommendation 601, "Encoding Parameters of Digital Television for Studios," 1982.

CCIR Recommendation 656, "Interfaces for Digital Component Video Signals in 525-Line and 625-Line Television Systems," 1982.

CCIR Recommendation 709, "Basic Parameter Values for the HDTV Standard for the Studio and for International Programme Exchange," 1990.

CCIR Recommendation 710, "Subjective Assessment Methods for Image Quality in High Definition Television," 1990.

CCIR Recommendation 714, "International Exchange of Programme Electronically Produced by means of High-Definition Television," 1990.

Chao, H. J. "Design of Leaky Bucket Access Control Schemes in ATM Networks," ICC '91, 1991, pp. 180-187.

Chemouil, P., M. Lebourges, and P. Gauthier. "Performance Evaluation of Adaptive Traffic Routing in a Metropolitan Network: A Case Study," GLOBECOM '89, 1989.

Chen, W-T., H-J. Liu, and Y-T. Tsay. "High-Throughput Cell Scheduling for Broadband Switching Systems," *IEEE Journal. Selected Areas in Communications,* Vol. 9, No. 9, December 1991.

Chiddix, James A. "Fiber backbone—Multichannel AM-Video Trunking," NCTA Technical papers, 1989.

Cicora, Walter S. "Cables' excellent position in HDTV," *ATC,* 1990.

Cidon, I. and I. Gopal. "Paris: An Approach to Integrated High-Speed Private Networks." *International Journal of Digital and Analog Cabled Systems,* Vol 1. No. 4.

Cidon, I., J. Derby, I. Gopal, and B. Kadaba. "A Critique of ATM from a Data Communications Perspective,". *Journal of High Speed Networks,* Vol. 1, No. 4.

Cisneros, A., and C. A. Brackett. "A Large ATM Switch Based on Memory Switches and Optical Star Couplers," *IEEE Journal of Selected Areas in Communications,* Vol. 9, No. 8, October 1991.

Cosmas, J. P., and A. Odinma-Okafor. "Characterization of Variable Rate Video Codecs in ATM to Geometrically Modulated Deterministic Process Model," ITC-13, 1991.

Cox, J. R., M. E. Gaddis, and J. S. Turner. "Project Zeus," *IEEE Network Magazine,* Vol. 7, No. 2, March 1993.

Crocetti, P., L. Fratta, et. al. "ATM Based SMDS for LANs/MANs Interconnection," *Proc.* XIV ISS, Yokohama, 1992.

D'Ambrosio, M., and R. Melen. "Performance Analysis of ATM Switching: A Review," *CSELT Tech. Rep.,* Vol. 20, No. 3.

Darcie, Thomas. "Subcarrier Multiplexing for Lightwave Networks and Video Distribution Systems," *IEEE JSAC,* Vol. 8 No. 7, Sep 1990, pp 1240–1248.

Data Communication Magazine (all 1993 issues).

Datapro's Broadband Networking, 1992.

Davie, B. S., J. M. Smith, and C. B. S. Traw. "Host Interfaces for ATM Networks," *High Performance Communications,* Kluwer Academic Publishers.

De Prycker, M., and J. Bauwens. "A Switching Exchange for an Asynchronous Time Division Based Network," ICC '87, Seattle, 1987.

DEC, Northern Telecom, Stratacom, Cisco, "Frame Relay Specification with Extensions," Revision 1.0, 1990.

Decina, M., P. Giacomazzi, and A. Pattavina, "Shuffle Interconnection Networks with Deflection Routing for ATM Switching: the Open-Loop Shuffleout," ITC-13, 1991.

DePrycker, M. *Asynchronous Transfer Mode: Solution for Broadband ISDN* (2nd edition), Ellis Horwood, Chichester, England.

Dittmann, L., and S. B. Jacobsen. "Statical Multiplexing of Identical Bursty Sources in an ATM network," GLOBECOM '88, 1988 p. 97.

Doshi, B. T., S. Dravida, P. K. Johri, and G. Ramamurthy. "Memory, Bandwidth, Processing and Fairness Considerations in Real Time Congestion Controls for Broadband Networks," ITC-13, 1991.

Doshi, B.T., and S. Dravida. "Congestion Controls for Bursty Data in High Speed Wide Area Networks: In Call Parameter Negotiations," ITC Seventh Specialist Seminar on Broadband Tech., 1990.

Dron, L. G., G. Ramamurthy, and B. Sengupta. "Delay Analysis of Continuous Bit-Rate Traffic over an ATM Network," *IEEE Journal on Selected Areas of Communication,* Vol. 9, No. 3, April 1991.

Dziong, Z., et al. "Bandwidth Management in ATM Networks," ITC-13, 1991.

Dziong, Z., K. Q. Liao, and L. Mason. "Flow Control Models for Multi-Service Networks with Delayed Call Set Up," INFOCOM '90, 1990.

Eckberg, A. E. "Generalized Peakedness of Teletraffic Processes," ITC-10, 1983. Eng, K., M. Hluchyj, Y. Yeh "Multicast and Broadcast Services in a Knockout Packet Switch," INFOCOM '88, 1A.4, 1988.

FCC Filings. Cox Cable and Continental Cable.

Fitzpatrick, G.J., et. al. "Analysis of Large Scale Three Stage Networks Serving Multirate Traffic," ITC '13, 1991.

Fujimoto, N., T. Ishihara, and K. Yamaguchi. "Broadband subscriber loop system using multi-gigabit intelligent optical shuttle nodes," *Proc.* GLOBECOM '87, paper 37.3.1.

Fujiyama, Y., et. al. "ATM Switching System Evolution and Implementation for B-ISDN," ICC '90, 1990.

Galassi, G., G. Rigolio, and L. Verri. "Resource Management and Dimensioning in ATM Networks," *IEEE Network Magazine*, May 1990.

Gallassi, G., G. Rigolio, and L. Fratta. "ATM: Bandwidth Assignment and Bandwidth Enforcement Policies," GLOBECOM '89, 1989.

Gerla, M. , J. A. S. Monteiro, and R. Pazos. "Topology Design and Bandwidth Allocation in ATM Nets," *IEEE Journal on Selected Areas of Communication*, Vol. 7, No. 8, Oct '89.

Ghanbari, M. "Two-Layer Coding of Video Signals for VBR Networks," *IEEE Journal on Selected Areas of Communication*, Vol. 7, No. 5, June 1989.

Giacopelli, J. N., J. J. Hickey, W. S. Marcus, W. D. Sincoskie, and M. Littlewood. "Sunshine : A High-Performance Self-Routing Broadband Packet Switch Architecture," *IEEE Journal of Selected Areas in Communications*, Vol. 9, No. 8, October 1991.

Gopal, I. S., I. Cidon, and H. Meleis. "PARIS: An Approach to Integrated Private Networks," ICC '87, 1987.

Griffin, John T. "Cost and performance comparison of fiber-optic CATV super trunks utilizing FM and Digital techniques," Jerrold Applied Media Lab, NCTA technical papers, 1989.

Gruber, J. G. "Delay Related Issues in Integrated Voice and Data Networks," *IEEE Journal on Selected Communication*, June 1981.

Gruber, J. G., and N. H. Le. "Performance Requirements for Integrated Voice and Data Networks," *IEEE Journal on Selected Areas of Communication*, December 1983.

Hajikano, K., et. al. "Asynchronous Transfer Mode Switching Architecture for Broadband ISDN-Multistage Self Routing Switching," ICC '88, 1988.

Handel, R., and M. N. Huber. *Integrated Broadband Networks; An Introduction to ATM-Based Networks*, Addison-Wesley, Reading, Mass.

Hart, George and Nick Hamitton Piercy, "Rogers Fiber Architecture, Rogers Engineering," 1989 NCTA technical papers.

Heldman, Robert. *Global Telecommunications—Layered Networks Layered Services*, McGraw-Hill Inc., 1992.

Henrion, M., et. al. "Switching Network Architecture for ATM Based Broadband Communications," ISS '90, Stockholm, 1990.

Hirano, M. , and N. Watanabi. "Characteristics of a Cell Multiplexer for Bursty ATM Traffic," ICC '89, 1989.

Hluchyj, M. G., et. al. "The Knockout Switch: a Simple Modular Architecture for High Performance Packet Switching," ISS '87,1987.

Huang, A., and S. Knauer. "Starlite, a Wideband Digital Switch," GLOBECOM '84, Atlanta,1984.

Hui, J. Y. *Switching and Traffic Theory for Integrated Broadband Networks*, Norwell, MA: Kluwer Academic Publishers, 1990.

Hui, J. Y. "Resource Allocation for Broadband Networks," *IEEE Journal on Selected Areas of Communication*, Vol. 6, No. 9, Dec '89.

Hui, J., and E. Authurs. "A Broadband Packet Switch for Integrated Transport," *IEEE Journal on Selected Areas of Communication*, Vol. 5, 1991.

IEEE Communication Magazine (all 1992 to April 1994 issues).

IEEE Network Magazine (all 1992, 1993 issues).

IEEE Standard: "DQDB Subnetwork of a Metropolitan Area Network," P802.6, 1991.

Irvin, D. R. "Making Broadband-ISDN Successful," *IEEE Network Magazine*, Vol. 7, No. 1, January 1993.

ISO-IEC/JTC1/CD 10918 (JPEG), "Digital Compression and Coding of Continuous-Tone Still Images," 1991.

ISO-IEC/JTC1/CD 11172 (MPEG I), "Coding of Moving Pictures and Associated Audio for Digital Storage Media at up to about 1.5 Mbps," 1990.

ISO-IEC/JTC1/SC2/WG9N 36/CD 11154 , "Progressive Bi-Level Image Compression," Revision 4.1, 1992.

ISO-IEC/JTC1/SC29/WG11 (MPEG II), "Coded Presentation of Picture and Audio Information," 1992.

ITU-T Draft Recommendation I.374, "Network Capability for the Support of Multimedia Services," 1992.

ITU-T Recommendation G.650, "Definition and Text Methods for the Relevant Parameters of Single Mode Fibers," 1992.

ITU-T Recommendation G.651, "Characteristics of a 50/125-µm Multimode Graded Index Optical Fiber Cable," 1992 (Rev).

ITU-T Recommendation G.652, "Characteristics of a Single-Mode Optical Fiber Cable," 1992 (Rev).

ITU-T Recommendation G.653, "Characteristics of a Dispersion-Shifted Single-Mode Optical Fiber Cable," 1992 (Rev).

ITU-T Recommendation G.654, "Characteristics of a 1550-nm Wavelength Loss-Minimized Single-Mode Optical Fiber Cable," 1992 (Rev).

ITU-T Recommendation G.703, "Physical/Electrical Characteristics of Hierarchical Digital Interfaces," 1991.

ITU-T Recommendation G.707, "Synchronous Digital Hierarchy Bit Rates," 1992 (Rev).

ITU-T Recommendation G.708," Network Node Interface for the Synchronous Digital Hierarchy," 1992 (Rev).

ITU-T Recommendation G.709, "Synchronous Multiplexing Structure," 1992 (Rev).

ITU-T Recommendation G.744, "Synchronous Digital Hierarchy (SDH) Management Information Model," (New; to be approved in 1993).

ITU-T Recommendation G.781, "Multiplexing Equipment for the SDH," 1990.

ITU-T Recommendation G.782, "Types and General Characteristics of Synchronous Digital Hierarchy (SDH) Multiplexing Equipment," 1990.

ITU-T Recommendation G.783, "Characteristics of Synchronous Digital Hierarchy (SDH) Multiplexing Equipment Functional Blocks," 1990.

ITU-T Recommendation G.784, "Synchronous Digital Hierarchy (SDH) Management," 1990.

ITU-T Recommendation G.7xx, "ATM Cell Mapping into Plesiochronous Digital Hierarchy (PDH)," 1992.

ITU-T Recommendation G.803, "Architecture of Transport Networks Based on the SDH," 1992.

ITU-T Recommendation G.825, "The Control of Jitter and Wander within Digital Network, which are Based on Synchronous Digital Hierarchy (SDH)," 1993.

ITU-T Recommendation G.831, "Performance and Management Capabilities of Transport Networks Based on the SDH," 1992.

ITU-T Recommendation G.957, "Optical Interfaces for Equipments and Relating to the Synchronous Digital Hierarchy," 1992.

ITU-T Recommendation G.958, "Digital Line Systems Based on the Synchronous Digital Hierarchy for Use on Optical Fiber Cables," 1990.

ITU-T Recommendation H.221, "Frame Structure for 64 to 1920 kbps Channel in Audiovisual Teleservices," 1990.

ITU-T Recommendation H.230, "Frame Synchronous Control and Indication Signals for Audiovisual Systems," 1990.

ITU-T Recommendation H.242, "System for Establishing Communication Between Audiovisual Terminals using Digital Channels up to 2 Mbps," 1990.

ITU-T Recommendation H.261, "Video Codec for Audiovisual Services at p × 64 kbps," 1990.

ITU-T Recommendation H.320, "Narrowband Visual Telephone Systems and Terminal Equipment," 1990.

ITU-T Recommendation H.32x, "Audiovisual Communication Terminal for BISDN," 1992.

ITU-T Recommendation I.113, "Vocabulary Terms for Broadband Aspects of ISDN," 1992 (Rev).

ITU-T Recommendation I.120, "Integrated Services Digital Network (ISDN)," 1992 (Rev).

ITU-T Recommendation I.121, "Broadband Aspects of ISDN," 1990.

ITU-T Recommendation I.122, "Framework for Providing Additional Packet Mode Bearer Service," 1991.

ITU-T Recommendation I.140, "Attribute Technique for the Characterization of the Telecommunication Services Supported by an ISDN and Network Capability of an ISDN," 1992 (Rev).

ITU-T Recommendation I.150, "BISDN ATM Functional Characteristics," 1992 (Rev).

ITU-T Recommendation I.211, "BISDN Service Aspects," 1992 (Rev).

ITU-T Recommendation I.233, "Frame Mode Bearer Services," 1992.

ITU-T Recommendation I.311, "BISDN General Network Aspects," 1992 (Rev).

ITU-T Recommendation I.321, "BISDN Protocol Reference Model and its Application," 1990.

ITU-T Recommendation I.327, "BISDN Functional Architecture Aspects," 1992.

ITU-T Recommendation I.35B, "BISDN ATM Cell Transfer Performance," 1992 (Draft).

ITU-T Recommendation I.361, "BISDN ATM Layer Specification," 1992.

ITU-T Recommendation I.362, "BISDN ATM Adaptation Layer (AAL) Functional Description," 1992.

ITU-T Recommendation I.363, "BISDN ATM Adaptation Layer (AAL) Specification," 1992.

ITU-T Recommendation I.364, "Support of Broadband Connectionless Data Service on BISDN," 1992.

ITU-T Recommendation I.370, "Congestion Management for the ISDN Frame Relaying Bearer Service," 1991.

ITU-T Recommendation I.371, "Traffic Control and Congestion Control in BISDN," 1992.

ITU-T Recommendation I.413, "BISDN User-Network Interface," 1992.

ITU-T Recommendation I.430, "ISDN Basic Rate User Network Interface Layer 1 Specification," 1992.

ITU-T Recommendation I.431, "ISDN Primary Rate User Network Interface Layer 1 Specification," 1992 (Rev).

ITU-T Recommendation I.432, "BISDN User-Network Interface—Physical Layer Specification," 1992.

ITU-T Recommendation I.610, "BISDN UNI Operations and Maintenance Principles," 1992.

ITU-T Recommendation Q.921 (I.441), "ISDN User Network Interface Data Layer Specification," 1988.

ITU-T Recommendation Q.922, "ISDN Data Link Layer Specification for Frame Mode Bearer Services," 1992.

ITU-T Recommendation Q.933, "DSS1 Signaling Specification for Frame Mode Bearer Service," 1992.

Joos, P., and W. Verbiest. "A Statistical Bandwidth Allocation and Usage Monitoring Algorithm for ATM Networks," ICC '89, 1989.

Kanayama, Y., Y. Maeda, and H. Ueda. "Virtual Path Management Functions for Broadband ATM Networks," GLOBECOM '91, 1991.

Kawashima, K., and H. Saito. "Teletraffic Issues in ATM Networks," *Computer Networks and ISDN Systems*, Vol. 20, 1990.

Kazovsky, Leonid. "Optical Signal Processing for Lightwave Communications Networks," *IEEE JSAC*, Vol. 8. No. 6, Aug 1990, pp. 973-981.

Kessler, Gary and David Train. *Metropolitan Area Networks—Concepts, Standards and Services,* McGraw-Hill, 1991

Kim, Y. M., and K.Y. Lee. "PR-Banyan: A Packet Switch with a Pseudo Randomizer for Nonuniform Traffic," ICC '91, 1991.

Kishino, F., K. Manabe, Y. Hayashi, and H. Yasuda. "Variable Bit-Rate Coding of Video Signals for ATM Networks," *IEEE Journal on Selected Areas of Communication,* Vol. 7, No. 5, June 1989.

Kleinrock, L. *Queuing Systems, Vol. 1: Theory,* New York: John Wiley & Sons, Inc, 1975.

Kleinrock, L. "ISDN-The Path to Broadband Networks," *IRE Proc.*, Vol. 79, 1991.

Kroner, H., P. J. Kuhn, and G. Willmann. "Performance Comparison for Resource Sharing Strategies between Lost-Call-Cleared and Reservation Traffic," ITC-13, 1991.

Kumar, Balaji, and N. Subramaniam, "Routing strategies for an hybrid network architecture," *Proceedings of the 1991 Symposium on Applied Computing*, April, 1991, Kansas City, MO.

Kuwahara, H., et.al. "Shared Buffer Memory Switch for an ATM Exchange," ICC '90, Boston, 1989.

Le Gall, D. J. "MPEG: A Video Compression Standard for Multimedia Applications," *Comm. ACM,* Vol. 34-4, 1991.

Lea, C. T. "The Load Sharing Banyan Network," *IEEE Trans. Comm.,* Vol. 35, No. 12, 1986.

Lee, D-S., K-H. Tzou, and S-Q. Li. "Control Analysis of Video Packet Loss in ATM Networks," *SPIE Visual Communications and Image Processing '90*, Vol. 1360, 1990.

Leland, W. E. "Window-Based Congestion Management in Broadband ATM Networks: The Performance of Three Access-Control Policies," GLOBECOM '89, 1989.

Leslie, I. "Fairisle: An ATM Network for the Local Area," *Proc.* ACM SIGCOMM '91, Zurich, September 3–6, 1991.

Louviom, J. R., J. Boyer, and J. B. Gravereaux. "Statistical Multiplexing of VBR Sources in ATM Networks," 3rd IEEE CAMAD Workshop, 1990.

Maglaris, B., D. Anastassiou, P. Sen, G. Karlsson, and J. Robins. "Performance Models of Statistical Multiplexing in Packet Video Communications," *IEEE Trans. on Communications,* Vol. 36, 1988.

Minzer, S. E. "Broadband ISDN and Asynchronous Transfer Mode (ATM)," *IEEE Comm. Magazine,* September 1989.

Miyao, Y. "A Call Admission Control Scheme in ATM Networks," ICC '91, 1991, pp. 391–396.

Network World, (articles in 1992, 1993 issues).

Nikolaidis, I., and I. F. Akyildiz. "Source Characterization and Statistical Multiplexing in ATM Networks," Technical Report GIT-CC-92/24 Georgia Tech.

Norgaard, K. "Evaluation of Output Traffic from an ATM Node," ITC-13, 1991, pp. 533–537.

Onvural, R. O., and I. Nikolaidis. "Routing in ATM Networks," *High Speed Networks,* H.G. Perros (ed.), New York: Plennum Pub., 1992.

Pangrac, David M. and Louis D. Williamson, "Fiber trunk and feeder—The continuing evolution," *ATC* NCTA technical papers, 1990.

Personick, S. D. *Fiber Optics : Technology and Applications,* Plenum Publishers.

Ransom, M. N. "The VISTAnet Gigabit Network Testbed," *Journal of High Speed Networks,* Vol. 1, No. 1.

Rice, W. O., and H. G. Perros. "What's New in B-ISDN Standards," *High Speed Networks,* Edited by H. G. Perros, Plennium Pub. Inc., 1992.

Russell, J. "Multimedia Networking Performance Requirements," *ATM Networks,* I. Viniotis and R. O. Onvural (eds.), New York: Plenum Pubs., 1993.

Saito, H., M. Kawarasaki, and H. Yamada. "An Analysis of Statistical Multiplexing in an ATM Transport Network," *IEEE Journal on Selected Areas of Communication,* Vol. 9, No. 3, Apr '91.

Schormans, J., J. Pitts, and E. Scharf. "Time Priorities in ATM Switches," ITC-13, 1991.

Schulzrinne, H., J. F. Kurose, and D. Towsley. "Congestion Control for Real-Time Traffic in High-Speed Networks," INFOCOM '90, 1990.

Schwartz, Mischa. *Computer—Communications Network Design and Analysis,* Prentice-Hall, Englewood Cliffs, NJ, 1977.

Schwartz, Mischa. *Telecommunications Networks,* Addison-Wesley Publishing Co., Reading, MA., 1987.

Sexton, M., and Reid. *Transmission Networking: SONET and the Synchronous Digital Hierarchy,* Artech House, Norwood, Mass.

Spohn, Darren L. *Data Network Design,* McGraw-Hill Series on Computer Communications, 1993.

Sriram, K., and D. M. Lucantoni. "Traffic Smoothing Effects of Bit Dropping in a Packet Voice Multiplexer," INFOCOM '88, 1988.

Sriram, K., R. S. McKinney, and M. H. Sherif. "Voice Packetization and Compression in Broadband ATM Networks," *IEEE Journal on Selected Areas of Communication,* Vol. 9, No. 3, April 1991.

Stallings, William. *ISDN—An introduction,* Macmillan Publishing Co., 1989

Sutherland, J., and L. Litteral. "Residential Video Services," *IEEE Communication Magazine,* Vol. 30, No. 7, 1992.

Suzuki, H., et. al. "Output Buffer Switch Architecture for Asynchronous Transfer Mode," ICC '89, Boston, 1989.

Tanenbaum, A. S. *Computer Networks* (2nd edition), Prentice Hall, Englewood Cliffs, N. J.

Telecom Data Report (all 1993 issues).

Telecommunication Magazine (all 1992, 1993 issues).

Tirtaatmadja, E., and R. A. Palmer. "The Application of Virtual Paths to the Interconnection of IEEE 802.6 Metropolitan Area Networks," *Proc.* XIII ISS, Stockholm, 1990.

Tseng, K. H., and M. T. Hsiao. "Admission Control of Voice/Data Integration in an ATM Network," ICC '91, 1991.

Turner, J. S. "Managing Bandwidth in ATM Networks with Bursty Traffic." *IEEE Network Magazine,* Vol. 6, No. 5, September 1992.

van Landegem, T., and P. Peschi. "Managing a Connectionless Virtual Overlay Network on Top of ATM," *Proc.* ICC '91, Denver, CO, 1991.

Verbeist, W., and L. Pinnoo. "A Variable Rate Video Codec for Asynchronous Transfer Mode Networks," *IEEE Journal on Selected Areas of Communication,* Vol. 7, No. 5, June 1989.

Virtamo, J. T, and J. W. Roberts. "Evaluating Buffer Requirements in an ATM Multiplexer," GLOBECOM '89, 1989.

Wang, Q., and V. S. Frost. "Efficient Estimation of Cell Blocking Probability for ATM Systems," ICC '91, 1991.

Woodworth, C. B., M. J. Karol, and R. D. Gillin. "A Flexible Broadband Packet Switch for a Multimedia Integrated Network," ICC '91, 1991.

Wulleman, R., and T. van Landegem. "Comparison of ATM Switching Architectures," *International Journal of Digital and Analog Cabled Systems,* Vol. 2, No. 4, 1989.

Yamada, H., and S. Sumita. "A Traffic Measurement Method and its Application for Cell Loss Probability Estimation in ATM Networks," *IEEE Journal on selected Areas of Communication,* Vol. 9, No. 3, April 1991.

Yamashita, H., H. G. Perros, and S. W. Hong. "Performance Modelling of a Shared Buffer ATM Switch Architecture," ITC-13, 1993.

Yasuda, Y., H. Yasuda, N. Ohta, and F. Kishino. "Packet Video Transmission through ATM Networks," GLOBECOM '89, 1989.

Yates, Robert K., Nolwen Mohé and Jorome Masson, *Fiber optics and CATV business strategy*, Artech House, Boston.

Zimmerman, H. "OSI Reference Model—The ISO Model of Architecture for Open Systems Interconnection," *IEEE Trans. on Communications,* Vol. 28, No. 4, April 1980.

Index

Boldface page numbers indicate illustrations

ABOUT THE AUTHOR

Balaji Kumar is a senior engineer at MCI in Dallas, Texas. He earned his M.B.A. at the University of Dallas and his M.S. in computer science at the University of Missouri. He publishes regularly in IEEE proceedings. As a senior network planning engineer at Bell Northern Research in Dallas, he was responsible for developing many internal BNR publications. Balaji lives in Allen, Texas.